Signal Processing Techniques for Knowledge Extraction and Information Fusion

Danilo Mandic • Martin Golz • Anthony Kuh •
Dragan Obradovic • Toshihisa Tanaka
Editors

Signal Processing Techniques for Knowledge Extraction and Information Fusion

Editors
Danilo Mandic
Imperial College London
London
UK

Martin Golz
University of Schmalkalden
Schmalkalden
Germany

Anthony Kuh
University of Hawaii
Manoa, HI
USA

Dragan Obradovic
Siemens AG
Munich
Germany

Toshihisa Tanaka
Tokyo University of Agriculture
 and Technology
Tokyo
Japan

ISBN: 978-0-387-74366-0   e-ISBN: 978-0-387-74367-7

Library of Congress Control Number: 2007941602

© 2008 Springer Science+Business Media, LLC
All rights reserved. This work may not be translated or copied in whole or in part without the written permission of the publisher (Springer Science+Business Media, LLC, 233 Spring Street, New York, NY 10013, USA), except for brief excerpts in connection with reviews or scholarly analysis. Use in connection with any form of information storage and retrieval, electronic adaptation, computer software, or by similar or dissimilar methodology now known or hereafter developed is forbidden.
The use in this publication of trade names, trademarks, service marks, and similar terms, even if they are not identified as such, is not to be taken as an expression of opinion as to whether or not they are subject to proprietary rights.

Printed on acid-free paper.

9 8 7 6 5 4 3 2 1

springer.com

Preface

This book emanated from many discussions about collaborative research among the editors. The discussions have focussed on using signal processing methods for knowledge extraction and information fusion in a number of applications from telecommunications to renewable energy and biomedical engineering. They have led to several successful collaborative efforts in organizing special sessions for international conferences and special issues of international journals. With the growing interest from researchers in different disciplines and encouragement from Springer editors Alex Greene and Katie Stanne, we were spurred to produce this book.

Knowledge extraction and information fusion have long been studied in various areas of computer science and engineering, and the number of applications for this class of techniques has been steadily growing. Features and other parameters that describe a process under consideration may be extracted directly from the data, and so it is natural to ask whether we can exploit digital signal processing (DSP) techniques for this purpose. Problems where noise, uncertainty, and complexity play major roles are naturally matched to DSP. This synergy of knowledge extraction and DSP is still under-explored, but has tremendous potential. It is the underlying theme of this book, which brings together the latest research in DSP-based knowledge extraction and information fusion, and proposes new directions for future research and applications. It is fitting, then, that this book touches on globally important applications, including sustainability (renewable energy), health care (understanding and interpreting biomedical signals) and communications (extraction and fusing of information from sensor networks).

The use of signal processing in data and sensor fusion is a rapidly growing research area, and we believe it will benefit from a work such as this, in which both background material and novel applications are presented. Some of the chapters come from extended papers originally presented at the special sessions in ICANN 2005 and KES 2006. We also asked active researchers in signal processing with specializations in machine learning and multimodal signal processing to make contributions to augment the scope of the book.

This book is divided in four parts with four chapters each.

Collaborative Signal Processing Algorithms

Chapter 1 by Jelfs et al. addresses hybrid adaptive filtering for signal modality characterization of real-world processes. This is achieved within a collaborative signal processing framework which quantifies in real-time, the presence of linearity and nonlinearity within a signal, with applications to the analysis of EEG data. This approach is then extended to the complex domain and the degree of nonlinearity in real-world wind measurements is assessed.

In Chap. 2, Hirata et al. extend the wind modelling approaches to address the control of wind farms. They provide an analysis of the wind features which are most relevant to the local forecasting of the wind profile. These are used as prior knowledge to enhance the forecasting model, which is then applied to the yaw control of a wind turbine.

A collaborative signal processing framework by means of hierarchical adaptive filters for the detection of sparseness in a system identification setting is presented in Chap. 3, by Boukis and Constantinides. This is supported by a thorough analysis with an emphasis on unbiasedness. It is shown that the unbiased solution corresponds to existence of a sparse sub-channel, and applications of this property are highlighted.

Chapter 4 by Zhang and Chambers addresses the estimation of the reverberation time, a difficult and important problem in room acoustics. This is achieved by blind source separation and adaptive noise cancellation, which in combination with the maximum likelihood principle yields excellent results in a simulated high noise environment. Applications and further developments of this strategy are discussed.

Signal Processing for Source Localization

Kuh and Zhu address the problem of sensor network localization in Chap. 5. Kernel methods are used to store signal strength information, and complex least squares kernel regression methods are employed to train the parameters for the support vector machine (SVM). The SVM is then used to estimate locations of sensors, and to track positions of mobile sensors. The chapter concludes by discussing distributed kernel regression methods to perform localization while saving on communication and energy costs.

Chapter 6, by Lenz et al., considers adaptive localization in wireless networks. They introduce an adaptive approach for simultaneous localization and learning based on theoretical propagation models and self-organizing maps, to demonstrate that it is possible to realize a self-calibrating positioning system with high accuracies. Results on real-world DECT and WLAN groups support the approach.

In Chap. 7, Host-Madsen et al. address signal processing methods for Doppler radar heart rate monitoring. This provides unobtrusive and ubiquitous detection of heart and respiration activity from distance. By leveraging recent advances in signal processing and wireless communication technologies, the authors explore robust radar monitoring techniques through MIMO signal processing. The applications of this method include health monitoring and surveillance.

Obradovic et al. present the fusion of onboard sensors and GPS for real-world car navigation in Chap. 8. The system is based on the position estimate obtained by Kalman filtering and GPS, and is aided by corrections provided by candidate trajectories on a digital map. In addition, fuzzy logic is applied to enhance guidance. This system is in operation in a number of car manufacturers.

Information Fusion in Imaging

In Chap. 9, Chumerin and Van Hulle consider the detection of independently moving objects as a component of the obstacle detection problem. They show that the fusion of information obtained from multiple heterogeneous sensors has the potential to outperform the vision-only description of driving scenes. In addition, the authors provide a high-level sensor fusion model for detection, classification, and tracking in this context.

Aghajan, Wu, and Kleihorst address distributed vision networks for human pose analysis in Chap. 10. This is achieved by collaborative processing and data fusion mechanisms, and under a low bandwidth communication constraint. The authors employ a 3D human body model as the convergence point of the spatiotemporal and feature fusion. This model also allows the cameras to interact and helps the evaluation of the relative values of the derived features.

The application of information fusion in E-cosmetics is addressed by Tsumura et al. in Chap. 11. The authors develop a practical skin color analysis and synthesis (fusion) technique which builds upon both the physical background and physiological understanding. The appearance of the reproduced skin features is analysed with respect to a number of practical constraints, including the imaging devices, illuminants, and environments.

Calhoun and Adalı consider the fusion of brain imaging data in Chap. 12. They utilize multiple image types to take advantage of the cross information. Unlike the standard approaches, where cross information is not taken into account, this approach is capable of detecting changes in functional magnetic resonance imaging (fMRI) activation maps. The benefits of the information fusion strategy are illustrated by real-world examples from neurophysiology.

Knowledge Extraction in Brain Science

Chapter 13, by Mandic et al. considers the "data fusion via fission" approach realized by empirical mode decomposition (EMD). Extension to the complex domain also helps to extract knowledge from processes which are strongly dependent on synchronization and phase alignment. Applications in real-world brain computer interfaces, e.g., in brain prosthetics and EEG artifact removal, illustrate the usefulness of this approach.

In Chap. 14, Rutkowski et al. consider some perceptual aspects of the fusion of information from multichannel EEG recordings. Time–frequency EMD features, together with the use of music theory, allow for a convenient and unique audio feedback in brain computer and brain machine (BCI/BMI) interfaces. This helps to ease the understanding of the notoriously difficult to analyse EEG data.

Cao and Chen consider the usefulness of knowledge extraction in brain death monitoring applications in Chap. 15. They combine robust principal factor analysis with independent component analysis to evaluate the statistical significance of the differences in EEG responses between quasi-brain-death and coma patients. The knowledge extraction principles here help to make a binary decision on the state of the consciousness of the patients.

Chapter 16, by Golz and Sommer, addresses a multimodal approach to the detection of extreme fatigue in car drivers. The signal processing framework is based on the fusion of linear (power spectrum) and nonlinear (delay vector variance) features, and knowledge extraction is performed via automatic input variable relevance detection. The analysis is supported by results from comprehensive experiments with a range of subjects.

London, *Danilo Mandic*
October 2007 *Martin Golz*
 Anthony Kuh
 Dragan Obradovic
 Toshihisa Tanaka

Acknowledgement

On behalf of the editors, I thank the authors for their contributions and for meeting such tight deadlines, and the reviewers for their valuable input.

The idea for this book arose from numerous discussions in international meetings and during the visits of several authors to Imperial College London. The visit of A. Kuh was made possible with the support of the Fulbright Commission; the Royal Society supported visits of M. Van Hulle and T. Tanaka; the Japan Society for the Promotion of Science (JSPS) also supported T. Tanaka.

The potential of signal processing for knowledge extraction and sensor, data, and information fusion has become clear through our special sessions in international conferences, such as ICANN 2005 and KES 2006, and in our special issue of the International Journal of VLSI Signal Processing Systems (Springer 2007). Perhaps the first gentle nudge to edit a publication in this area came from S.Y. Kung, who encouraged us to organise a special issue of his journal dedicated to this field. Simon Haykin made me aware of the need for a book covering this area and has been inspirational throughout.

I also thank the members of the IEEE Signal Processing Society Technical Committee on Machine Learning for Signal Processing for their vision and stimulating discussions. In particular, Tülay Adalı, David Miller, Jan Larsen, and Marc Van Hulle have been extremely supportive. I am also grateful to the organisers of MLSP 2005, KES 2006, MLSP 2007, and ICASSP 2007 for giving me the opportunity to give tutorial and keynote speeches related to the theme of this book. The feedback from these lectures has been most valuable.

It is not possible to mention all the colleagues and friends who have helped towards this book. For more than a decade, Tony Constantinides has been reminding me of the importance of fixed point theory in this area, and Kazuyuki Aihara and Jonathon Chambers have helped to realise the potential of information fusion for heterogeneous measurements. Maria Petrou has been influential in promoting data fusion concepts at Imperial. Andrzej Cichocki and his team from RIKEN have provided invigorating discussions and continuing support.

A special thanks to my students who have been extremely supportive and helpful. Beth Jelfs took on the painstaking job of going through every chapter and ensuring the book compiles. A less dedicated and resolute person would have given up long before the end of this project. Soroush Javidi has created and maintained our book website, David Looney has undertaken a number of editing jobs, and Ling Li has always been around to help.

Henry Goldstein has helped to edit and make this book more readable. Finally, I express my appreciation to the signal processing tradition and vibrant research atmosphere at Imperial, which have made delving into this area so rewarding.

Imperial College London, *Danilo Mandic*
October 2007

Contents

Part I Collaborative Signal Processing Algorithms

1 Collaborative Adaptive Filters for Online Knowledge Extraction and Information Fusion
Beth Jelfs, Phebe Vayanos, Soroush Javidi, Vanessa Su Lee Goh, and Danilo P. Mandic 3
1.1 Introduction 3
    1.1.1 Previous Online Approaches 5
    1.1.2 Collaborative Adaptive Filters 6
1.2 Derivation of The Hybrid Filter 7
1.3 Detection of the Nature of Signals: Nonlinearity 8
    1.3.1 Tracking Changes in Nonlinearity of Signals 10
1.4 Detection of the Nature of Signals: Complex Domain 12
    1.4.1 Split-Complex vs. Fully-Complex 13
    1.4.2 Complex Nature of Wind 17
1.5 Conclusions 19
References 20

2 Wind Modelling and its Possible Application to Control of Wind Farms
Yoshito Hirata, Hideyuki Suzuki, and Kazuyuki Aihara 23
2.1 Formulating Yaw Control for a Wind Turbine 23
2.2 Characteristics for Time Series of the Wind 25
    2.2.1 Surrogate Data 25
    2.2.2 Results 25
2.3 Modelling and Predicting the Wind 27
    2.3.1 Multivariate Embedding 27
    2.3.2 Radial Basis Functions 28
    2.3.3 Possible Coordinate Systems 30
    2.3.4 Direct vs. Iterative Methods 30
    2.3.5 Measurements of the Wind 30
    2.3.6 Results 32

2.4	Applying the Wind Prediction to the Yaw Control	34
2.5	Conclusions	34
References		35

3 Hierarchical Filters in a Collaborative Filtering Framework for System Identification and Knowledge Retrieval
Christos Boukis and Anthony G. Constantinides 37

3.1	Introduction	37
3.2	Hierarchical Structures	39
	3.2.1 Generalised Structures	40
	3.2.2 Equivalence with FIR	41
3.3	Multilayer Adaptive Algorithms	43
	3.3.1 The Hierarchical Least Mean Square Algorithm	43
	3.3.2 Evaluation of the Performance of HLMS	44
	3.3.3 The Hierarchical Gradient Descent Algorithm	45
3.4	Applications	46
	3.4.1 Standard Filtering Applications	46
	3.4.2 Knowledge Extraction	47
3.5	Conclusions	49
A	Mathematical Analysis of the HLMS	50
References		53

4 Acoustic Parameter Extraction From Occupied Rooms Utilizing Blind Source Separation
Yonggang Zhang and Jonathon A. Chambers 55

4.1	Introduction	55
4.2	Blind Estimation of Room RT in Occupied Rooms	57
	4.2.1 MLE-Based RT Estimation Method	57
	4.2.2 Proposed Noise Reducing Preprocessing	59
4.3	A Demonstrative Study	60
	4.3.1 Blind Source Separation	62
	4.3.2 Adaptive Noise Cancellation	65
4.4	Simulation Results	67
4.5	Discussion	72
4.6	Conclusion	73
References		74

Part II Signal Processing for Source Localization

5 Sensor Network Localization Using Least Squares Kernel Regression
Anthony Kuh and Chaopin Zhu 77

5.1	Introduction	77
5.2	Sensor Network Model	80
5.3	Localization Using Classification Methods	81

5.4	Least Squares Subspace Kernel Regression Algorithm		82
	5.4.1	Least Squares Kernel Subspace Algorithm	82
	5.4.2	Recursive Kernel Subspace Least Squares Algorithm	84
5.5	Localization Using Kernel Regression Algorithms		85
	5.5.1	Centralized Kernel Regression	85
	5.5.2	Kernel Regression for Mobile Sensors	86
	5.5.3	Distributed Kernel Regression	87
5.6	Simulations		89
	5.6.1	Stationary Motes	90
	5.6.2	Mobile Motes	91
	5.6.3	Distributed Algorithm	92
5.7	Summary and Further Directions		93
References			94

6 Adaptive Localization in Wireless Networks
Henning Lenz, Bruno Betoni Parodi, Hui Wang, Andrei Szabo, Joachim Bamberger, Dragan Obradovic, Joachim Horn, and Uwe D. Hanebeck 97

6.1	Introduction		97
6.2	RF Propagation Modelling		98
	6.2.1	Characteristics of the Indoor Propagation Channel	99
	6.2.2	Parametric Channel Models	99
	6.2.3	Geo Map-Based Models	100
	6.2.4	Non-Parametric Models	102
6.3	Localization Solution		103
6.4	Simultaneous Localization and Learning		104
	6.4.1	Kohonen SOM	105
	6.4.2	Main Algorithm	106
	6.4.3	Comparison Between SOM and SLL	107
	6.4.4	Convergence Properties of SLL	107
	6.4.5	Statistical Conditions for SLL	113
6.5	Results on 2D Real-World Scenarios		116
6.6	Conclusions		118
References			119

7 Signal Processing Methods for Doppler Radar Heart Rate Monitoring
Anders Høst-Madsen, Nicolas Petrochilos, Olga Boric-Lubecke, Victor M. Lubecke, Byung-Kwon Park, and Qin Zhou 121

7.1	Introduction		121
7.2	Signal Model		123
	7.2.1	Physiological Signal Model	125
7.3	Single Person Signal Processing		126
	7.3.1	Demodulation	126
	7.3.2	Detection of Heartbeat and Estimation of Heart Rate	127

7.4 Multiple People Signal Processing.............................132
    7.4.1 Heartbeat Signal......................................133
    7.4.2 Algorithm 133
    7.4.3 Results 134
7.5 Conclusion 138
References 139

8 Multimodal Fusion for Car Navigation Systems
*Dragan Obradovic, Henning Lenz, Markus Schupfner,
and Kai Heesche*..141
8.1 Introduction 141
8.2 Kalman Filter-Based Sensor Fusion for Dead Reckoning
    Improvement 143
8.3 Map Matching Improvement by Pattern Recognition 146
    8.3.1 Generation of Feature Vectors by State Machines 147
    8.3.2 Evaluation of Certainties of Road Alternatives Based
          on Feature Vector Comparison 150
8.4 Fuzzy Guidance 154
8.5 Conclusions 157
References 157

Part III Information Fusion in Imaging

9 Cue and Sensor Fusion for Independent Moving Objects Detection and Description in Driving Scenes
N. Chumerin and M.M. Van Hulle 161
9.1 Introduction 161
9.2 Vision Sensor Data Processing................................164
    9.2.1 Vision Sensor Setup 164
    9.2.2 Independent Motion Stream............................165
    9.2.3 Recognition Stream 167
    9.2.4 Training 168
    9.2.5 Visual Streams Fusion 170
9.3 IMO Detection and Tracking 171
9.4 Classification and Description of the IMOs 171
9.5 LIDAR Sensor Data Processing................................172
    9.5.1 LIDAR Sensor Setup 172
    9.5.2 Ground Plane Estimation 173
    9.5.3 LIDAR Obstacles Projection 175
9.6 Vision and LIDAR Fusion 175
9.7 Results 176
9.8 Conclusions and Future Steps 177
References 178

10 Distributed Vision Networks for Human Pose Analysis
Hamid Aghajan, Chen Wu, and Richard Kleihorst 181
10.1 Introduction 181
10.2 A Unifying Framework 183
10.3 Smart Camera Networks 184
10.4 Opportunistic Fusion Mechanisms 185
10.5 Human Posture Estimation 187
10.6 The 3D Human Body Model 189
10.7 In-Node Feature Extraction 190
10.8 Collaborative Posture Estimation 192
10.9 Towards Behavior Interpretation 195
10.10 Conclusions 198
References 199

11 Skin Color Separation and Synthesis for E-Cosmetics
*Norimichi Tsumura, Nobutoshi Ojima, Toshiya Nakaguchi,
and Yoichi Miyake* 201
11.1 Introduction 201
11.2 Image-Based Skin Color Analysis and Synthesis 203
11.3 Shading Removal by Color Vector Space Analysis:
     Simple Inverse Lighting Technique 205
     11.3.1 Imaging Model 205
     11.3.2 Finding the Skin Color Plane in the Face and Projection
            Technique for Shading Removal 208
11.4 Validation of the Analysis 210
11.5 Image-Based Skin Color and Texture Analysis/Synthesis 211
11.6 Data-Driven Physiologically Based Skin
     Texture Control 212
11.7 Conclusion and Discussion 218
References 219

12 ICA for Fusion of Brain Imaging Data
Vince D. Calhoun and Tülay Adalı 221
12.1 Introduction 221
12.2 An Overview of Different Approaches for Fusion 223
12.3 A Brief Description of Imaging Modalities
     and Feature Generation 224
     12.3.1 Functional Magnetic Resonance Imaging 224
     12.3.2 Structural Magnetic Resonance Imaging 226
     12.3.3 Diffusion Tensor Imaging 226
     12.3.4 Electroencephalogram 227
12.4 Brain Imaging Feature Generation 228
12.5 Feature-Based Fusion Framework Using ICA 228
12.6 Application of the Fusion Framework 230
     12.6.1 Multitask fMRI 231
     12.6.2 Functional Magnetic Resonance Imaging–Structural
            Functional Magnetic Resonance Imaging 231

XVI Contents

      12.6.3 Functional Magnetic Resonance Imaging–Event-Related Potential 233
      12.6.4 Structural Magnetic Resonance Imaging–Diffusion Tensor Imaging 233
      12.6.5 Parallel Independent Component Analysis 235
12.7 Selection of Joint Components 235
12.8 Conclusion 237
References 237

Part IV Knowledge Extraction in Brain Science

13 Complex Empirical Mode Decomposition for Multichannel Information Fusion
Danilo P. Mandic, George Souretis, Wai Yie Leong, David Looney, Marc M. Van Hulle, and Toshihisa Tanaka 243
13.1 Introduction 243
      13.1.1 Data Fusion Principles 244
13.2 Empirical Mode Decomposition 244
13.3 Ensemble Empirical Mode Decomposition 247
13.4 Extending EMD to the Complex Domain 249
      13.4.1 Complex Empirical Mode Decomposition 251
      13.4.2 Rotation Invariant Empirical Mode Decomposition 254
      13.4.3 Complex EMD as Knowledge Extraction Tool for Brain Prosthetics 254
13.5 Empirical Mode Decomposition as a Fixed Point Iteration 257
13.6 Discussion and Conclusions 258
References 259

14 Information Fusion for Perceptual Feedback: A Brain Activity Sonification Approach
Tomasz M. Rutkowski, Andrzej Cichocki, and Danilo P. Mandic 261
14.1 Introduction 261
14.2 Principles of Brain Sonification 263
14.3 Empirical Mode Decomposition 264
      14.3.1 EEG and EMD: A Match Made in Heaven? 265
      14.3.2 Time–Frequency Analysis of EEG and MIDI Representation 269
14.4 Experiments 271
14.5 Conclusions 272
References 273

15 Advanced EEG Signal Processing in Brain Death Diagnosis
Jianting Cao and Zhe Chen 275
15.1 Introduction 275

15.2　Background and EEG Recordings 276
　　　15.2.1 Diagnosis of Brain Death 276
　　　15.2.2 EEG Preliminary Examination and Diagnosis System 276
　　　15.2.3 EEG Recordings 278
15.3　EEG Signal Processing 279
　　　15.3.1 A Model of EEG Signal Analysis 280
　　　15.3.2 A Robust Prewhitening Method for Noise Reduction 280
　　　15.3.3 Independent Component Analysis 283
　　　15.3.4 Fourier Analysis and Time–Frequency Analysis 285
15.4　EEG Preliminary Examination with ICA 285
　　　15.4.1 Extracted EEG Brain Activity from Comatose Patients.... 286
　　　15.4.2 The Patients Without EEG Brain Activities 287
15.5　Quantitative EEG Analysis with Complexity Measures 288
　　　15.5.1 The Approximate Entropy 289
　　　15.5.2 The Normalized Singular Spectrum Entropy............. 290
　　　15.5.3 The C_0 Complexity 291
　　　15.5.4 Detrended Fluctuation Analysis 292
　　　15.5.5 Quantitative Comparison Results 292
　　　15.5.6 Classification 295
15.6　Conclusion and Future Study.................................. 296
References 297

16 Automatic Knowledge Extraction: Fusion of Human Expert Ratings and Biosignal Features for Fatigue Monitoring Applications

Martin Golz and David Sommer 299

16.1　Introduction 299
16.2　Fatigue Monitoring 301
　　　16.2.1 Problem 301
　　　16.2.2 Human Expert Ratings 302
　　　16.2.3 Experiments 303
　　　16.2.4 Feature Extraction 305
16.3　Feature Fusion and Classification 306
　　　16.3.1 Learning Vector Quantization 307
　　　16.3.2 Automatic Relevance Determination 308
　　　16.3.3 Support Vector Machines 309
16.4　Results 310
　　　16.4.1 Feature Fusion 310
　　　16.4.2 Feature Relevance 312
　　　16.4.3 Intra-Subject and Inter-Subject Variability 313
16.5　Conclusions and Future Work 314
References 315

Index 317

Contributors

Tülay Adalı
University of Maryland Baltimore
County, Baltimore
MD 21250, USA
adali@umbc.edu

Hamid Aghajan
Department of Electrical Engineering
Stanford University
CA, USA
hamid@wsnl.stanford.edu

Kazuyuki Aihara
Institute of Industrial Science
The University of Tokyo
153-8505, Japan
aihara@sat.t.u-tokyo.ac.jp

Joachim Bamberger
Siemens AG
Otto-Hahn-Ring 6
81730 Munich, Germany
joachim.bamberger@siemens.com

Bruno Betoni Parodi
Siemens AG
Otto-Hahn-Ring 6
81730 Munich, Germany
betoni@gmail.com

Olga Boric-Lubecke
Kai Sensors, Inc./University
of Hawaii, Honolulu
HI 96822, USA
olga@ieee.org

Christos Boukis
Athens Information Technology
Peania/Athens
19002, Greece
cbou@ait.edu.gr

Vince Calhoun
The MIND Institute/University
of New Mexico, 1101 Yale Boulevard
Albuquerque, NM 87131, USA
vcalhoun@unm.edu

Jianting Cao
Saitama Institute of Technology
Saitama
369-0293, Japan
cao@sit.ac.jp

Jonathon A. Chambers
Advanced Signal Processing Group
Loughborough University
Loughborough, UK
j.a.chambers@lboro.ac.uk

Zhe Chen
Massachusetts General Hospital
Harvard Medical School
Boston, MA 02114, USA
zhechen@neurostat.mgh.harvard.edu

Nikolay Chumerin
Katholieke Universiteit Leuven
Herestraat 49, bus 1021
B-3000 Leuven, Belgium
nikolay.chumerin@med.kuleuven.be

Andrzej Cichocki
Brain Science Institute
RIKEN, Saitama
351-0198, Japan
cia@brain.riken.jp

Anthony Constantinides
Imperial College London
Exhibition Road, London
SW7 2BT, UK
agc@imperial.ac.uk

Vanessa Su Lee Goh
Nederlandse Aardolie Maatschappij
B.V., PO Box 28000
9400 HH Assen, The Netherlands
Vanessa.Goh@shell.com

Martin Golz
University of Applied Sciences
Schmalkalden
Germany
m.golz@fh-sm.de

Uwe Hanebeck
Universitt Karlsruhe (TH)
Kaiserstr. 12
76131 Karlsruhe, Germany
uwe.hanebeck@ieee.org

Kai Heesche
Siemens AG
Otto-Hahn-Ring 6
81730 Munich, Germany
kai.heesche@siemens.com

Yoshito Hirata
Institute of Industrial Science
The University of Tokyo
153-8505, Japan
yoshito@sat.t.u-tokyo.ac.jp

Joachim Horn
Helmut-Schmidt-University/
University of the Federal Armed
Forces, 22043 Hamburg, Germany
joachim.horn@hsu-hh.de

Anders Høst-Madsen
Kai Sensors, Inc./University
of Hawaii, Honolulu
HI 96822, USA
ahm@hawaii.edu

Soroush Javidi
Imperial College London
Exhibition Road, London
SW7 2BT, UK
soroush.javidi@imperial.ac.uk

Beth Jelfs
Imperial College London
Exhibition Road, London
SW7 2BT, UK
beth.jelfs@imperial.ac.uk

Richard Kleihorst
NXP Semiconductor Research
Eindhoven
The Netherlands
Richard.Kleihorst@nxp.com

Anthony Kuh
University of Hawaii
Honolulu
HI 96822, USA
kuh@spectra.eng.hawaii.edu

Henning Lenz
Siemens AG
Oestliche Rheinbrueckenstr. 50
76187 Karlsruhe, Germany
henning.lenz@siemens.com

Wai Yie Leong
Agency for Science, Technology
and Research, (A*STAR) SIMTech
71 Nanyang Drive, Singapore 638075
waiyie@ieee.org

David Looney
Imperial College London
Exhibition Road, London
SW7 2BT, UK
david.looney06@imperial.ac.uk

Victor M. Lubecke
Kai Sensors, Inc./University of
Hawaii, Honolulu
HI 96822, USA
lubecke@ieee.org

Danilo P. Mandic
Imperial College London
Exhibition Road, London
SW7 2BT, UK
d.mandic@imperial.ac.uk

Yoichi Miyake
Graduate School of Advanced
Integration Science, Chiba University
263-8522, Japan
miyake@faculty.chiba-u.jp

Toshiya Nakaguchi
Graduate School of Advanced
Integration Science, Chiba University
263-8522, Japan
nakaguchi@faculty.chiba-u.jp

Dragan Obradovic
Siemens AG
Otto-Hahn-Ring 6
81730 Munich, Germany
dragan.obradovic@siemens.com

Nobutoshi Ojima
Kao Corporation
Japan
ojima.nobutashi@kao.co.jp

Byung-Kwon Park
University of Hawaii
Honolulu
HI 96822, USA
byungp@hawaii.edu

Nicolas Petrochilos
University of Hawaii
Honolulu
HI 96822, USA
petro@ahi.eng.hawaii.edu

Tomasz Rutkowski
Brain Science Institute
RIKEN, Saitama
351-0198, Japan
tomek@brain.riken.jp

Markus Schupfner
Harman/Becker Automotive Systems
GmbH, Moosacherstr. 48
80809 Munich, Germany
MSchupfner@harmanbecker.com

David Sommer
University of Applied Sciences
Schmalkalden
Germany
d.sommer@fh-sm.de

George Souretis
Imperial College London
Exhibition Road, London
SW7 2BT, UK
g.souretis@imperial.ac.uk

Hideyuki Suzuki
Institute of Industrial Science
The University of Tokyo
153-8505, Japan
hideyuki@sat.t.u-tokyo.ac.jp

Andrei Szabo
Siemens AG
Otto-Hahn-Ring 6
81730 Munich, Germany
andrei.szabo@siemens.com

Toshihisa Tanaka
Tokyo University of Agriculture
and Technology
Japan
tanakat@cc.tuat.ac.jp

Norimichi Tsumura
Graduate School of Advanced
Integration Science, Chiba University
263-8522, Japan
tsumura@faculty.chiba-u.jp

Marc M. Van Hulle
Katholieke Universiteit Leuven
Herestraat 49, bus 1021
B-3000 Leuven, Belgium
marc.vanhulle@med.kuleuven.be

Phebe Vayanos
Imperial College London
Exhibition Road, London
SW7 2BT, UK
foivi.vayanos@imperial.ac.uk

Hui Wang
Siemens AG
Otto-Hahn-Ring 6
81730 Munich, Germany
hui.wang.ext@siemens.com

Chen Wu
Department of Electrical
Engineering
Stanford University
CA, USA
chenwu@stanford.edu

Yonggang Zhang
Advanced Signal Processing
Group
Loughborough University
Loughborough, UK
Y.Zhang5@lboro.ac.uk

Qin Zhou
Broadcom Inc.
USA
lucy.qinzhou@gmail.com

Chaopin Zhu
Juniper Networks
Sunnyvale
CA 94089, USA
czhu@juniper.net

Part I

Collaborative Signal Processing Algorithms

1

Collaborative Adaptive Filters for Online Knowledge Extraction and Information Fusion

Beth Jelfs, Phebe Vayanos, Soroush Javidi, Vanessa Su Lee Goh, and Danilo P. Mandic

We present a method for extracting information (or knowledge) about the nature of a signal. This is achieved by employing recent developments in signal characterisation for online analysis of the changes in signal modality. We show that it is possible to use the fusion of the outputs of adaptive filters to produce a single collaborative hybrid filter and that by tracking the dynamics of the mixing parameter of this filter rather than the actual filter performance, a clear indication as to the nature of the signal is given. Implementations of the proposed hybrid filter in both the real $\mathbb{R}$ and the complex $\mathbb{C}$ domains are analysed and the potential of such a scheme for tracking signal nonlinearity in both domains is highlighted. Simulations on linear and nonlinear signals in a prediction configuration support the analysis; real world applications of the approach have been illustrated on electroencephalogram (EEG), radar and wind data.

1.1 Introduction

Signal modality characterisation is becoming an increasingly important area of multidisciplinary research and large effort has been put into devising efficient algorithms for this purpose. Research in this area started in mid-1990s but its applications in machine learning and signal processing are only recently becoming apparent. Before discussing characterisation of signal modalities certain key properties for defining the nature of a signal should be outlined [8, 21]:

1. Linear (strict definition) – A linear signal is generated by a linear time-invariant system, driven by white Gaussian noise.
2. Linear (commonly adopted) – Definition 1 is relaxed somewhat by allowing the distribution of the signal to deviate from the Gaussian one, which can be interpreted as a linear signal from 1. measured by a static (possibly nonlinear) observation function.

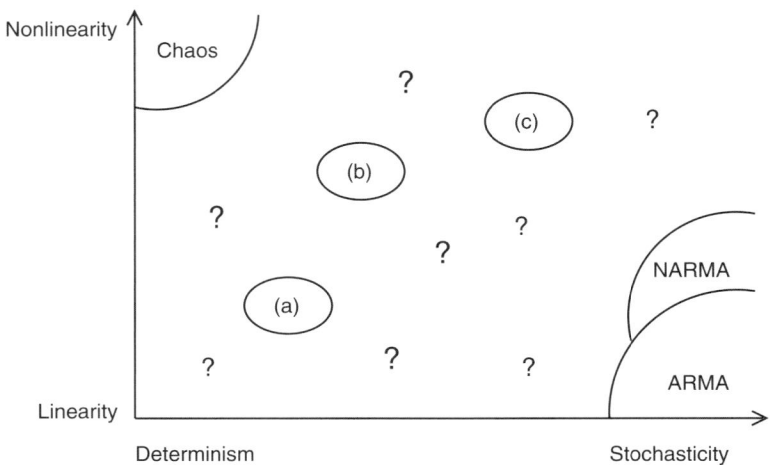

Fig. 1.1. Deterministic vs. stochastic nature or linear vs. nonlinear nature

3. Nonlinear – A signal that cannot be generated in the above way is considered nonlinear.
4. Deterministic (predictable) – A signal is considered deterministic if it can be precisely described by a set of equations.
5. Stochastic – A signal that is not deterministic.

Figure 1.1 (modified from [19]) illustrates the range of signals spanned by the characteristics of nonlinearity and stochasticity. While signals with certain characteristics are well defined, for instance chaotic signals (nonlinear and deterministic) or those produced by autoregressive moving average (ARMA) models (linear and stochastic signals), these represent only the extremes in signal nature and do not highlight the majority of signals which do not fit into such classifications. Due to the presence of such factors as noise or uncertainty, any real world signals are represented in the areas (a), (b), (c) or '?'; these are significant areas about which we know little or nothing. As changes in the signal nature between linear and nonlinear and deterministic and stochastic can reveal information (knowledge) which is critical in certain applications (e.g., health conditions) the accurate characterisation of the nature of signals is a key prerequisite prior to choosing a signal processing framework.

The existing algorithms in this area are based on hypothesis testing [6, 7, 20] and describe the signal changes in a statistical manner. However, there are very few online algorithms which are suitable for this purpose. The purpose of the approach described in this chapter is to introduce a class of online algorithms which can be used not only to identify, but also to track changes in the nature of the signal (signal modality detection).

One intuitive method to determine the nature of a signal has been to present the signal as input to two adaptive filters with different characteristics, one nonlinear and the other linear. By comparing the responses of each filter, this can be used to identify whether the input signal is linear or not. While this is a very useful simple test for signal nonlinearity, it does not provide an online solution. There are additional ambiguities due to the need to choose many parameters of the corresponding filters and this approach does not rely on the "synergy" between the filters considered.

1.1.1 Previous Online Approaches

In [17] an online approach is considered which successfully tracks the degree of nonlinearity of a signal using adaptive algorithms, but relies on a parametric model to effectively model the system to provide a true indication of the degree of nonlinearity. Figure 1.2 shows an implementation of this method using a third-order Volterra filter and the normalised least mean square (NLMS) algorithm with a step size $\mu = 0.008$ to update the system parameters. The system input and output can be described by

$$u[k] = \sum_{i=0}^{I} a_i x[k-i] \text{ where } I = 2 \text{ and } a_0 = 0.5, \ a_1 = 0.25, \ a_2 = 0.125,$$
$$y[k] = F(u[k]; k) + \eta[k], \tag{1.1}$$

where $x[k]$ are i.i.d uniformly distributed over the range $[-0.5, 0.5]$ and $\eta[k] \sim \mathcal{N}(0, 0.0026)$. The function $F(u[k]; k)$ varies with k

$$F(u[k]; k) = \begin{cases} u^3[k] & \text{for } 10{,}000 < k \leq 20{,}000, \\ u^2[k] & \text{for } 30{,}000 < k \leq 40{,}000, \\ u[k] & \text{at all other times}. \end{cases} \tag{1.2}$$

The output $y[k]$ can be seen in the first trace of Fig. 1.2, the second and third traces show the residual estimation errors of the optimal linear system and

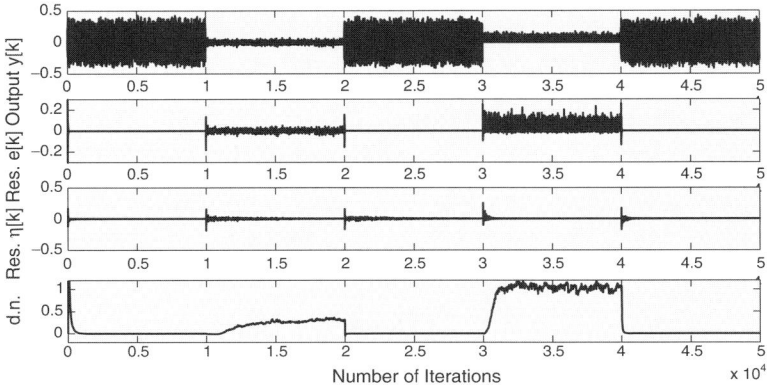

Fig. 1.2. Estimated degree of signal nonlinearity for an input alternating from linear to nonlinear

Volterra system, respectively, the final trace is the estimated degree of signal nonlinearity. While these results show that this approach can detect changes in nonlinearity and is not affected by the presence of noise, this may be largely due to the nature of the input signal in question being particularly suited to the Volterra model.

This type of method relies on the nature of the nonlinearity under observation being suited to the actual signal model; in real world situations it is not always possible to know the nonlinearity in advance, therefore their application is limited. To overcome these limitations, we propose a much more flexible method based on collaborative adaptive filtering.

1.1.2 Collaborative Adaptive Filters

Developing on the well-established tracking capabilities of adaptive filters using combinations of adaptive subfilters in a more natural way produces a single hybrid filter without the need for any knowledge of underlying signal generation models. Hybrid filters consist of multiple individual adaptive subfilters operating in parallel and all feeding into a mixing algorithm which produces the single output of the filter [4, 13]. The mixing algorithms are also adaptive and combine the outputs of each subfilter based on the estimate of their current performance on the input signal from their instantaneous output error.

Many previous applications of hybrid filters have focused mainly on the improved performance they can offer over the individual constituent filters. Our aim is to focus on one additional effect of the mixing algorithm, that is, to show whether it can give an indication of which filter is currently responding to the input signal most effectively. Therefore, intuitively by selecting algorithms which are particularly suited to one type of input signals, it is possible to cause the mixing algorithm to adapt according to fundamental properties of the input signal.

A simple form of mixing algorithm for two adaptive filters is a convex combination. Convexity can be described as [5]

$$\lambda x + (1 - \lambda)y \quad \text{where} \quad \lambda \in [0, 1]. \tag{1.3}$$

For x and y being two points on a line, as shown in Fig. 1.3, their convex mixture (1.3) will lie on the same line between x and y.

For convex mixing of the outputs of adaptive filters, it is intuitively clear that initially λ will adapt to favour the faster filter (that is the filter with faster learning rate) and following convergence it will favour the filter with

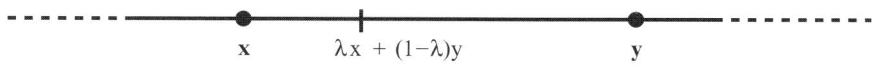

Fig. 1.3. Convexity

better steady-state properties[1]; should one of the subfilters fail to converge, the values of λ adapt such that the hybrid filter follows the stable subfilter [16]. The approach in this chapter focuses on observing the dynamics of mixing parameter λ, to allow conclusions to be drawn about the current nature of the input signal.

1.2 Derivation of The Hybrid Filter

Unlike the existing approaches to hybrid adaptive filters which focus on the quantitative performance of such filters, in this case the design of the hybrid filters is such that it should combine the characteristics of two distinctly different adaptive filters. Signal modality characterisation is achieved by making the value of the "mixing" parameter λ adapt according to the fundamental dynamics of the input signal. In this chapter we illustrate applications of this method for characterisation of nonlinearity and complexity on both synthetic and real world data, but this method can be equally well applied to any other signal characteristics. With that in mind we start from the general derivation of the convex hybrid filter before moving on to specific implementations.

Figure 1.4 shows the block diagram of a hybrid filter consisting of two adaptive filters combined in a convex manner. At every time instant k, the output of the hybrid filter, $y(k)$, is an adaptive convex combination of the output of the first subfilter $y_1(k)$ and the output of the second subfilter $y_2(k)$, and is given by

$$y(k) = \lambda(k)y_1(k) + (1 - \lambda(k))\, y_2(k), \qquad (1.4)$$

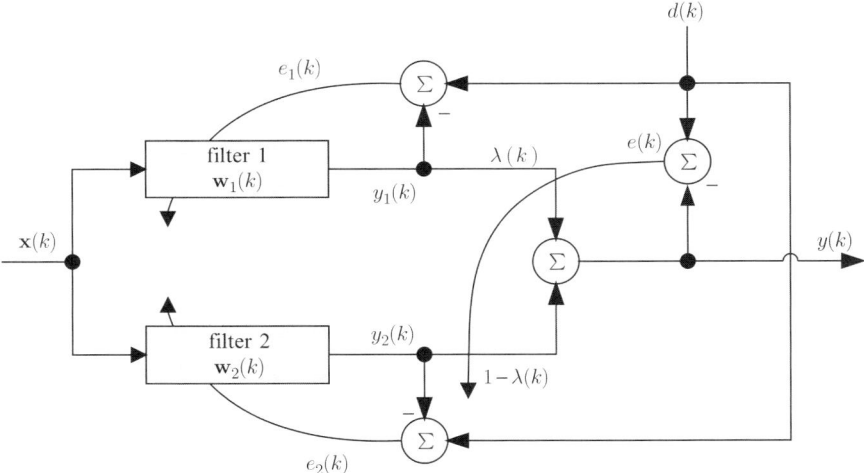

Fig. 1.4. Convex combination of adaptive filters (hybrid filter)

---

[1] Unlike traditional search then converge approaches this method allows for potentially nonstationary data.

where $y_1(k) = \mathbf{x}^\mathrm{T}(k)\mathbf{w}_1(k)$ and $y_2(k) = \mathbf{x}^\mathrm{T}(k)\mathbf{w}_2(k)$ are the outputs of the two subfilters with corresponding weight vectors $\mathbf{w}_1(k) = [w_{1,1}(k), \ldots, w_{1,N}(k)]^\mathrm{T}$ and $\mathbf{w}_2(k) = [w_{2,1}(k), \ldots, w_{2,N}(k)]^\mathrm{T}$ which are dependent on the algorithms used to train the subfilters based on the common input vector $\mathbf{x}(k) = [x_1(k), \ldots, x_N(k)]^\mathrm{T}$ for filters of length N.

To preserve the inherent characteristics of the subfilters, which are the basis of our approach, the constituent subfilters are updated by their own errors $e_1(k)$ and $e_2(k)$, using a common desired signal $d(k)$, whereas the parameter λ is updated based on the overall error $e(k)$. The convex mixing parameter $\lambda(k)$ is updated based on minimisation of the quadratic cost function $E(k) = \frac{1}{2}e^2(k)$ using the following gradient adaptation:

$$\lambda(k+1) = \lambda(k) - \mu_\lambda \nabla_\lambda E(k)_{|\lambda=\lambda(k)}, \qquad (1.5)$$

where μ_λ is the adaptation step-size. From (1.4) and (1.5), using an LMS type adaptation, the λ update can be obtained as

$$\lambda(k+1) = \lambda(k) - \frac{\mu_\lambda}{2}\frac{\partial e^2(k)}{\partial \lambda(k)} = \lambda(k) + \mu_\lambda e(k)(y_1(k) - y_2(k)). \qquad (1.6)$$

To ensure the combination of adaptive filters remains a convex function, it is critical that λ remains within the range $0 \leq \lambda(k) \leq 1$. In [4] the authors obtained this through the use of a sigmoid function as a post-nonlinearity to bound $\lambda(k)$. Since, to determine the changes in the modality of a signal, we are not interested in the overall performance of the filter but in the dynamics of parameter λ, the use of a sigmoid function would interfere with true values of $\lambda(k)$ and was therefore not appropriate. In this case a hard limit on the set of allowed values for $\lambda(k)$ was therefore implemented.

1.3 Detection of the Nature of Signals: Nonlinearity

Implementations of the hybrid filter described above using the LMS algorithm [23] to train one of the subfilters and the generalised normalised gradient descent (GNGD) algorithm [15] for the other, have been used to distinguish the linearity/nonlinearity of a signal [11]. The LMS algorithm was chosen as it is widely used, known for its robustness and excellent steady-state properties whereas the GNGD algorithm has a faster convergence speed and better tracking capabilities. By exploiting these properties it is possible to show that due to the synergy and simultaneous mode of operation, the hybrid filter has excellent tracking capabilities for signals with extrema in their inherent linearity and nonlinearity characteristics.

The output of the LMS trained subfilter y_LMS is generated from [23]

$$\begin{aligned} y_\mathrm{LMS}(k) &= \mathbf{x}^\mathrm{T}(k)\mathbf{w}_\mathrm{LMS}(k), \\ e_\mathrm{LMS}(k) &= d(k) - y_\mathrm{LMS}(k), \\ \mathbf{w}_\mathrm{LMS}(k+1) &= \mathbf{w}_\mathrm{LMS}(k) + \mu_\mathrm{LMS} e_\mathrm{LMS}(k)\mathbf{x}(k) \end{aligned} \qquad (1.7)$$

and y_{GNGD} is the corresponding output of the GNGD trained subfilter given by [15]

$$y_{\text{GNGD}}(k) = \mathbf{x}^{\text{T}}(k)\mathbf{w}_{\text{GNGD}}(k) \qquad (1.8)$$
$$e_{\text{GNGD}}(k) = d(k) - y_{\text{GNGD}}(k)$$
$$\mathbf{w}_{\text{GNGD}}(k+1) = \mathbf{w}_{\text{GNGD}}(k) + \frac{\mu_{\text{GNGD}}}{\|\mathbf{x}(k)\|_2^2 + \varepsilon(k)} e_{\text{GNGD}}(k)\mathbf{x}(k)$$
$$\varepsilon(k+1) = \varepsilon(k) - \rho\mu_{\text{GNGD}} \frac{e_{\text{GNGD}}(k)e_{\text{GNGD}}(k-1)\mathbf{x}^{\text{T}}(k)\mathbf{x}(k-1)}{\left(\|\mathbf{x}(k-1)\|_2^2 + \varepsilon(k-1)\right)^2}$$

where the step-size parameters of the filters are μ_{LMS} and μ_{GNGD}, and in the case of the GNGD ρ is the step-size adaptation parameter and ε the regularisation term.

By evaluating the resultant hybrid filter in an adaptive one-step ahead prediction setting with the length of the adaptive filters set to $N = 10$, it is possible to illustrate the ability of the hybrid filter to identify the modality of a signal of interest. The behaviour of λ has been investigated for benchmark synthetic linear and nonlinear inputs. Values of λ were averaged over a set of 1,000 independent simulation runs, for the inputs described by a stable linear AR(4) process:

$$x(k) = 1.79x(k-1) - 1.85x(k-2) + 1.27x(k-3) - 0.41x(k-4) + n(k) \quad (1.9)$$

and a benchmark nonlinear signal [18]:

$$x(k+1) = \frac{x(k)}{1+x^2(k)} + n^3(k), \qquad (1.10)$$

where $n(k)$ is a zero mean, unit variance white Gaussian process. The values of the step-sizes used were $\mu_{\text{LMS}} = 0.01$ and $\mu_{\text{GNGD}} = 0.6$. For the GNGD filter $\rho = 0.15$ and the initial value of the regularisation parameter was $\varepsilon(0) = 0.1$. Within the convex combination of the filters, filter 1 corresponds to the GNGD trained subfilter and filter 2 to the LMS trained subfilter, the step-size for the adaptation of $\lambda(k)$ was $\mu_\lambda = 0.05$ and the initial value[2] of $\lambda(0) = 1$.

From the curves shown in Fig. 1.5 it can be seen the value of $\lambda(k)$ for both inputs moves towards zero as the adaptation progresses. As expected, the output of the convex combination of adaptive filters approaches the output of the LMS filter y_{LMS} predominately. This is due to the better steady-state properties of the LMS filter when compared to the GNGD filter, which due to its constantly 'alert' state does not settle in the steady state as well as the LMS. In the early stages of adaptation, the nonlinear input (1.10) adapts to become dominated by the LMS filter much faster than the linear input and

---

[2] Since GNGD exhibits much faster convergence than LMS, it is natural to start the adaptation with $\lambda(0) = 1$. This way, we avoid possible artefacts that may arise due to the slow initial response to the changes in signal modality.

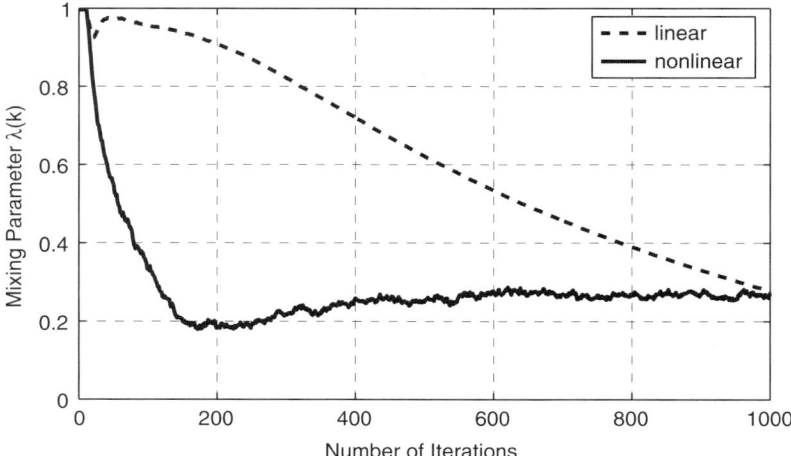

Fig. 1.5. Comparison of the mixing parameter λ for linear and nonlinear inputs

rapidly converges, whereas the linear input (1.9) changes much more gradually between the two filters.[3]

1.3.1 Tracking Changes in Nonlinearity of Signals

It is also possible to use changes in λ along the adaptation to track the changes in signal modality. Since the behaviour of λ as a response to the different inputs is clearly distinct, especially in the earliest stages of adaptation, the convex combination was presented with an input signal which alternated between linear (1.9) and nonlinear (1.10). The input signal was alternated every 200 samples and the corresponding dynamics of the mixing parameter $\lambda(k)$ are shown in Fig. 1.6. From Fig. 1.6 it is clear that the value of $\lambda(k)$ adapts in a way which ensures that the output of the convex combination is dominated by the filter most appropriate for the input signal characteristics.

To illustrate the discrimination ability of the proposed approach, the next set of simulations shows the results of the same experiment as in Fig. 1.6, but for a decreased number of samples between the alternating segments of data. Figure 1.7 shows the response of $\lambda(k)$ to the input signal alternating every 100 and 50 samples, respectively. There is a small anomaly in the values of λ immediately following the change in input signal from nonlinear to linear, which can be clearly seen in Fig. 1.7 around sample numbers $100i$, $i = 1, 2, \ldots$, where the value of λ exhibits a small dip before it increases. This is due to the fact that the input to both the current AR process (1.9) and the tap inputs to both filters use previous nonlinear samples where we are in fact predicting the

---

[3] Both filters perform well on a linear input and are competing along the adaptation.

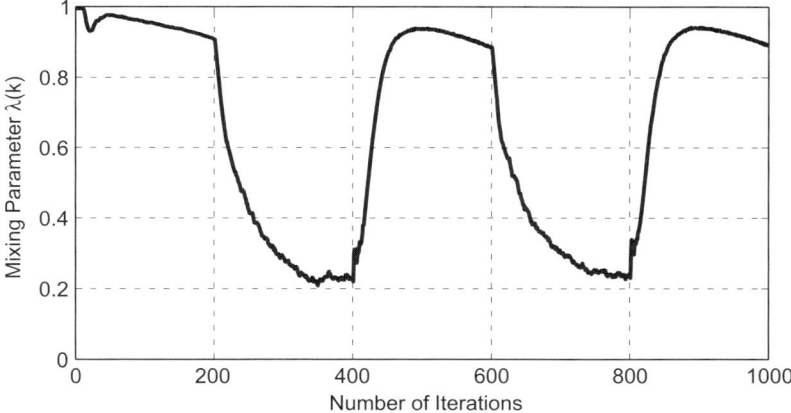

Fig. 1.6. Evolution of the mixing parameter λ for input nature alternating every 200 samples

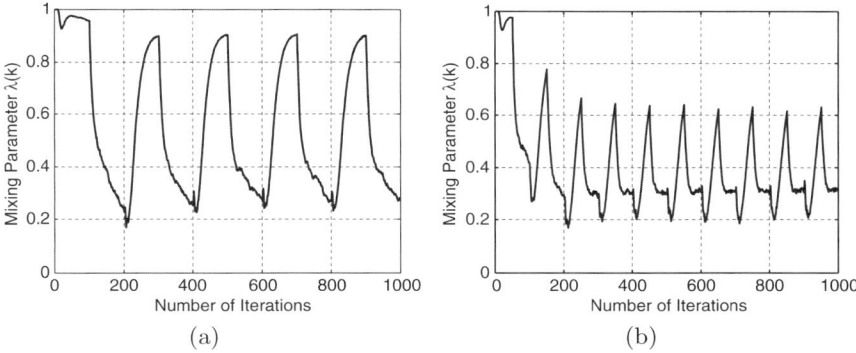

Fig. 1.7. Evolution of the mixing parameter λ for a signal with input nature alternating between linear to nonlinear. (**a**) Input signal nature alternating every 100 samples and (**b**) input signal nature alternating every 50 samples

first few "linear" samples. This does not become an issue when alternations between the input signals occur less regularly or if there is a more natural progression from "linear" to "nonlinear" in the the input signal.

Real World Applications

To examine the usefulness of this approach for the processing of real world signals, a set of EEG signals has been analysed. Following the standard practice, the EEG sensor signals were averaged across all the channels and any trends in the data were removed. Figure 1.8 shows the response of λ when applied to two different sets of EEG data from epileptic patients, both showing the

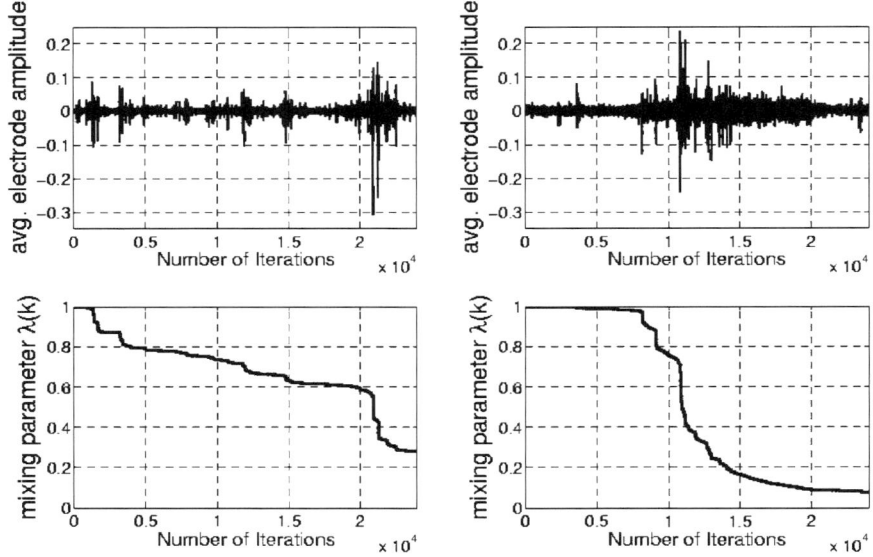

Fig. 1.8. *Top panel*: EEG signals for two patients showing epileptic seizures. *Bottom panel*: corresponding adaptations of λ

onset of a seizure as indicated by a sudden change in the value of λ. These results show that this approach can effectively detect changes in the nature of the EEG signals which can be very difficult to achieve otherwise.

1.4 Detection of the Nature of Signals: Complex Domain

For generality, building upon identification and tracking of nonlinearity in the real domain $\mathbb{R}$, we shall extend this to the complex domain $\mathbb{C}$. To facilitate this, the update of λ (1.6) was extended to the complex domain, resulting in

$$\begin{aligned}\lambda(k+1) &= \lambda(k) - \frac{\mu_\lambda}{2}\left\{e(k)\frac{\partial e^*(k)}{\partial \lambda(k)} + e^*(k)\frac{\partial e(k)}{\partial \lambda(k)}\right\} \\ &= \lambda(k) + \mu_\lambda \Re\left\{e(k)(y_1(k) - y_2(k))^*\right\},\end{aligned} \quad (1.11)$$

where $(\cdot)^*$ denotes the complex conjugation operator. For the complex version of the hybrid convex combination, the subfilters previously discussed were substituted with the complex NLMS and complex normalised nonlinear gradient descent (NNGD) [14], in this case the normalised versions were used as opposed to the standard complex LMS (CLMS) and NGD (CNGD) to overcome problems with the convergence of the individual subfilters and hence dependence on the combination of input signal statistics. The CLMS update is given by [22]

$$\mathbf{w}_{\mathrm{CLMS}}(k+1) = \mathbf{w}_{\mathrm{CLMS}}(k) + \eta_{\mathrm{CLMS}}(k)e_{\mathrm{CLMS}}(k)\mathbf{x}^*(k), \quad (1.12)$$

where η denotes the learning rate which for the CLMS is $\eta_{\text{CLMS}}(k) = \mu_{\text{CLMS}}$ and for the CNLMS and $\eta_{\text{CLMS}}(k) = \mu_{\text{CLMS}}/(\|\mathbf{x}(k)\|_2^2 + \varepsilon)$.

The CNGD is described by

$$e_{\text{CNGD}}(k) = d(k) - \text{net}(k),$$
$$\text{net}(k) = \Phi\left(\mathbf{x}^{\text{T}}(k)\mathbf{w}_{\text{CNGD}}(k)\right),$$
$$\mathbf{w}_{\text{CNGD}}(k+1) = \mathbf{w}_{\text{CNGD}}(k) - \eta_{\text{CNGD}}(k)\nabla_{\mathbf{w}}E(k), \quad (1.13)$$

where $\text{net}(k)$ is the net input, $\Phi(\cdot)$ denotes the complex nonlinearity and $E(k)$ is the cost function given by

$$E(k) = \frac{1}{2}|e(k)|^2. \quad (1.14)$$

Following the standard complex LMS derivation [22] for a fully complex nonlinear activation function (AF), Φ, the weight update is expressed as

$$\mathbf{w}_{\text{CNGD}}(k+1) = \mathbf{w}_{\text{CNGD}}(k) + \eta_{\text{CNGD}}(k)e_{\text{CNGD}}(k)\left(\Phi'\left[\text{net}(k)\right]\right)^*\mathbf{x}^*(k) \quad (1.15)$$

where $(\cdot)'$ denotes the complex differentiation operator and $\eta_{\text{CNGD}} = \mu_{\text{CNGD}}$ the CNGD and $\eta_{\text{CNGD}} = \mu_{\text{CNGD}}/(C + [\Phi'(\text{net}(k))]^2\|\mathbf{x}\|_2^2)$ denotes CNNGD.

For the purposes of tracking changes in the nonlinearity of signals, the hybrid filter was again presented with an input signal alternating between linear and nonlinear. The process $n(k)$ in the linear AR(4) signal (1.9) was replaced with a complex white Gaussian process again with zero mean and unit variance,

$$n(k) = n_{\text{r}}(k) + \text{j}n_{\text{i}}(k),$$

where the real and imaginary components of n are mutually independent sequences having equal variances so that $\sigma_n^2 = \sigma_{n_{\text{r}}}^2 + \sigma_{n_{\text{i}}}^2$. The complex benchmark nonlinear signal [18] was

$$x(k) = \frac{x^2(k-1)\left(x(k-1)+2.5\right)}{1+x^2(k-1)+x^2(k-2)} + n(k-1) \quad (1.16)$$

and the nonlinearity used was the sigmoid function

$$\Phi(z) = \frac{1}{1+e^{-z}}, \quad \text{where } z \in \mathbb{C} \quad (1.17)$$

Figure 1.9 shows the response of λ to the input signal alternating every 200 and every 100 samples and again the hybrid filter was clearly capable of tracking such changes in the nonlinearity of the input signal.

1.4.1 Split-Complex vs. Fully-Complex

Whilst being able to identify the nonlinearity of a signal is important and can give key knowledge about the signal under observation, within nonlinear

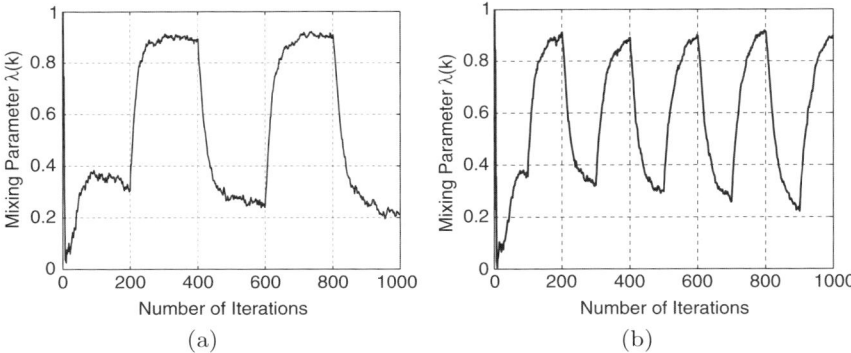

Fig. 1.9. Evolution of the mixing parameter λ for a signal with input nature alternating between linear to nonlinear. (**a**) Input signal nature alternating every 200 samples and (**b**) input signal nature alternating every 100 samples

adaptive filtering in $\mathbb{C}$ one of the biggest problems is the choice of nonlinear complex AF. There are three main methods to deal with this:

- Processing the real and imaginary components separately using a real nonlinearity
- Processing in the complex domain using a so-called "split-complex" nonlinearity
- Or using a so-called "fully-complex" nonlinearity

A fully-complex nonlinearity is a function $f : \mathbb{C} \to \mathbb{C}$ and are the most efficient in using higher order statistics within a signal [12]. For a split-complex function the real and imaginary components of the input are separated and fed through the dual real valued AF $f_R(x) = f_I(x)$, $x \in \mathbb{R}$. A split complex AF can be represented as

$$\Phi_{\text{split}}(z) = f_R(z^r) + jf_I(z^i) = u(z^r) + jv(z^i) \tag{1.18}$$

Algorithms using split complex AFs have been shown to give good results. However due to their reliance on the real and imaginary weight updates being mutually exclusive, they are not suitable when the real and imaginary components of a signal are strongly correlated.

Consider the Ikeda map, a well-known benchmark signal in chaos theory [2], given by

$$\begin{aligned} x(k+1) &= 1 + u\left[x(k)\cos t(k) - y(k)\sin t(k)\right], \\ y(k+1) &= u\left[x(k)\sin t(k) + y(k)\cos t(k)\right], \end{aligned} \tag{1.19}$$

where u is a parameter and

$$t(k) = 0.4 - \frac{6}{1 + x^2(k) + y^2(k)}. \tag{1.20}$$

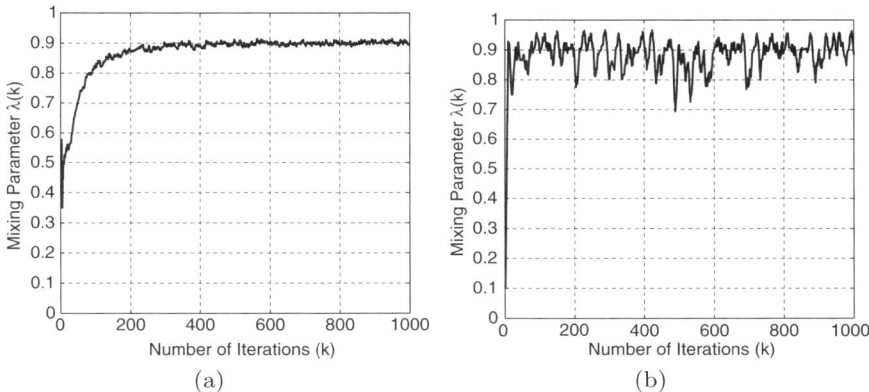

Fig. 1.10. Evolution of the mixing parameter λ for prediction of Ikeda map. (**a**) Nonlinear vs. linear and (**b**) fully- vs. split-complex

Figure 1.10a illustrates that the hybrid filter can clearly identify the Ikeda map as nonlinear when presented with it as an input. It is natural, however, to expect that as the signal generation mechanism is in the form of coupled difference equations, by representing the pair $[x(n), y(n)]$ as a vector in $\mathbb{C}$, the Ikeda map (1.19) will represent a fully complex signal. This is indeed confirmed by the simulation results shown in Fig. 1.10b where to test the application of the hybrid filter method for detection of the nature of nonlinear complex signals, the hybrid filter consisted of a combination of a fully-complex and a split-complex subfilter trained by the CNGD algorithm with $\lambda = 1$ corresponding to the fully-complex subfilter. As expected (by design), from Fig. 1.10, the Ikeda map is a nonlinear signal (Fig. 1.10a) which exhibits fully-complex nonlinear properties (Fig. 1.10b).

To illustrate this further, Fig. 1.11 shows the performance of the complex real time recurrent learning (CRTRL) algorithm [9] for both split- and fully-complex learning on the prediction of the Ikeda map; observe that the split-complex CRTRL did not respond as well as the fully-complex version to prediction of the Ikeda map.

Knowledge of the complex nonlinearity that describes a real-world complex signal is critical, as it can help us to understand the nature of the dynamics of the system under observation (radar, sonar, vector fields). To illustrate the tracking capabilities of this hybrid filter, the filter was presented with real world radar data. The radar data comes from a maritime radar (IPIX, publicly available from [1]), for different sea states, "low" (calm sea) and "high" (turbulent) states. While there are off-line statistical tests for radar data [10] and it has been shown that radar data is predominantly fully complex in nature when the target is in the beam [8], on-line estimation algorithms are lacking and it is clear that it is important to track the onset of changes in the nature while recording. Figure 1.12 shows the evolution of λ when predicting radar data that was alternated every 50 samples from the low sea state to

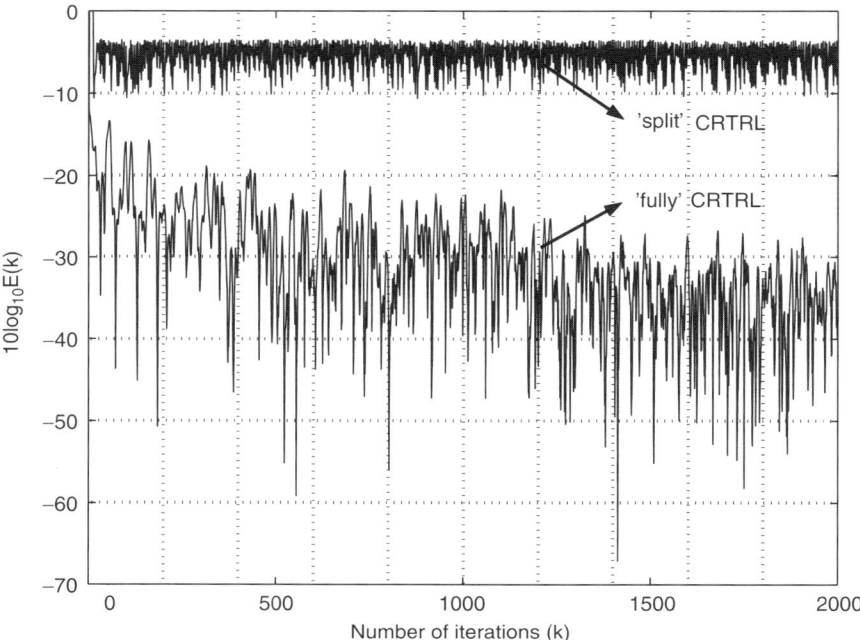

Fig. 1.11. Learning curves for fully-complex vs. split-complex prediction of Ikeda map

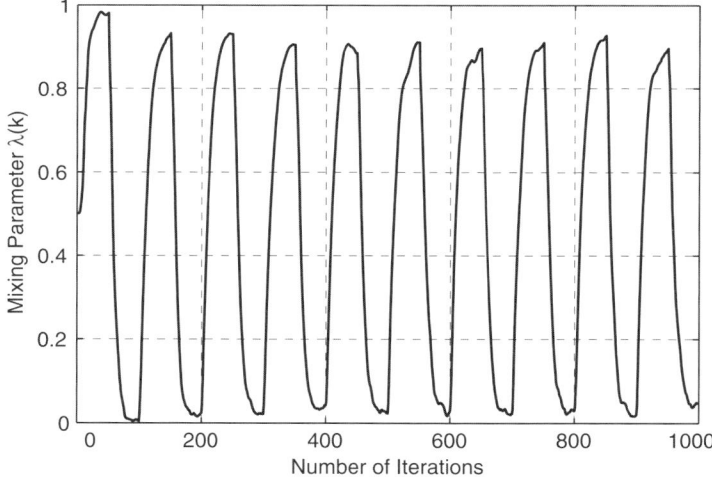

Fig. 1.12. Evolution of the mixing parameter λ for alternating blocks of 400 data samples of radar data from the "low" to "high" sea state

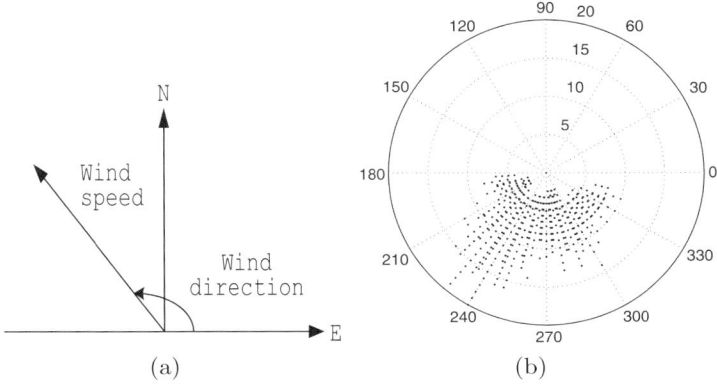

Fig. 1.13. Wind recordings as complex [speed,direction] quantities. (**a**) Wind vector representation and (**b**) wind rose representation

the high sea state. All the data sets were standardised, so all the magnitudes were in the range $[-1, 1]$. The initial weight vectors were set to zero and the filter order $N = 10$. Figure 1.12 shows that the modality of the high sea state was predominantly fully complex and similarly the low sea state was predominantly split complex.

1.4.2 Complex Nature of Wind

Wind modelling is an illustration for the need for complex valued techniques; Fig. 1.13a represents a wind rose plot of direction vs. magnitude and shows the need for wind to be modelled based on both direction and speed. Wind is normally measured either as a bivariate process of these measurements [3] or, despite the clear interdependence between the components, only the speed component in taken into account. From Fig. 1.13b it is clear that wind could also be represented as a vector of speed and direction components in the North–East coordinate system. Following this, the wind vector $\mathbf{v}(k)$ can be represented in the complex domain $\mathbb{C}$, as

$$\mathbf{v}(k) = |\mathbf{v}(k)|e^{j\theta(k)} = v_{\mathrm{E}}(k) + jv_{\mathrm{N}}(k), \tag{1.21}$$

where v is the speed component and θ the direction, modelled as a single complex quantity.

Complex Surrogate Data Testing for the Nature of Wind

Following the approach from [8], to support the complex-valued modelling of wind we first employ a statistical test for the complex nature of wind. The test is based on the complex-valued surrogate data analysis and is set within

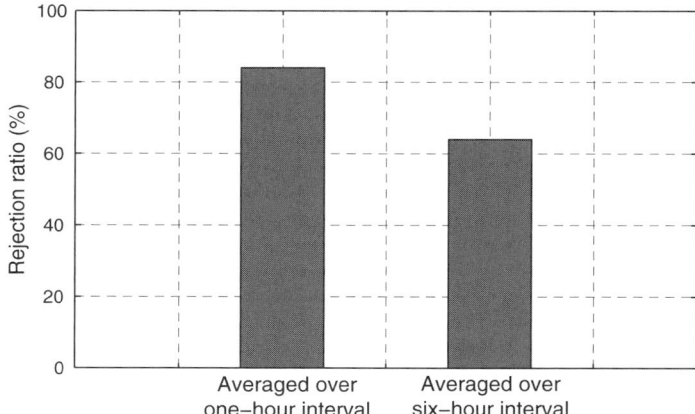

Fig. 1.14. Complex valued surrogate data test for the complex nature of wind signal (speed and direction)

the framework of hypothesis testing and the statistical testing methodology from [7] is adopted. The signals are characterised by the delay vector variance (DVV) method and the null hypothesis was that the original signal is complex-valued. Figure 1.14 shows the results of this test indicating there is a significant component dependence within the complex-valued wind signal representation. This is indicated by the rejection ratio of the null hypothesis of fully complex wind data being significantly greater than zero.

The results from Fig. 1.14 show the proposed test repeated 100 times, with the number of times the null hypothesis was rejected computed. The wind signal was averaged over either 1-h intervals or 6-h intervals and as can be seen from the rejection ratios, the signal averaged over 1 h showed a stronger indication of having a complex nature than those averaged over 6 h. Therefore, the components of complex-valued wind signals become more dual univariate and linear when averaged over longer intervals, this is in line with results from probability theory where a random signal becomes more Gaussian (and therefore linear) with the increase in the degree of averaging.

Tracking Modality of Wind Data

As it has been shown that wind can be considered a complex valued quantity and that it is possible to track the nature of complex signals using the hybrid filter combination of a fully complex and a split complex adaptive filter, the hybrid filter was used to predict a set of wind data. The data used was measurements of the wind in an urban area over one day. The filter length was set to $N = 10$, the learning rates of the split and fully complex NGD algorithms were $\mu_{\text{split}} = 0.01$ and $\mu_{\text{fully}} = 0.01$ and the step size of the learning parameter was $\mu_\lambda = 0.5$. The results of this can be seen in Fig. 1.15, as for the majority

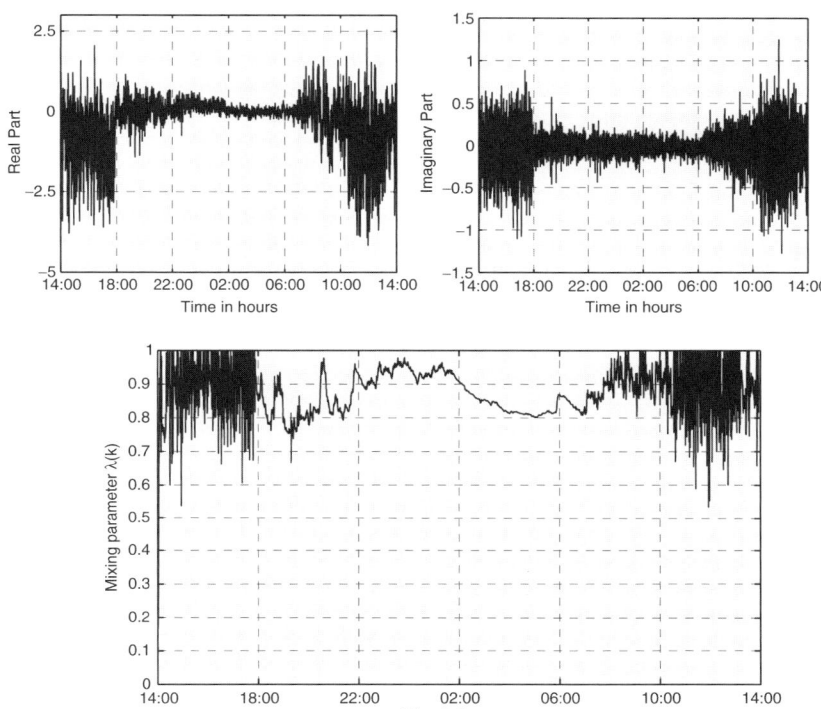

Fig. 1.15. Evolution of the mixing parameter λ for the prediction of wind

of the time the value of λ is around 0.9 the wind signal can be considered mainly fully complex. It is also clear that the first and last measurements are more unstable in nature as λ oscillated in the range $[0.5, 0.9]$, this indicates the intermittent wind nature was mainly fully complex but does at times become more split complex. Since these measurements were taken from a 24 h period starting from 14:00 these sections correspond to the recordings taken between 14:00–18:00 and 08:00–14:00 the next day, this is to be expected as during these times the wind is changing rapidly compared to the "calm" period in the late evening and the early morning.

1.5 Conclusions

We have proposed a novel approach to identify changes in the modality of a signal. This is achieved by a convex combination of two adaptive filters for which the transient responses are significantly different. By training the two filters with different algorithms, it is possible to exploit the difference in the performance capabilities of each. The evolution of the adaptive convex mixing parameter λ, helps determine which filter is more suited to the current input signal dynamics, and thereby gain information about the nature of the signal. This way, information fusion is achieved by collaborative modular learning, suitable for the online mode of operation. The analysis and simulations illustrate that there is significant potential for the use of this method for online tracking of some fundamental properties of the input signal. Both synthetic and real world examples on EEG, radar and wind data support the analysis. The extension to the simultaneous tracking of several parameters follows naturally; this can be achieved by a hierarchical and distributed structure of hybrid filters.

Acknowledgements

We wish to thank Prof. Kazuyuki Aihara and Dr. Yoshito Hirata from the Institute of Industrial Science, University of Tokyo, Japan for providing the wind data sets. We also thank Dr. Mo Chen from Imperial College London for helping with the surrogate data testing.

References

1. The McMaster IPIX radar sea clutter database. http://soma.crl.mcmaster.ca/ipix/
2. Aihara, K. (ed.): Applied Chaos and Applicable Chaos. Tokyo: Science–Sha (1994)
3. Alexiadis, M.C., Dokopoulos, P.S., Sahsamanoglou, H.S., Manousaridis, I.M.: Short-term forecasting of wind speed and related electrical power. Solar Energy **63**(1), 61–68 (1998)
4. Arenas-Garcia, J., Figueiras-Vidal, A., Sayed, A.: Steady state performance of convex combinations of adaptive filters. In: Proceedings IEEE International Conference on Acoustics, Speech and Signal Processing (ICASSP '05), vol. 4, pp. 33–36 (2005)
5. Cichocki, A., Unbehauen, R.: Neural Networks for Optimisation and Signal Processing. Wiley, New York (1993)
6. Gautama, T., Mandic, D., Hulle, M.V.: Signal nonlinearity in fMRI: a comparison between BOLD and MION. IEEE Transactions on Medical Imaging, **22**(5), 636–644 (2003)
7. Gautama, T., Mandic, D., Hulle, M.V.: The delay vector variance method for detecting determinism and nonlinearity in time series. Physica D **190**(3–4), 167–176 (2004)
8. Gautama, T., Mandic, D., Hulle, M.V.: On the characterisaion of the deterministic/stochastic and linear/nonlinear nature of time series. Technical report dpm-04-5, Imperial College (2004)
9. Goh, S.L., Mandic, D.P.: A general complex valued RTRL algorithm for nonlinear adaptive filters. Neural Computation **16**(12), 2699–2731 (2004)
10. Haykin, S., Principe, J.: Making sense of a complex world [chaotic events modeling]. IEEE Signal Processing Magazine **15**(3), 66–81 (1998)

11. Jelfs, B., Vayanos, P., Chen, M., Goh, S.L., Boukis, C., Gautama, T., Rutkowski, T., Kuh, T., Mandic, D.: An online method for detecting nonlinearity within a signal. In: Proceedings 10th International Conference on Knowledge-Based and Intelligent Information and Engineering Systems KES2006, vol. 3, pp. 1216–1223 (2006)
12. Kim, T., Adali, T.: Approximation by fully complex multilayer perceptrons. Neural Computation **15**(7), 1641–1666 (2003)
13. Kozat, S., Singer, A.: Multi-stage adaptive signal processing algorithms. In: Proceedings of the IEEE Sensor Array and Multichannel Signal Processing Workshop, pp. 380–384 (2000)
14. Mandic, D.: NNGD algorithm for neural adaptive filters. Electronics Letters **36**(9), 845–846 (2000)
15. Mandic, D.: A generalized normalized gradient descent algorithm. IEEE Signal Processing Letters **11**(2), 115–118 (2004)
16. Mandic, D., Vayanos, P., Boukis, C., Jelfs, B., Goh, S., Gautama, T., Rutkowski, T.: Collaborative adaptive learning using hybrid filters. In: Proceedings of the IEEE International Conference on Acoustics, Speech and Signal Processing, ICASSP 2007, vol. 3, pp. 921–924 (2007)
17. Mizuta, H., Jibu, M., Yana, K.: Adaptive estimation of the degree of system nonlinearity. In: Proceedings IEEE Adaptive Systems for Signal Processing and Control Symposium (AS-SPCC), pp. 352–356 (2000)
18. Narendra, K., Parthasarathy, K.: Identification and control of dynamical systems using neural networks. IEEE Transactions on Neural Networks **1**(1), 4–27 (1990)
19. Schreiber, T.: Interdisciplinary application of nonlinear time series methods. Physics Reports **308**(1), 1–64 (1999)
20. Schreiber, T., Schmitz, A.: Discrimination power of measures for nonlinearity in a time series. Physical Review E **55**(5), 5443–5447 (1997)
21. Schreiber, T., Schmitz, A.: Surrogate time series. Physica D **142**, 346–382 (2000)
22. Widrow, B., McCool, J., Ball, M.: The complex LMS algorithm. Proceedings of the IEEE **63**(4), 719–720 (1975)
23. Widrow, B., Stearns, S.D.: Adaptive Signal Processing. Prentice-Hall, Englewood Cliffs, NJ (1985)

2

Wind Modelling and its Possible Application to Control of Wind Farms

Yoshito Hirata, Hideyuki Suzuki, and Kazuyuki Aihara

Because of global warming and oil depletion, the number of wind turbines is increasing. Wind turbines commonly have a horizontal axis, but some wind turbines have a vertical axis. A problem of wind turbines with horizontal axis is that we need to face them towards the wind for maximising energy production. If the geography around a wind turbine is complicated, then the wind direction keeps changing. Thus, if one can predict the wind direction to some extent, then one may generate more electricity by controlling wind turbines according to the prediction.

In this chapter, we discuss how to model the wind. First, we formulate a problem and clarify which properties of the wind we need to predict. Second, we discuss the characteristics for time series of the wind. Third, we model the wind based on the knowledge and predict it. We prepare different models for predicting wind direction and absolute wind speed. Finally, we apply the prediction and simulate control of a wind turbine. Since we integrate predictions for wind direction and absolute wind speed to obtain an optimal control, the whole scheme can be regarded as heterogeneous fusion.

2.1 Formulating Yaw Control for a Wind Turbine

There are three purposes for predicting the wind in applications of wind turbine. The first purpose is to estimate energy production by wind turbines in the future. By estimating the energy production, we can prepare other sources of energy production to meet the demand. The time scale of the prediction for this purpose ranges from 30 min to some days and the prediction can be done by either a medium-range weather forecasts [20] or time series prediction [2–4]. The second purpose is to avoid gusts. For this purpose, the time scale of prediction is on the order of seconds. A Markov chain [12–14, 18] is effective for the prediction of this purpose. The third purpose is to generate more electricity. On this chapter, we focus on this third purpose.

There are two types of wind turbines. The major one has a horizontal axis, while the minor one has a vertical axis. The horizontal axis is more popular because it can attain better efficacy.

There are several factors one may control in a wind turbine with a horizontal axis. The examples include the direction of wind turbine (yaw angle), pitch angles, and speeds of rotations. Although mechanical details could be different, the problems for controlling these factors can be written mathematically in a similar way. As an example, we formulate a problem of yaw control.

In yaw control, the net benefits we can get are divided into two parts: the first part is benefits we may obtain; the second part is costs for controls. The unit of the net benefits is a currency: yen, US dollars, UK pounds, or other currencies. The benefits we may obtain from the energy production correlates with the amount of energy production. Thus, they are shown by using a power curve, a function $f(u)$ that returns the amount of energy produced given a wind speed u. Here the wind speed is perpendicular to the face of the rotor. Let v_t be the absolute wind speed at time t, θ_t, the wind direction, and $\varphi(t)$, the direction of the wind turbine. We also denote a set of v_t, θ_t, $\varphi(t)$ by v, θ and φ, respectively. Then, the benefits $b(\varphi, v, \theta)$ from time k to $k+K$ can be written as

$$\sum_{t=k+1}^{k+K} f(v_t \cos(\theta_t - \varphi(t))). \tag{2.1}$$

Since we have freedom for choosing the scale of the currency, we scale the currency in a way that the maximum electricity generated within 2 s yields a unit of the currency.

The costs for the yaw control include the electricity we need, replacements of devices due to stress, fatigue, and friction, and others coming from unknown sources. The term of the costs is a function of the orbit φ, wind speed v, and wind direction θ. Therefore, we denote it by $c(\varphi, v, \theta)$. In the simulations shown later, we assume that $c(\varphi, v, \theta)$ can be written as

$$C_1 \sum_{t=k+1}^{k+K} |\varphi(t) - \varphi(t-1)|$$
$$+ C_2 \sum_{t=k+1}^{k+K} |(\varphi(t) - \varphi(t-1)) - (\varphi(t-1) - \varphi(t-2))| \tag{2.2}$$
$$+ C_3 \sum_{t=k+1}^{k+K} |\varphi(t) - \theta_t|.$$

The first term depends on how much we turn a wind turbine. The second term depends on how much we change the angular velocity of wind turbine. The third term depends on the difference between the directions of the wind and the wind turbine.

The net benefits can be written as $b(\varphi, v, \theta) - c(\varphi, v, \theta)$. By replacing v_t and θ_t with their predictions $\hat{v}_t$ and $\hat{\theta}_t$, a problem of yaw control can be written as
$$\min_{\varphi} b(\varphi, \hat{v}, \hat{\theta}) - c(\varphi, \hat{v}, \hat{\theta}). \tag{2.3}$$

The above problem means that for yaw control, we need to predict the wind direction and absolute wind speed for the duration of K.

2.2 Characteristics for Time Series of the Wind

To predict the future wind, we need to know its properties. In this section, we discuss the characteristics of the wind.

2.2.1 Surrogate Data

To characterise the data sets, we used surrogate data analysis [21–25]. Surrogate data analysis is hypothesis testing. At the beginning of the analysis, one decides a null-hypothesis. Then we generate a set of random data that are consistent with the null-hypothesis. These random data are used for obtaining a confidence interval of a test statistic. If the value of the test statistic obtained from the original data is within the confidence interval, then one cannot reject the null-hypothesis. Otherwise, one rejects the null-hypothesis.

Common null-hypotheses are non-serial dependence [21], linearity [22, 25], and periodic orbits with uncorrelated noise [24]. In this chapter, we use the first two since the third surrogate has meaning when a time series is pseudo-periodic.

2.2.2 Results

In this section, we used a data set of the wind measured on 25 August 2004 at the top of a building in Institute of Industrial Science, The University of Tokyo. The measurement was of 50 Hz and it has three components, namely, the east wind, the north wind, and the upward wind. Part of data set is shown in Fig. 2.1.

First we preprocessed the data set. We divided a data set of the east and north winds into 36 segments of scalar time series of length 10,000. Then we matched the beginning and the end of time series [23] to avoid artefacts their mismatch may cause. By applying the method of Kennel [15] with four-dimensional embedding space with a uniform delay of 2 s, we confirmed the stationarity for 10 segments out of 36 segments. We used the delay of 2 s because this is a time scale we are interested in the later part of this chapter. We used the four-dimensional embedding space because in most cases we tested, the false nearest neighbour method [16] showed that with four-dimensional embedding space, the ratio of the false nearest neighbours is less

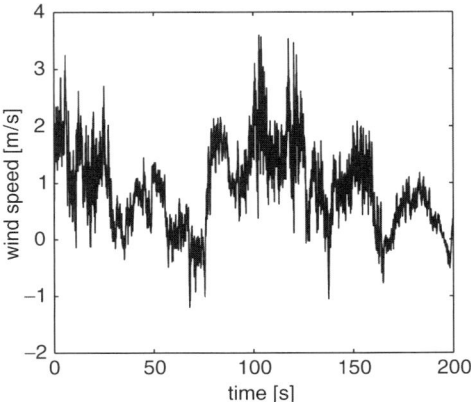

Fig. 2.1. Part of data set used in surrogate data analysis

than 1%. We had to exclude nonstationary segments since nonstationarity might cause a spurious rejection [26].

Second, we used random shuffle surrogates [21] for testing serial dependence. We used the Wayland statistic [27] for the test statistic. We used the embedding dimensions between 1 and 12. We generated 39 random shuffle surrogates and thus the significant level for each embedding dimension was 5%. To make the significant level of the multiple tests 1%, we need to have more than or equal to three rejections out of 12 embedding dimensions. We found that all the ten stationary segments have serial dependence. But a peculiar feature is that the Wayland statistic for the original data is greater than that for the random shuffle surrogates [5] (see Fig. 2.2). This may be interpreted

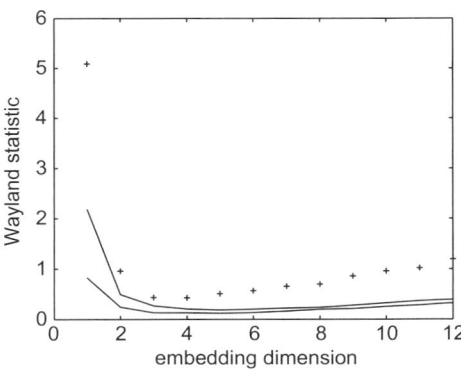

Fig. 2.2. Surrogate test for serial dependence. The *crosses* show the values obtained from the actual data set and the *solid lines*, the minimum and maximum for 39 random shuffle surrogates

that the original data are less deterministic than the random shuffle surrogates while this interpretation should be wrong. After testing some models, we found that this strange phenomenon could happen if an original data set is contaminated with either observational noise or dynamical noise [5]. The common characteristic between these two possible causes is that a time series has small fluctuations around trends. Thus, the wind may be considered as small fluctuations with long-term trends.

Third, we tested the nonlinearity. We generated iterative amplitude adjusted Fourier transform surrogates [22] and compared the original data sets and their surrogates with the Wayland statistic of the original data. We found a rejection in a multiple test for three segments out of 10 stationary segments. The number of rejections is significantly higher than the chance level. Thus the wind may be assumed to be a nonlinear process.

2.3 Modelling and Predicting the Wind

In the previous section, we discussed that a nonlinear model is appropriate for modelling the wind. In this section, we build a model and evaluate its performance by a prediction error.

2.3.1 Multivariate Embedding

In most cases, nonlinear time series analysis starts with delay embedding. Let $x_i(t)$ be the ith observable at time t. Let $x(t)$ represent whole the state at time t. Then, we choose

$$(x_i(t), x_i(t-\tau), x_i(t-2\tau), \cdots, x_i(t-(\kappa-1)\tau)). \tag{2.4}$$

If κ is sufficiently large, then this vector gives an embedding, meaning that the vector and $x(t)$ is one-to-one and the associated tangent space at each point is also one-to-one. For the comparison later, we call this vector uniform embedding.

In practice, Judd and Mees [10] extended the notion of delay embedding and defined the following vector:

$$(x_i(t-\tau_1), x_i(t-\tau_2), \cdots, x_i(t-\tau_\kappa)). \tag{2.5}$$

They call this vector non-uniform embedding. The method of Judd and Mees [10] fits a linear model and chooses the vector that minimises description length, an information criterion.

The non-uniform embedding was extended to multivariate time series by Garcia and Almeida [1]. Namely, one tries to construct a vector which looks like

$$(x_{i_1}(t-\tau_1), x_{i_2}(t-\tau_2), \ldots, x_{i_\kappa}(t-\tau_\kappa)). \tag{2.6}$$

Their method is based on the extension of false nearest neighbour method [16], a standard method for estimating the embedding dimension. In this chapter, we used the extension [8] of Judd and Mees [10], another method for obtaining an non-uniform embedding from a multivariate time series, since the method gives, within a shorter time, a non-uniform embedding which is as reasonable as the method of Garcia and Almeida.

2.3.2 Radial Basis Functions

The model we built was a radial basis function model since it can approximate any continuous function relatively well. We use $x^i(t)$ for showing a reconstruction for the ith coordinate at time t. Then a radial basis function model can be written as

$$x_i(t+1) = \delta_i + \alpha_i \cdot x^i(t) + \sum_l \beta_{i,l} \exp\left[-\frac{\|x^i(t) - \gamma_{i,l}\|^2}{2\nu^2}\right], \quad (2.7)$$

where δ_i, α_i, $\beta_{i,l}$, $\gamma_{i,l}$, and ν are parameters for the model. The centres of the radial basis functions were chosen using the technique of chaperons [9]. In this technique, one chooses a set of points from observations and adds Gaussian noise to them, whose standard deviation is 30% of that of the data. We prepared 100 radial basis functions using this technique. We also set ν to the standard deviation of the time series.

Since we have already decided $\gamma_{i,k}$ and ν, now our model is a pseudo-linear model and the remaining parameters may be decided by the least squares solution. But a simple application of the least squares solution will face the problem of overfitting. To avoid overfitting, we decided the remaining parameters for the radial basis function model as prescribed by Judd and Mees [9] using the Normalised Maximum Likelihood [19] for a model selection criterion. To explain the details, first let us define the Normalised Maximum Likelihood.

First we need to define a set of basis functions. In this chapter, the set is made of the 100 radial basis functions and linear terms. Suppose that now we have a number P of basis functions $\{\zeta_j : j \in F\}$, where $F = \{1, 2, \ldots, P\}$ is the set of indices for all the basis functions. Let D be the biggest delay among $\{\tau_i\}$. The length of given time series is denoted by N. Define $\tilde{N} = N - D$. Let V be the $\tilde{N} \times P$ matrix whose (q, r) component is the value for the rth basis function at time q, i.e.,

$$V = \begin{pmatrix} \zeta_1(x^i(D)) & \zeta_2(x^i(D)) & \cdots & \zeta_P(x^i(D)) \\ \vdots & & & \vdots \\ \zeta_1(x^i(t)) & \zeta_2(x^i(t)) & \cdots & \zeta_P(x^i(t)) \\ \vdots & & & \vdots \\ \zeta_1(x^i(N-1)) & \zeta_2(x^i(N-1)) & \cdots & \zeta_P(x^i(N-1)) \end{pmatrix}. \quad (2.8)$$

Denote a set of indices for basis functions by B. Thus, it holds that $B \subset F$. Let us denote by V_B the matrix that is formed from the columns of V with indices in $B = \{j_1, \ldots, j_J\}$, i.e.,

$$V_B = \begin{pmatrix} \zeta_{j_1}(x^i(D)) & \zeta_{j_2}(x^i(D)) & \cdots & \zeta_{j_J}(x^i(D)) \\ \vdots & \vdots & & \vdots \\ \zeta_{j_1}(x^i(t)) & \zeta_{j_2}(x^i(t)) & \cdots & \zeta_{j_J}(x^i(t)) \\ \vdots & \vdots & & \vdots \\ \zeta_{j_1}(x^i(N-1)) & \zeta_{j_2}(x^i(N-1)) & \cdots & \zeta_{j_J}(x^i(N-1)) \end{pmatrix}. \quad (2.9)$$

Let us also define ξ as

$$\xi = (x_i(D+1), \ldots, x_i(t+1), \ldots, x_i(N))^\mathrm{T}. \quad (2.10)$$

Here 'T' shows the transposition of matrix.

Let λ_B be the least squares solution for $\xi \approx V_B \lambda_B$. Then the prediction error e_B can be written as $e_B = \xi - V_B \lambda_B$. When minimising the squared error $e_B^\mathrm{T} e_B$, we can find the solution by

$$\lambda_B = (V_B^\mathrm{T} V_B)^{-1} V_B^\mathrm{T} \xi. \quad (2.11)$$

Define R by

$$R = \frac{1}{\tilde{N}} (V_B \lambda_B)^\mathrm{T} V_B \lambda_B. \quad (2.12)$$

Then the Normalised Maximum Likelihood [19] for a set of p basis functions B is given by

$$\mathcal{L}(B) = \frac{\tilde{N} - p}{2} \ln \frac{e_B^\mathrm{T} e_B}{\tilde{N}} + \frac{p}{2} \ln R - \ln \Gamma \left(\frac{\tilde{N} - p}{2} \right) - \ln \Gamma \left(\frac{p}{2} \right) - \ln p. \quad (2.13)$$

Here Γ shows the Gamma function.

Using the Normalised Maximum Likelihood and these notations, we can write down the algorithm proposed in [9] for selecting an optimal set of basis functions in the following way:

1. Normalise V so that each column has unit length.
2. Let B and B' be empty sets.
3. Obtain the prediction error. If B' is empty, then $e_{B'} = \xi$. If not, then $e_{B'} = \xi - V_{B'} \lambda_{B'}$. Here $\lambda_{B'}$ can be obtained using the formula of (2.11).
4. Find the basis function that matches the prediction error best. Let $\mu = V^\mathrm{T} e_{B'}$. Then the biggest component of μ corresponds to the best basis function matching to the prediction error. Let g be the index for the best matching basis function. Let $B' \leftarrow B' \cup \{g\}$.
5. Find the basis function in B' that least contributes to making the error small. Let h be the index for the basis function whose corresponding $\lambda_{B'}$ is the smallest. If $g \neq h$, then $B' \leftarrow B' \backslash \{h\}$ and go to Step 3.

6. If B is empty or $\mathcal{L}(B') < \mathcal{L}(B)$, then set $B \leftarrow B'$ and go back to Step 3.
7. The chosen B is the optimal set of basis functions.

We used the above algorithm for deciding δ_i, α_i, and $\beta_{i,l}$. Namely, for basis functions which are not selected, we set the corresponding components to 0. For selected basis functions, we use values obtained from the above fitting.

2.3.3 Possible Coordinate Systems

There are two possible coordinate systems for representing a wind vector. The first one is the polar coordinate system. The second one is the rectangular coordinate system. The problem of the polar coordinate system is how to represent the wind direction. The wind direction has a value between 0° and 360°, which is on a ring. If one tries to show the wind direction using a value in an interval, then one has to split the ring somewhere and it causes discontinuity. As a result, the wind direction cannot be predicted well [6]. Thus, the rectangular coordinate system is better than the polar coordinate system for predicting the wind direction.

2.3.4 Direct vs. Iterative Methods

The model now at our hands predicts a step ahead. However, as discussed in Sect. 2.1, we need to predict the wind direction and absolute wind speed for a certain range of time. There are two options for it: The first option is to prepare several models each of which has a different prediction step. This approach is called a direct method. The second option is to use a model for predicting a step ahead several times and realise a multiple step prediction. This approach is called an iterative method.

There are some discussions on which model is better on a certain occasion [11, 17]. Generally speaking, when we have a good one-step prediction, then an iterative method is better. Otherwise, a direct method might be better. Judd and Small [11] proposed a mixture of direct and iterative methods. In their method, one first uses an iterative method and predicts the future values. This step is called ϕ-step since the one-step model is shown using ϕ. Then one uses a corrector ψ that makes the prediction more accurate. This step is called ψ-step due to the notation of the corrector. As a total, the method is called the $\psi\phi$ method.

2.3.5 Measurements of the Wind

The data sets we used were measured at about 1 m high on the campus of the Institute of Industrial Science, The University of Tokyo. We used two identical anemometers which record the east, the north, and the upward winds with 50 Hz. We measured the wind for nine different days between September 2005 and May 2006. On some days, we placed the two anemometers 5 m apart in

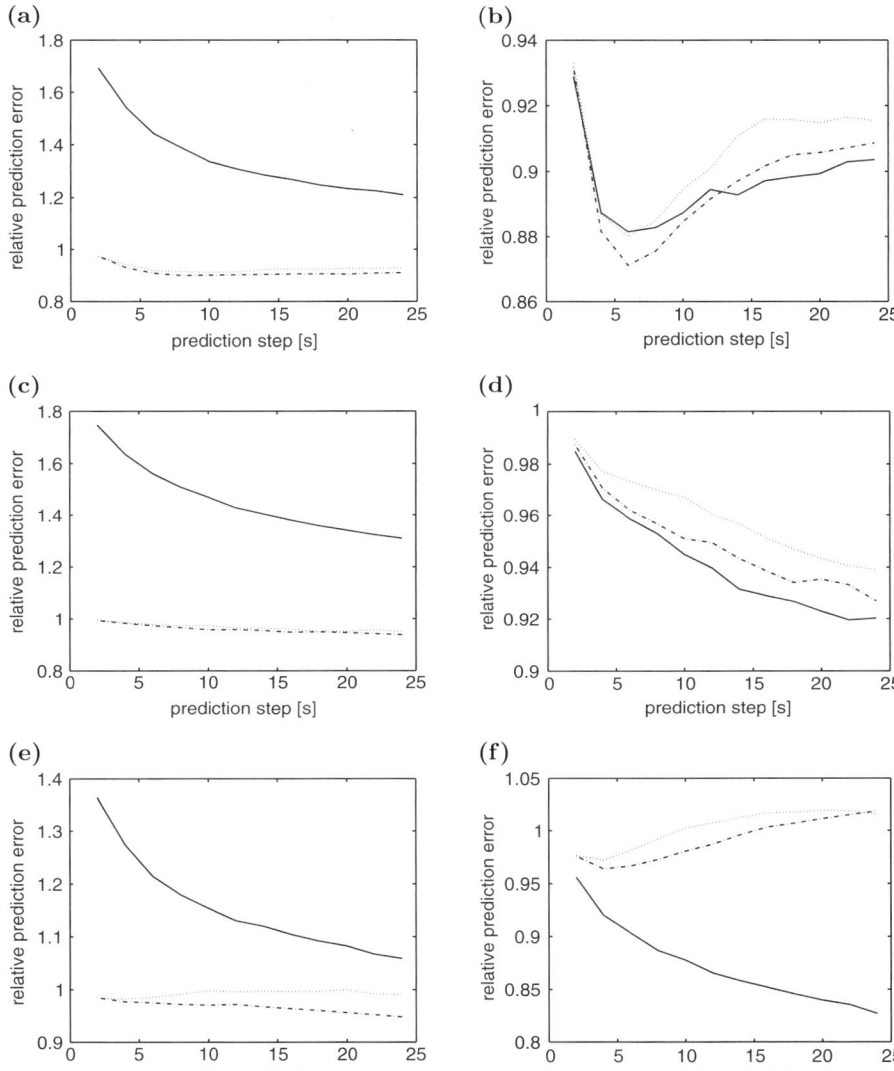

Fig. 2.3. Means of root mean square errors for the compared combinations of methods relative to those of persistent prediction. (**a**) and (**b**) are cases where the additional observation was at upstream. (**c**) and (**d**) are cases where the additional observation was at downstream. (**e**) and (**f**) are cases where the additional observation was nearby but at neither downstream nor upstream. (**a**), (**c**), and (**e**) show root mean square errors for the wind direction, and (**b**), (**d**), and (**f**), those for the absolute wind speed. In each part, the *solid line* shows the case of polar coordinates with the direct method, the *dotted line*, that of rectangular coordinates with the direct method, and the *broken line*, that of rectangular coordinates with the $\psi\phi$ method

the north–south direction. On the other days, we placed them 5 m apart in the east–west direction. Since we are interested in the dynamics of 2 s order, we took the moving average of 2 s and resampled it by 2 s.

We used measurements taken from the two points for predicting the wind at one of them as we can expect better prediction by doing this [7].

2.3.6 Results

In the previous sections, we argued that there are two coordinate systems and two possible methods for the prediction. Thus, theoretically there are four possible combinations. In this section, however, we compare three of them, namely, the polar coordinates with the direct method, the rectangular coordinates with the direct method, and the rectangular coordinates with the $\psi\phi$ method, since the polar coordinates with an iterative method does not provide good prediction since they have to use the prediction of the wind direction several times.

In Fig. 2.3, we compared the performances of the three combinations of the methods by the prediction error for each method relative to that for the persistent prediction, where we used the value of 2 s before as a prediction.

When we predicted the wind direction, the rectangular coordinates with the $\psi\phi$ method achieved smaller mean prediction errors than the other two combinations (Fig. 2.3a, c, e). We interpret this results as follows: Because the wind direction takes a value on the ring, the polar coordinates are not a good idea for representing the wind direction. By using the $\psi\phi$ method, we can make the prediction errors smaller compared to the direct prediction.

When we predicted the absolute wind speed (Fig. 2.3b, d, f), the polar coordinates with the direct method gave the smallest mean prediction error for all cases except for short prediction steps of the case where the additional observation was at upstream. We think that the polar coordinates were better

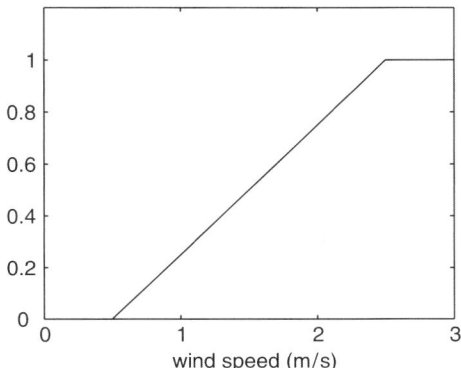

Fig. 2.4. Power curve used for the simulation

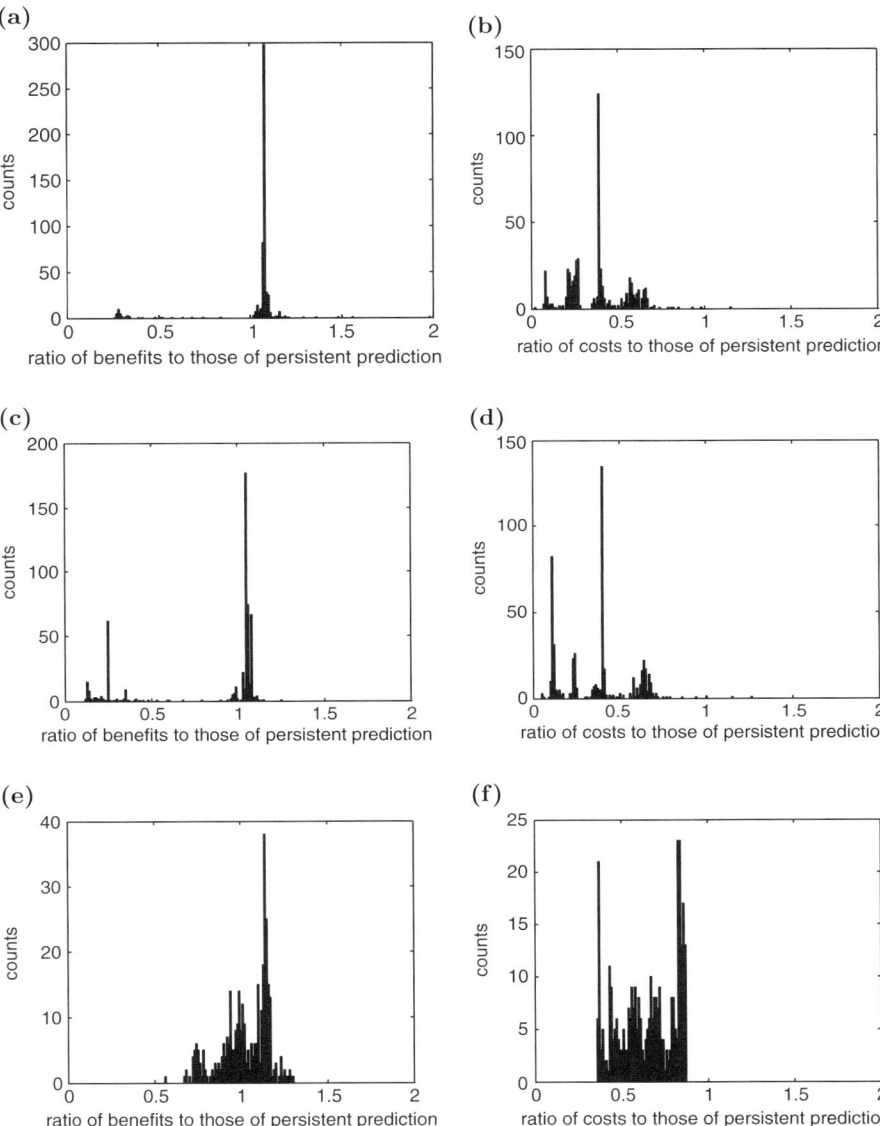

Fig. 2.5. Results of simulations for energy production. (**a**), (**c**), and (**e**) show the histograms of ratios of expected net benefits for the combination of the proposed prediction methods to those for the persistent prediction. (**b**), (**d**), and (**f**) show the histograms of ratios of expected costs for controlling a wind turbine based on the proposed prediction to those based on the persistent prediction. (**a**) and (**b**) are cases where the additional observation was at upstream. (**c**) and (**d**) are cases where the additional observation was at downstream. (**e**) and (**f**) are cases where the additional observation was nearby but at neither downstream nor upstream

for predicting the absolute wind speed because the polar coordinates represent what we want to predict directly.

2.4 Applying the Wind Prediction to the Yaw Control

In Sect. 2.3, we observed that the wind direction was predicted well when the rectangular coordinates with the $\psi\phi$ method was used, while the absolute wind speed was predicted well when the polar coordinates and the direct method were used. In this section, we compared the combination of these methods with the persistent prediction in expected energy.

From the actual data, we predicted the wind directions and absolute wind speeds for prediction steps between 2 and 24 s. Then we fed the predictions to the model of wind turbine presented in Sect. 2.1 and decided the control policy. For estimating energy production, we used the actual wind data.

We used a power curve shown in Fig. 2.4 for the simulation. We set $C_1 = C_2 = C_3 = 0.01$.

We used the polar coordinates with the direct method and the rectangular coordinates with the $\psi\phi$ method for predicting the absolute wind speed and the wind direction, respectively. This combination is the best, based on the results of Sect. 2.3. We compared the results with the case where we used the persistent prediction for both the absolute wind speed and the wind direction.

The results were shown in Fig. 2.5. The median of ratio of the expected net benefits for the proposed prediction to those for the persistent prediction is greater than 1. The median of ratio of the expected control costs for the proposed prediction to those for the persistent prediction is smaller than 1. These results mean that the proposed prediction works well for controlling a wind turbine. The above observations were always true and do not depend on the place of the additional observation point much.

2.5 Conclusions

In this chapter, we showed it is possible that wind modelling may help to increase energy production. To show this, we first formulated the problem for controlling the yaw angle of a wind turbine and found that we need to predict the wind direction and the absolute wind speed for a certain range of the future. Next, we investigated the properties of time series of the wind using surrogate data analysis. We showed that the time series always have serial dependence and they sometimes have nonlinearity. We also argued that times series of the wind may be characterised by small fluctuations with trends. These findings are positive signs for the prediction. Then, we modelled the wind and found that the wind direction is predicted well by the rectangular coordinates and the $\psi\phi$ method, and the absolute wind speed is predicted well if one uses the polar coordinates and the direct method. Finally, we applied the prediction of the wind to the simulation of wind turbine control. We found that the nonlinear prediction can potentially increase the energy production. An important point is that although our gain of information is small in terms of the root mean square error, it is big enough for controlling a wind turbine optimally.

We feel that the methods discussed here could be useful for controlling a small wind turbine since a small wind turbine can be directed with a small force very quickly, and in the scale of a small wind turbine, the wind at the edges of the blades is not so much different from that at the centre.

The focus of this chapter has been that a heterogeneous data fusion strategy is useful for controlling a wind turbine, and therefore the models of turbine, stress, fatigue, and friction used are synthetic. Although these models are not guaranteed to provide a feasible and safe solution for a real wind turbine, we believe that they are sufficiently realistic to illustrate the proposed framework.

Acknowledgement

This study was partially supported by the Industrial Technology Research Grant Program in 2003, from the New Energy and Industrial Technology Development Organization (NEDO) of Japan.

References

1. Garcia, S.P., Almeida, J.S.: Multivariate phase space reconstruction by nearest neighbor embedding with different time delays. Physical Review E **72**(2), 027205 (2005)
2. Goh, S.L., Chen, M., Popvic, H., Aihara, K., Obradovic, D., Mandic, D.P.: Complex-valued forecasting of wind profile. Renewable Energy **31**(11), 1733–1750 (2006)
3. Goh, S.L., Mandic, D.P.: A complex-valued RTRL algorithm for recurrent neural networks. Neural Computation **16**(12), 2699–2713 (2004)
4. Goh, S.L., Mandic, D.P.: Nonlinear adaptive prediction of complex-valued signals by complex-valued PRNN. IEEE Transactions on Signal Processing **53**(5), 1827–1836 (2005)
5. Hirata, Y., Horai, S., Suzuki, H., Aihara, K.: Testing serial dependence by Random-shuffle surrogates and the wayland method. Physics Letters A (2007). DOI 10.1016/j.physleta.2007.05.061
6. Hirata, Y., Mandic, D.P., Suzuki, H., Aihara, K.: Wind direction modelling using multiple observation points. Philosophical Transacations of the Royal Society A (2007). DOI 10.1098/rsta.2007.2112
7. Hirata, Y., Suzuki, H., Aihara, K.: Predicting the wind using spatial correlation. In: Proceedings of 2005 International Symposium on Nonlinear Theory and its Applications (NOLTA 2005) (2005)

8. Hirata, Y., Suzuki, H., Aihara, K.: Reconstructing state spaces from multivariate data using variable delays. Physical Review E **74**(2), 026202 (2006)
9. Judd, K., Mees, A.: On selecting models for nonlinear time-series. Physica D **82**(4), 426–444 (1995)
10. Judd, K., Mees, A.: Embedding as a modeling problem. Physica D **120**(3–4), 273–286 (1998)
11. Judd, K., Small, M.: Towards long-term prediction. Physica D **136**(1–2), 31–44 (2000)
12. Kantz, H., Holstein, D., Ragwitz, M., Vitanov, N.K.: Extreme events in surface wind: Predicting turbulent gusts. In: S. Bocaletti, B.J. Gluckman, J. Kurths, L.M. Pecora, R. Meucci, O. Yordanov (eds.) Proceedings of the 8th Experimental Chaos Conference, no. 742 in AIP Conference Proceedings. American Institute of Physics, New York (2004)
13. Kantz, H., Holstein, D., Ragwitz, M., Vitanov, N.K.: Markov chain model for turbulent wind speed data. Physica A **342**(1–2), 315–321 (2004)
14. Kantz, H., Holstein, D., Ragwitz, M., Vitanov, N.K.: Short time prediction of wind speeds from local measurements. In: J. Peinke, P. Schaumann, S. Barth (eds.) Wind Energy: Proceedings of the EUROMECH Colloquium. Springer, Berlin Heidelberg New York (2006)
15. Kennel, M.B.: Statistical test for dynamical nonstationarity in observed time-series data. Physical Review E **56**(1), 316–321 (1997)
16. Kennel, M.B., Brown, R., Abarbanel, H.D.I.: Determining embedding dimension for phase-space reconstruction using a geometrical construction. Physical Review A **45**(6), 3403–3411 (1992)
17. McNames, J.: A nearest trajectory strategy for time series prediction. In: Proceedings of the International Workshop on Advanced Black-Box Techniques for Nonlinear Modeling (1998)
18. Ragwitz, M., Kantz, H.: Detecting non-linear structure and predicting turbulent gusts in surface wind velocities. Europhysics Letters **51**(6), 595–601 (2000)
19. Rissanen, J.: MDL denoising. IEEE Transactions on Information Theory **46**(7), 2537–2543 (2000)
20. Roulston, M.S., Kaplan, D.T., Hardenberg, J., Smith, L.A.: Using medium-range weather forecasts to improve the value of wind energy production. Renewable Energy **28**(4), 585–602 (2003)
21. Scheinkman, J.A., LeBaron, B.: Nonlinear dynamics and stock returns. Journal of Business **62**(3), 311–337 (1989)
22. Schreiber, T., Schmitz, A.: Improved surrogate data for nonlinearity tests. Physical Review Letters **77**(4), 635–638 (1996)
23. Schreiber, T., Schmitz, A.: Surrogate time series. Physica D **142**(3–4), 346–382 (2000)
24. Small, M., Yu, D., Harrison, R.G.: Surrogate test for pseudoperiodic time series. Physical Review Letters **87**(18), 188101 (2001)
25. Theiler, J., Eubank, S., Longtin, A., Galdrikian, B., Farmer, J.D.: Testing for nonlinearity in time-series: the method of surrogate data. Physica D **58**(1–4), 77–94 (1992)
26. Timmer, J.: Power of surrogate data testing with respect to nonstationarity. Physical Review E **58**(4), 5153–5156 (1998)
27. Wayland, R., Bromley, D., Pickett, D., Passamante, A.: Recognizing determinism in a time-series. Physical Review Letters **70**(5), 580–582 (1993)

3

Hierarchical Filters in a Collaborative Filtering Framework for System Identification and Knowledge Retrieval

Christos Boukis and Anthony G. Constantinides

This chapter provides a critical review of hierarchical filters and the associated adaptive learning algorithms. Hierarchical filters are collaborative adaptive filtering architectures where short-length adaptive transversal filters are combined into layers, which are then combined into a multilayered structures. These structures offer potentially faster speed of convergence compared to the standard finite impulse response (FIR) filters, which is due to the small order of their constituting sub-filters. Several approaches can be used to adapt the coefficients of hierarchical filters. These include the use of the standard least mean square (LMS) algorithm for every sub-filter, via a variant of linear back-propagation, through to using a different algorithm for every layer and every sub-filter within the layer. Unless the input signal is white or the unknown channel is sparse, hierarchical filters converge to biased solutions. We make use of this property to propose a collaborative approach to the identification of sparse channels. The performances of these algorithms are evaluated for a variety of applications, including system identification and sparsity detection. The benefits and limitations of hierarchical adaptive filtering in this context are highlighted.

3.1 Introduction

Conventional adaptive filtering systems rely on the use of digital filters the weights of which are updated through adaptive algorithms to meet some appropriate and predefined criteria. There is a vast variety of filters that can be used for this purpose [18] including finite impulse response (FIR), infinite impulse response (IIR), gamma [15], Laguerre [17] and Kautz [10]. The algorithms that are used for the adaptation of their weights have a recursive nature so as to be able to perform in an online fashion, and their aim is the minimisation of a unimodal, convex cost function. Typical examples are Gradient Descent algorithms [7] like the least mean square (LMS) and

Raphson–Newton [9] algorithms like the recursive least squares (RLS), which attempt to minimise the square of the adaptation error.

Due to its simplicity in analysis, reduced computational complexity and in robustness the dominant approach in adaptive signal processing applications is the use of FIR filters that update their coefficients through stochastic gradient descent algorithms that minimise a quadratic cost function. These structures have been successfully applied to a plethora of real-time systems like noise cancellation, acoustic echo cancellation, feedback control and channel equalisation [7]. However, their performance, in terms of convergence speed and steady-state misadjustment, depends strongly on the length of the adaptive filter. Hence, they are not appropriate in applications where filters of large order are required. To anticipate these problems several solutions have been proposed in the literature like the use of filters that have some form of feedback, or adaptation of the coefficients in an appropriate transform domain [2].

We consider the use of multilayer adaptive structure organised in a hierarchical manner [14], called hierarchical filters (HFs). Their objective is the combination of several adaptive filters to produce more robust and more accurate estimates than transversal filters in shorter time intervals. These sub-filters are usually tap-delay lines [4], although the use of other types of filters is not prevented, and they can be of any order. Several policies can be applied for the adaptation of the weights of their sub-filters. Typical examples are the Hierarchical LMS [19] that attempts to minimise the local error within every sub-filter, the hierarchical gradient descent (HGD) that is a member of the class of the back-propagation algorithms [3] and it aims at the minimisation of the global error. Moreover, algorithms that use a different optimisation criterion for every sub-filter can be developed. Each one of these algorithms has different characteristics: HLMS converges to biased solutions unless the unknown channel is sparse or the input signal is white, HGD might converge to local minima and so on.

The performance of HFs in a variety of adaptive filtering applications, like prediction and system identification has been examined thoroughly in the literature [11, 13, 16]. The short length of their sub-filters results in faster convergence than typical tap-delay lines, although they often produce biased estimates. To justify this behaviour, we provide a rigorous mathematical analysis. We also show that every HF has an equivalent FIR filter. Another interesting field of application of HFs is the domain of knowledge extraction [5]. Embarking upon the properties of the algorithms that have been developed for hierarchical structures, HFs can be used for sparsity detection within unknown channels. Moreover, since HFs behave as integrated structures that fuse the estimates of the sub-filters of the input layer, they can be used to detect the modality of a signal by using a different even-moment of the adaptation error for every input-layer sub-filter as cost function.

This chapter is organised as follows: In Sect. 3.2 the various types of HFs are presented and their parameters are discussed. In Sect. 3.3 the basic algorithms that are used for the adaptation of the coefficients of these filters are presented. Section 3.4 deals with the application of HFs to standard filtering applications and with their use for knowledge extraction. Finally Sect. 3.5 concludes this chapter.

3.2 Hierarchical Structures

The fundamental idea of HFs is to combine multiple short-length filters instead of a single one of high order in adaptive filtering applications so as to provide more robust estimates [14]. These filters perform in parallel forming the input layer of the HF. Their outputs are combined in the superjacent layer by linear combiners. The outputs of these linear combiners are then fed into the combiners of the next layer and so on. In general, the input of a tap at level l is the output of a sub-filter of the $(l-1)$th level. If the HF consists of sub-filters of length β then the input of the jth tap of the ith sub-filter at level l, is the output of the $((i-1)\beta + j)$th sub-filter of the $(l-1)$th level. The top layer, called the output layer, combines the output signals of the underlying layer and produces a single global estimate. In Fig. 3.1 an HF with three layers consisting of sub-filters of length 2 is presented (a (3,2) structure).[1] The output of this structure is given by [4]

$$y(n) = \sum_{i=1}^{2} \theta_{1,i}^{(3)}(n) \sum_{j=1}^{2} \theta_{i,j}^{(2)}(n) \sum_{m=1}^{2} \theta_{(i-1)2+j,m}^{(1)}(n) x(n - d_{i,j,m}), \qquad (3.1)$$

where $d_{i,j,m} = ((i-1)2 + j - 1)2 + (m-1)$, $\theta_{i,j}^{(l)}$ is the jth coefficient of the ith sub-filter at the lth layer and $x(n)$ is the input signal. The order of an HF is the number of the delays included in its input layer.

Another way to assess the performance of HF is the following: Assuming that the set of optimal coefficients is $\mathcal{F} = \{f_0, f_1, \ldots, f_{(N-1)}\}$, and using an HF of α layers, with sub-filters of length $\beta = \sqrt[\alpha]{M}$, then each sub-filter at the input layer estimates a subset of $\mathcal{F}$. More specifically, the first sub-filter estimates the subset of coefficients $\mathcal{F}_1 = \{f_0, f_1, \ldots, f_{(\beta-1)}\}$, the second sub-filter estimates the subset $\mathcal{F}_2 = \{f_\beta, \ldots, f_{(2\beta-1)}\}$, and in general the ith sub-filter of the input layer estimates the subset $\mathcal{F}_i = \{f_{(i-1)\beta}, \ldots, f_{i\beta-1}\}$ of the set of the unknown coefficients. Since the inputs to the sub-filters of the structure are mutually disjoint, it holds that $\mathcal{F}_1 \cap \mathcal{F}_2 \cap \cdots \cap \mathcal{F}_{(M/\beta)} = \emptyset$ and $\mathcal{F} = \mathcal{F}_1 \cup \mathcal{F}_2 \cup \cdots \cup \mathcal{F}_{(M/\beta)}$. These estimates are merged subsequently into larger sets in the subsequent levels of the structure, and finally, at the output level, an estimate of the whole set of the optimal coefficients is produced.

For our convenience, in the rest of the chapter we will deal with HFs with two layers, sub-filters of length β in the input layer and a linear combiner of length M/β in the output layer.

---

[1] An HF with constant sub-filter length β, α layers will be denoted as an (α, β) structure.

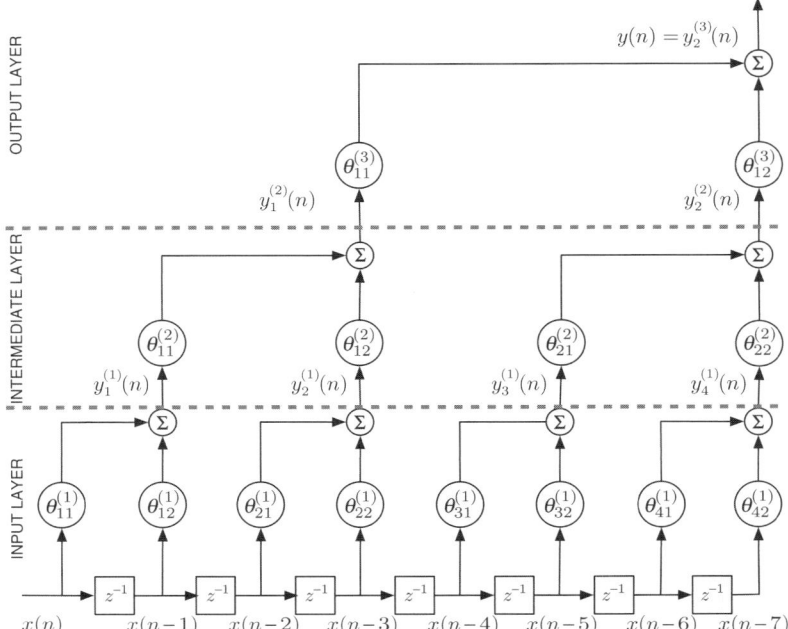

Fig. 3.1. A hierarchical filter with three layers, consisting of four sub-filters of length $\beta = 2$ each at the input layer, two sub-filters at the intermediate layer and one sub-filter at the output layer

3.2.1 Generalised Structures

In Fig. 3.2, an alternative representation of an HF with two layers is depicted. From this it is clearly observed that the input layer consists of transversal filters that operate in parallel. In the case where the delay Δ is equal to the length of the sub-filters, i.e. when $\Delta = \beta$, the inputs of the sub-filters are mutually disjoint. This is the case we have dealt with so far (Fig. 3.1). However, when $\Delta < \beta$ the input vectors of sub-filters that lie on the same layer overlap. In this case, the input sample $x(n-d)$ can be used by two or more sub-filters. A typical example of an HF with three layers and where the input vectors of neighbouring sub-filters overlap is presented in Fig. 3.3.

When $\Delta = 0$, which results in fully connected structures, all the sub-filters of a layer use the same regressor vector. The output is a combination of $\prod_{l=1}^{\alpha} N^{(l)}$ terms, where α is the number of levels of the hierarchy and $N^{(l)}$ is the length of the sub-filters of the lth level. In this case, the intersection of the subsets of the set of the unknown coefficients, that are estimated from the transversal filters of the input layer, is a nonempty set, i.e. $\mathcal{F}_1 \cap \mathcal{F}_2 \cap \cdots \cap \mathcal{F}_{(M/\beta)} \neq \emptyset$, although it still holds that $\mathcal{F} = \mathcal{F}_1 \cup \mathcal{F}_2 \cup \cdots \cup \mathcal{F}_{(M/\beta)}$.

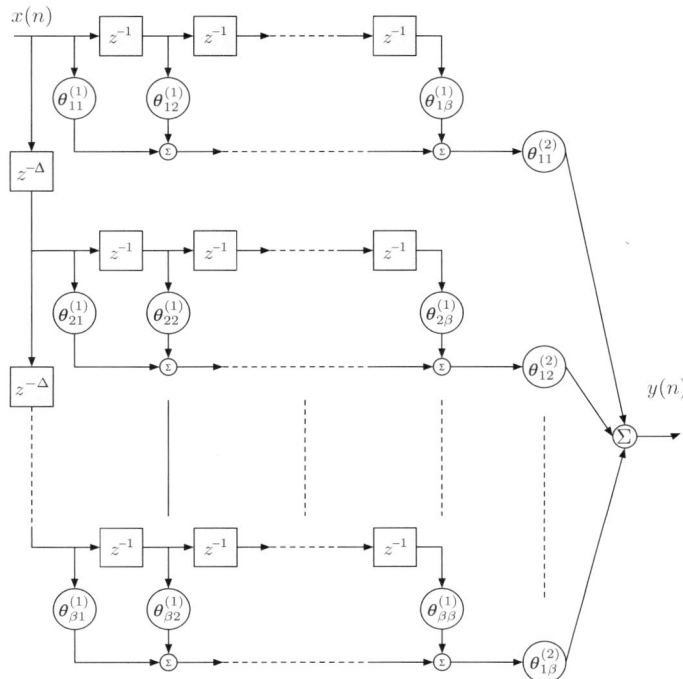

Fig. 3.2. An alternative representation of a hierarchical filter with two layers

This happens because, due to overlapping the subsets of the coefficients that adjacent sub-filters attempt to estimate have common elements.

3.2.2 Equivalence with FIR

The number of the past samples of the input signal, that are used for the computation of the output signal of an HF (i.e., the memory of the system), depends only on the number of the delay elements at the input layer [13]. This can also be seen from (3.1), which represents the output of a (3,2) HF.

Based on this observation, it is concluded that HFs are functionally equivalent to single-layered FIR filters, since their outputs are of the form $y(n) = \mathbf{f}^t\mathbf{x}(n)$, with $\mathbf{f} = [f_0, f_1, \ldots, f_{N-1}]^\mathrm{T}$ the parameter vector and $\mathbf{x}(n) = [x(n), x(n-1), \ldots, x(n-M+1)]^t$ the regressor vector, respectively. In the case of sub-filters that use mutually disjoint inputs, the coefficients of the equivalent FIR filter are products of the coefficients of the HF. When overlapping between the inputs of the sub-filters of the same layer occurs, the total coefficient of every input sample $x(n-d)$, where $d = 0, 1, \ldots, M-1$ is a sum of the products of the coefficients of the HF. To illustrate this, Table 3.1 presents the coefficients of the equivalent FIR filters for the HFs of Figs. 3.1 and 3.3.

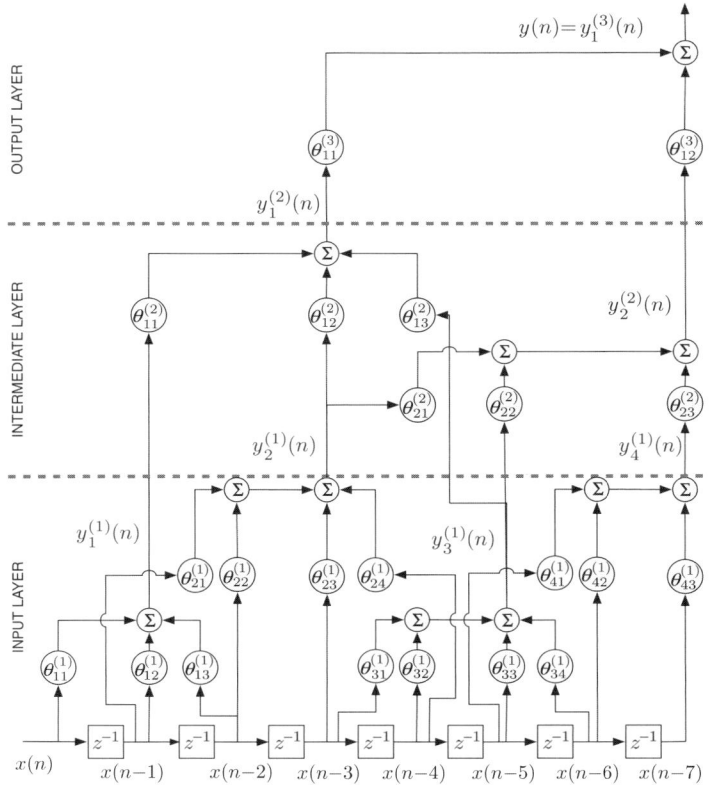

Fig. 3.3. A hierarchical filter with three layers, where the input vectors of neighbouring sub-filters share common elements

Table 3.1. The coefficients of the equivalent FIR filters of the HFs of Figs. 3.1 and 3.3

FIR Coefs.	HF of Fig. 3.1	HF of Fig. 3.3
f_0	$\theta_{11}^{(3)}\theta_{11}^{(2)}\theta_{11}^{(1)}$	$\theta_{11}^{(3)}\theta_{11}^{(2)}\theta_{11}^{(1)}$
f_1	$\theta_{11}^{(3)}\theta_{11}^{(2)}\theta_{12}^{(1)}$	$\theta_{11}^{(3)}\theta_{11}^{(2)}\theta_{12}^{(1)} + (\theta_{11}^{(3)}w_{12}^{(2)} + \theta_{12}^{(3)}\theta_{21}^{(2)})\theta_{21}^{(1)}$
f_2	$\theta_{11}^{(3)}\theta_{12}^{(2)}\theta_{21}^{(1)}$	$\theta_{11}^{(3)}\theta_{11}^{(2)}\theta_{13}^{(1)} + (\theta_{11}^{(3)}\theta_{12}^{(2)} + \theta_{12}^{(3)}\theta_{21}^{(2)})\theta_{22}^{(1)}$
f_3	$\theta_{11}^{(3)}\theta_{12}^{(2)}\theta_{22}^{(1)}$	$(\theta_{11}^{(3)}\theta_{12}^{(2)} + \theta_{12}^{(3)}\theta_{21}^{(2)})\theta_{23}^{(1)} + (\theta_{11}^{(3)}\theta_{13}^{(2)} + \theta_{12}^{(3)}\theta_{22}^{(2)})\theta_{31}^{(1)}$
f_4	$\theta_{12}^{(3)}\theta_{21}^{(2)}\theta_{31}^{(1)}$	$(\theta_{11}^{(3)}\theta_{12}^{(2)} + \theta_{12}^{(3)}\theta_{21}^{(2)})\theta_{24}^{(1)} + (\theta_{11}^{(3)}\theta_{13}^{(2)} + \theta_{12}^{(3)}\theta_{22}^{(2)})\theta_{32}^{(1)}$
f_5	$\theta_{12}^{(3)}\theta_{21}^{(2)}\theta_{32}^{(1)}$	$(\theta_{11}^{(3)}\theta_{13}^{(2)} + \theta_{12}^{(3)}\theta_{22}^{(2)})\theta_{33}^{(1)} + \theta_{12}^{(3)}\theta_{23}^{(2)}\theta_{41}^{(1)}$
f_6	$\theta_{12}^{(3)}\theta_{22}^{(2)}\theta_{41}^{(1)}$	$(\theta_{11}^{(3)}\theta_{13}^{(2)} + \theta_{12}^{(3)}\theta_{22}^{(2)})\theta_{34}^{(1)} + \theta_{12}^{(3)}\theta_{23}^{(2)}\theta_{42}^{(1)}$
f_7	$\theta_{12}^{(3)}\theta_{22}^{(2)}\theta_{42}^{(1)}$	$\theta_{12}^{(3)}\theta_{23}^{(2)}\theta_{43}^{(1)}$

3.3 Multilayer Adaptive Algorithms

For the adaptation of the coefficients of HFs several recursive algorithms have been presented in the literature. Typical examples are the hierarchical least mean square (HLMS) [19], the hierarchical gradient descent (HGD) [4] and the hierarchical recursive least squares (HRLS) [20] algorithms. The first two are members of the class of stochastic gradient descent (SGD) algorithms, since in their adaptation they approximate the expectation of the gradient of a predefined objective function with its instantaneous value, while the latter is a Newton–Raphson algorithm.

3.3.1 The Hierarchical Least Mean Square Algorithm

The objective of the HLMS is the minimisation of the local cost function within each sub-filter, defined as the square of the difference between the desired response and the output signal of this specific sub-filter. This is achieved by applying the LMS algorithm to every sub-filter [19]. In the subsequence, we will present the HLMS algorithm for an HF with two layers, having M/β sub-filters of length β at the input layer and a sub-filter of order $(M/\beta - 1)$ at the output layer (Fig. 3.1).

The objective function of the ith sub-filter of the lth layer of an HF is given by

$$J_i^{(l)}(n) = \frac{1}{2}\left(d(n) - [\boldsymbol{\theta}_i^{(l)}]^t \mathbf{x}_i^{(l)}(n)\right)^2, \qquad (3.2)$$

where $\mathbf{x}_i^{(l)}(n)$ the corresponding input vector defined as

$$\mathbf{x}_i^{(l)}(n) = \begin{cases} [x(n-(i-1)\beta), \ldots, x(n-(i-1)\beta-\beta+1)]^t & \text{for } l=1 \\ [y_1^{(1)}(n), y_2^{(1)}(n), \ldots, y_{M/\beta}^{(1)}(n)]^t & \text{for } l=2 \end{cases} \qquad (3.3)$$

and

$$\boldsymbol{\theta}_i^{(l)} = \begin{cases} [\theta_{i,1}^{(1)}, \theta_{i,2}^{(1)}, \ldots, \theta_{i,\beta}^{(1)}]^t & \text{for } l=1 \\ [\theta_{1,1}^{(2)}, \theta_{1,2}^{(2)}, \ldots, \theta_{1,M/\beta}^{(2)}]^t & \text{for } l=2 \end{cases} \qquad (3.4)$$

is the coefficient vector ($i = 1, 2, \ldots, M/\beta$). Hence, the coefficients of each sub-filter are updated according to the recursive equations

$$\boldsymbol{\theta}_i^{(l)}(n+1) = \boldsymbol{\theta}_i^{(l)}(n) + \mu e_i^{(l)}(n)\mathbf{x}_i^{(l)}(n), \qquad (3.5)$$

where $e_i^{(l)}(n)$ is the adaptation error given by $e_i^{(l)}(n) = d(n) - [\boldsymbol{\theta}_i^{(l)}]^t \mathbf{x}_i^{(l)}(n)$ for the ith sub-filter of the lth layer of the HF. The computational complexity of the HLMS depends on the number of layers and the order of the sub-filters and for structures that use constant sub-filter length throughout all layers is given by $\mathcal{O}(\beta)(\beta^\alpha - 1)/(\beta - 1)$, where β is the sub-filter length and α the number of layers.

The Wiener solution of this structure towards which its coefficients converge is given by (Appendix A)

$$\boldsymbol{\theta}_i^{(1)} = \boldsymbol{\theta}_i^\circ + \sum_{j=1, j \neq i}^{M/\beta} R_{\beta \times \beta}^{(1)}{}^{-1}(0) R_{\beta \times \beta}^{(1)}((i-j)\beta) \boldsymbol{\theta}_j^\circ \qquad (3.6)$$

for the coefficients of the input layer, when the desired signal is produced by a moving average (MA) process described by $d(n) = [\boldsymbol{\theta}^\circ]^t \mathbf{x}(n) = \sum_{j=1}^{\beta} \boldsymbol{\theta}_j^\circ \mathbf{x}_j^{(1)}(n)$ and by

$$\boldsymbol{\Theta}_1^{(2)} = \left[\boldsymbol{\Theta}^{(1)t} R_{M \times M}^{(1)}(0) \boldsymbol{\Theta}^{(1)} \right]^{-1} \boldsymbol{\Theta}^{(1)t} R_{M \times M}^{(1)}(0) \boldsymbol{\Theta}^\circ \qquad (3.7)$$

for the coefficients of the second layer, where

$$\boldsymbol{\Theta}^{(1)} = \begin{bmatrix} \boldsymbol{\theta}_1^{(1)}(n) & \mathbf{0}_{\beta \times 1} & \cdots & \mathbf{0}_{\beta \times 1} \\ \mathbf{0}_{\beta \times 1} & \boldsymbol{\theta}_2^{(1)}(n) & \cdots & \mathbf{0}_{\beta \times 1} \\ \vdots & \vdots & \ddots & \vdots \\ \mathbf{0}_{\beta \times 1} & \mathbf{0}_{\beta \times 1} & \cdots & \boldsymbol{\theta}_{M/\beta}^{(1)}(n) \end{bmatrix}, \qquad (3.8)$$

$R_{\beta \times \beta}^{(1)}$ the $\beta \times \beta$ and $R_{M \times M}^{(1)}$ the $M \times M$ autocorrelation matrices of the input signal of the first layer.

3.3.2 Evaluation of the Performance of HLMS

Equations (3.6) and (3.7) are very important since they describe the behaviour of the HLMS, and they explain most of the attributes that have been observed and reported in previous published works.

From (3.6), which illustrates the dependence of the estimates $\boldsymbol{\theta}_i^{(1)}$ on the statistical properties of the input signal $x(n)$ and the optimum coefficient values $\boldsymbol{\theta}_j^\circ$, $i, j = 1, 2, \ldots, \beta$, it is apparent that, the estimates of the unknown coefficients derived from the sub-filters of the input layer of an HF are usually biased [16]. Indeed, the bias term $\sum_{j=1, j \neq i}^{\beta} R_{\beta \times \beta}^{(1)}{}^{-1}(0) R_{\beta \times \beta}^{(1)}((i-j)\beta) \boldsymbol{\theta}_j^\circ$ is minimised when the input signal is white since in this case the autocorrelation $R_{\beta \times \beta}^{(1)}((i-j)\beta)$ is zero for $j \neq i$. Furthermore, if only a few of the optimal coefficients are nonzero forming a group into a vector $\boldsymbol{\theta}_{i_o}^\circ$ and $\boldsymbol{\theta}_i^\circ = 0$ for every $i \neq i_o$ then the estimates of these coefficients are unbiased, and the bias in the estimation of the rest zero coefficients is usually negligible, since from (3.6) and (3.7), it is derived that for $i = i_o$, $\boldsymbol{\theta}_i^{(1)} = \boldsymbol{\theta}_{i_o}^\circ$ and $\theta_i^{(2)} = 1$, while $\boldsymbol{\theta}_i^{(1)} = R_{\beta \times \beta}^{(1)}{}^{-1}(0) R_{\beta \times \beta}^{(1)}((i_o - i)\beta) \boldsymbol{\theta}_{i_o}^\circ$ and $\theta_{1i}^{(2)} = 0$ for $i \neq i_o$. In this case any failure of the first level of the structure to identify successfully the optimum coefficients is corrected at the second level.

Writing (3.6) in matrix form yields

$$\boldsymbol{\Theta}^{(1)} = \mathbb{R}^{(1)}{}^{-1}(0) R_{M \times M}^{(1)}(0) \boldsymbol{\Theta}^\circ, \qquad (3.9)$$

where

$$\mathbb{R}^{(1)^{-1}}(0) = \begin{bmatrix} R^{(1)}_{\beta \times \beta}(0)^{-1} & 0_{\beta \times \beta} & \cdots & 0_{\beta \times \beta} \\ \vdots & \vdots & \ddots & \vdots \\ 0_{\beta \times \beta} & 0_{\beta \times \beta} & \cdots & R^{(1)}_{\beta \times \beta}(0)^{-1} \end{bmatrix}. \quad (3.10)$$

From (3.9) it is observed that the bias decreases as β increases and when $\beta = M$ then $\mathbf{\Theta}^{(1)} = \mathbf{\Theta}^{\circ}$ since in this case $\mathbb{R}^{(1)^{-1}}(0) = R^{(1)}_{N \times N}{}^{-1}(0)$.

The fact that the estimated coefficients of the output level of the structures employed by these hierarchical algorithms depend on the statistical characteristics of the input signal as well is highlighted in (3.7). Moreover, this mathematical expression illustrates the effect of the estimated coefficients of the input level on the estimates of the output layer. Obviously, an error in the estimation of the coefficients of the lower level is propagated towards the higher levels of the structure.

3.3.3 The Hierarchical Gradient Descent Algorithm

The HGD algorithm [4] adapts the coefficients of the HF so as to minimise the total adaptation error, defined as the difference between the desired response and the output of the HF. It is a member of the class of back-propagation algorithms and it consists of two passes: in the forward pass the output of the structure and the gradient of the total error with respect to the coefficients of the only sub-filter of this structure is computed, while in the backward pass this gradient value is propagated towards the lower layers of this structure [9]. For the case of an HF with two layers and no overlapping, where the length of the input layer sub-filters is β while that of the unique output layer sub-filter is M/β, HGD can be summarised in the following set of equations:

```
Forward pass: Output Gradient computation
```

$$y_1^{(2)}(n) = \sum_{i=1}^{M/\beta} \theta_{1i}^{(2)} \sum_{j=1}^{\beta} \theta_{ij}^{(1)}(n) \cdot x(n - (i-1)\beta - (j-1)) \quad (3.11a)$$

$$\gamma_1^{(2)}(n) = -(d(n) - y_1^{(2)}(n)) = -e(n) \quad (3.11b)$$

$$\boldsymbol{\theta}_1^{(2)}(n+1) = \boldsymbol{\theta}_1^{(2)}(n) + \mu \gamma_1^{(2)}(n) \mathbf{x}_1^{(2)}(n) \quad (3.11c)$$

```
Backward pass: Gradient propagation
```

$$\gamma_i^{(1)}(n) = -\theta_{1i}^{(2)}(n) \gamma_1^{(2)}(n) \quad (3.12a)$$

$$\boldsymbol{\theta}_i^{(1)}(n+1) = \boldsymbol{\theta}_i^{(1)}(n) + \mu \gamma_i^{(1)}(n) \mathbf{x}_i^{(1)}(n) \quad (3.12b)$$

The HGD algorithm, contrary to the HLMS, converges to unbiased solutions, like all the back-propagation algorithms. It suffers from all the drawbacks that accompany back-propagation algorithms, like slow convergence and multimodal error surfaces. Moreover, its mathematical analysis is difficult enough, and it is similar to that of the standard back-propagation algorithm. Introducing a momentum term in the coefficients updating equation can reduce the settling time and prevent the convergence to local minima [3].

3.4 Applications

Like all adaptive filters, HFs can be applied to a wide range of applications like system identification, prediction, channel equalisation and noise cancellation among others. In these systems, the use of HFs is similar to that of any other adaptive filter: given an input signal they attempt to follow a specific desired response by training their coefficients with an appropriate learning algorithm. Moreover, HFs can be used for knowledge retrieval. More analytically:

- Relying on the fact that the HLMS returns extremely biased estimates when performing in a system identification context, unless the major part of the energy of the unknown channel is contained in just a few coefficients of the unknown channel, sparsity detection can be performed without requiring the identification of the unknown channel [5].
- Using a different even moment of the adaptation error for every sub-filter of the input layer as cost function, it is possible to develop a system for the identification of the dynamics of the desired response.

To illustrate the potential of HF in the subsequence, initially we will discuss their performance in a system identification setting. Subsequently, we will present a system that detects sparsity within an unknown channel, by simply comparing the performance of an HF that updates its weight values with the HLMS to that of a transversal filter whose coefficients are trained with the LMS algorithm.

3.4.1 Standard Filtering Applications

The performance of the HGD and the HLMS algorithm in a system identification context is depicted in Fig. 3.4. The HF that was used had two layers. The input layer had four sub-filters of length 10, while the output layer had a unique sub-filter of length 4. Two unknown channels were employed (a) an FIR filter of order 39 whose coefficients were generated by a Gaussian process $\mathcal{N}(0,1)$ (*complete channel*) and (b) an FIR filter with 40 coefficients that were produced by a Gaussian process $\mathcal{N}(0, 10^{-3})$ apart from $\theta^o_{11}, \theta^o_{12}, \ldots, \theta^o_{15}$, which were produced by a process $\mathcal{N}(0,1)$ (*sparse channel*). This is a typical case in many telecommunication channels [1, 12] and in underwater communications [6]. The input signal was either white with zero mean an unit variance, or coloured noise that was produced by feeding a stable autoregressive model with transfer function $F(z) = 1/(1 - 1.79z^{-1} + 1.85z^{-2} - 1.27z^{-3} + 0.41z^{-4})$ with white Gaussian noise $w(k) \sim \mathcal{N}(0,1)$. The MSE curves of a transversal filter of length 40, whose coefficients were updated with the LMS algorithm are also provided as a benchmark. The desired response was contaminated by 40 dB additive white Gaussian noise. The MSE curves illustrated in Fig. 3.4 have been produced by averaging the results of 100 independent simulation runs. From these it is concluded that

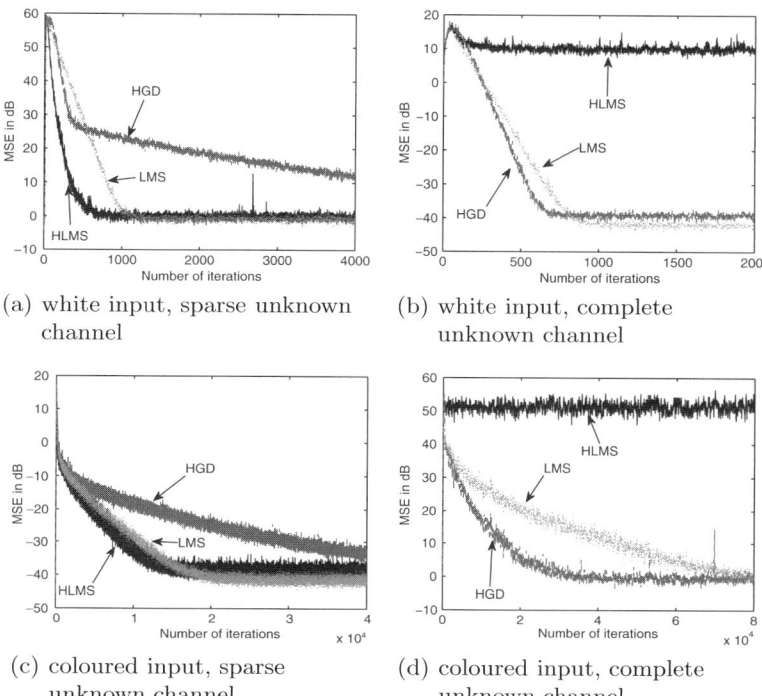

Fig. 3.4. The performance of the HLMS and the HGD algorithm for a two-layered HF with four sub-filters of length 10 at the output layer and a sub-filter of length 4 at the output layer compared to that of the LMS applied to an FIR filter of length 40

- HLMS outperforms HGD and LMS, in terms of convergence speed, when the unknown channel is *sparse*; however, when the energy of the unknown channel is distributed to all of its coefficients (complete channel) the estimates that HLMS produces are biased, resulting in significantly increased steady state error.
- When the unknown channel is *complete* HGD reaches steady state faster than HLMS and LMS; this is not the case though when the unknown is sparse. Moreover, HGD converges to unbiased solutions irrespectively of the characteristics of the unknown channel.

Similar results are obtained from the application of HFs to other adaptive filtering applications, like prediction and channel equalisation.

3.4.2 Knowledge Extraction

In this section, a system that compares the outputs of an HF and an FIR filter to decide whether an unknown channel is sparse is presented. The fundamental

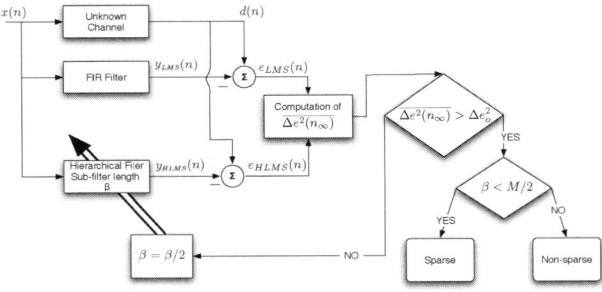

Fig. 3.5. Diagram of the experimental set-up used for sparsity detection

idea of this system is to compare the output of an HF to that of an FIR filter to decide whether a channel is sparse or not (Fig. 3.5). Its functioning can be described as follows: Initially a proper order $M-1$ for the adaptive filters is chosen so as to avoid under-modelling of the unknown channel. A value for n_∞, which denotes the number of iterations required by the LMS algorithm to reach steady state, is also chosen. Notice that n_∞ depends on the order of the adaptive filter and the employed learning rate. The hierarchical filter has two layers. The sub-filter length is initially set to $\beta = M/2$. At time instant $n = 0$ the adaptation of the coefficients of the FIR and the hierarchical filter commences. The quantity that is observed so as to make a decision concerning the sparsity of the channel is

$$\Delta e^2(n) = e_{\text{HLMS}}^2(n) - e_{\text{LMS}}^2(n).$$

To reduce the *randomness* of this signal and to be able to get more robust decisions a momentum term was introduced [8] resulting in

$$\overline{\Delta e^2(n)} = (1-\lambda)\overline{\Delta e^2(n-1)} + \lambda \Delta e^2(n). \qquad (3.13)$$

When the value of $\overline{\Delta e^2(n)}$ tends to zero as n increases, then it is concluded that the unknown channel is sparse. On the contrary, when there is no sparseness $\Delta e^2(n)$ progressively increases. Thus, observing the value of $\overline{\Delta e^2(n)}$ at a time instant n_∞ after the commencement of the adaptation process and comparing it to a hard-bound Δe_o^2 a decision can be made on the sparsity of the channel. If $\overline{\Delta e^2(n)} < \Delta e_o^2$ the sub-filter length is decreased and the whole procedure is repeated for this new value. When $\overline{\Delta e^2(n_\infty)}$ is found to have a large positive value then the final length of the sub-filter β_f is checked. If this is smaller than $M/2$, i.e. its initial value, then the channel is sparse. The nonzero coefficients in this case are located in a neighbourhood of width $2\beta_f$ in the impulse response of the unknown channel. Their exact location can be found by checking which one of the coefficients of the output layer converges to a nonzero value when $\beta = 2\beta_f$.

To evaluate the performance of proposed method, the determination of the sparseness of an unknown channel of length 256 was attempted. Its impulse

Table 3.2. Decisions of the consecutive steps proposed algorithm algorithm for sparsity detection

Step	β	$10\log(\overline{\Delta e^2(n_\infty)})$	$\overline{\Delta e^2(n_\infty)} < \Delta e_o^2$
1	64	−283.3	Yes
2	32	−285.44	Yes
3	**16**	**−287.36**	**Yes**
4	8	10.43	No

The decision threshold Δe_o^2 was set to 10^{-5}

response had only eight nonzero samples located at positions $k = 85, 86, \ldots, 92$ and their values were given by $\theta_k^o = \exp(-(k-k_0)/\tau)$, where $k = k_o, k_o + 1, \ldots, k_o + d$.

To enhance adaptation, the learning rate of the LMS algorithm was normalised with respect to the power of its input vector. For the HLMS algorithm the step-size of every sub-filter was normalised with respect to the power of its regressor vector. The order of both adaptive filters was set to 256 as well, to avoid under-modelling. For this length a proper value for n_∞ was 3×10^3. The decision on the sparsity of the unknown channel was based on the comparison of $\overline{\Delta e^2(n_\infty)}$ to the threshold $\Delta e_o^2 = 10^{-5}$. The input signal was white noise of zero mean and unit variance and for the computation of $\overline{\Delta e^2(n_\infty)}$ the momentum term λ was set to 0.8.

Initially the sub-filter length was set to $\beta = M/2$, i.e. 64. Since $\Delta e^2(n)$ fell bellow the predefined threshold within a few iterations the sub-filter length was reduced to $\beta = M/4 = 32$. The value of $\Delta e^2(n)$ became again smaller than Δe_o^2 and the β was reduced again. The same procedure was repeated several times. After some sub-filter order reductions a value β_f was found for which $\overline{\Delta e^2(n_\infty)} > \Delta e_o^2$. This value in our case was $\beta_f = 8$. Since $\beta_f < M/2$, it was concluded that the unknown channel was sparse (Table 3.2). Moreover, by observing that $\theta_{1,5}^{(2)} >> \theta_{1,i}^{(2)}$ for $i = 1, \ldots, 4, 6, \ldots, 8$ when $\beta = 16$ it was concluded that the energy of the impulse response of the unknown channel was concentrated in the coefficients $[\theta_{i_o\beta}^\star, \ldots, \theta_{(i_o+1)\beta-1}^\star] = [\theta_{80}^\star, \ldots, \theta_{95}^\star]$.

The squared error difference $\overline{\Delta e^2(n)}$ for the different stages of the proposed method is illustrated in Fig. 3.6 as a function of time. It is observed that when $\overline{\Delta e^2(n)}$ is decreasing with time the coefficient $\theta_{1,i_o}^{(2)}$ of the second layer that converges to 1 indicates that the nonzero coefficients $\theta_k^\star$, of the unknown channel, have indexes $k = i_o\beta, \ldots, (i_o+1)\beta - 1$. This does not hold when $\overline{\Delta e^2(n_\infty)} > \Delta e_o^2$.

3.5 Conclusions

This chapter has addressed the use of hierarchical filters, which consist of several filters organised in a hierarchical structure, for the detection of sparsity in the system identification setting. The factors that affect the performance of

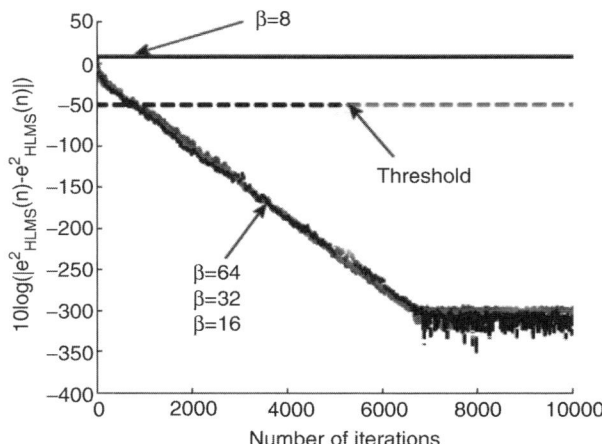

Fig. 3.6. The value of $\Delta e^2(n)$ as a function of time for several phases of the proposed sparsity detection method

these collaborative (and possibly distributed) adaptive filtering architectures, that is, the number of layer of the structure, the length of the sub-filters and the amount of overlapping between the input vectors of neighbouring sub-filters, have been thoroughly examined. The adaptation of the coefficients of hierarchical filters with least squares algorithms, including the hierarchical least mean square and the hierarchical gradient descent, has been discussed and the relation between hierarchical and FIR filters has been illustrated.

We have illustrated the use of hierarchical filters for knowledge extraction, in particular, this has been achieved for the assessment of the degree of sparsity of an unknown channel. This approach exploits the fact that the HLMS has significantly higher steady-state error than the LMS algorithm, unless the unknown channel is sparse and its nonzero coefficients lie in the vicinity of a single sub-filter within the input layer. It has been shown that by careful design of the input layer of the hierarchical filter, efficient sparsity detection can be achieved.

A Mathematical Analysis of the HLMS

When HLMS is employed for the adaptation of the coefficients of an HF, the objective function that the ith sub-filter of the first layer attempts to minimise is

$$J_i^{(1)}(n) = \frac{1}{2}[e_i^{(1)}(n)]^2, \qquad (3.14)$$

where $e_i^{(1)}(n)$ the *local* adaptation error. Taking the gradient of (3.14) with respect to $\boldsymbol{\theta}_i^{(1)}$ the optimal solution in the MSE sense for the coefficients of this sub-filter is found to be

$$\boldsymbol{\theta}_i^{(1)} = R_i^{(1)^{-1}} r_i^{(1)}, \tag{3.15}$$

where $R_i^{(1)}$ the autocorrelation matrix $E\left\{\mathbf{x}_i^{(1)^\mathrm{T}}(n)\mathbf{x}_i^{(1)}(n)\right\}$ of the input signal of the corresponding sub-filter and $r_i^{(1)} = E\{d(n)\mathbf{x}_i^{(1)}(n)\}$ the cross-correlation vector of the output with the input. For wide sense stationary processes, it holds that $R_1^{(1)} = R_2^{(1)} = \cdots = R_i^{(1)} = \cdots = R_\beta^{(1)} = R_{\beta \times \beta}^{(1)}$. Without loss of generality we assume that the desired response is an MA process of order $(M-1)$ (there is no measurement noise and the order of the process is the same with that of the HF), described by

$$d(n) = [\boldsymbol{\Theta}^\circ]^t \mathbf{x}(n) = \sum_{i=1}^{\beta} \boldsymbol{\theta}_i^\circ \mathbf{x}_i^{(1)}(n), \tag{3.16}$$

where $\boldsymbol{\Theta}^\circ = [\boldsymbol{\theta}_1^{\circ t}, \boldsymbol{\theta}_2^{\circ t}, \ldots, \boldsymbol{\theta}_\beta^{\circ t}]^t$ and $\boldsymbol{\theta}_i^\circ = [\theta_{(i-1)\beta+1}^\circ, \ldots, \theta_{i\beta}^\circ]^t$ the coefficient vectors and $\mathbf{x}(n) = [\mathbf{x}_1^{(1)^t}(n), \mathbf{x}_2^{(1)^t}(n), \ldots, \mathbf{x}_\beta^{(1)^\mathrm{T}}(n)]^t$ and $\mathbf{x}_i^{(1)}(n) = [x(n-(i-1)\beta), \ldots, x(n-i\beta+1)]^\mathrm{T}$ the input vectors. Thus the cross-correlation vector $r_i^{(1)}$ is given by

$$r_i^{(1)} = E\left\{\sum_{j=1}^{M/\beta} R_{\beta \times \beta}^{(1)}((i-j)\beta)\,\boldsymbol{\theta}_j^\circ\right\}. \tag{3.17}$$

Combining (3.15) and (3.17) results in the following expression that illustrates the relation between the coefficients $\boldsymbol{\theta}_i^{(1)}$ computed by the sub-filters of the input layer and the unknown coefficients θ_i°

$$\boldsymbol{\theta}_i^{(1)} = \boldsymbol{\theta}_i^\circ + \sum_{j=1, j \neq i}^{M/\beta} R_{\beta \times \beta}^{(1)^{-1}}(0) R_{\beta \times \beta}^{(1)}((i-j)\beta)\,\boldsymbol{\theta}_j^\circ. \tag{3.18}$$

Equation (3.18) can be re-written in matrix form as follows:

$$\boldsymbol{\Theta}^{(1)} = \mathbb{R}^{(1)^{-1}}(0) R_{M \times M}^{(1)}(0) \boldsymbol{\Theta}^\circ, \tag{3.19}$$

where $\mathbb{R}^{(1)^{-1}}(0)$ is an $M \times M$ matrix defined as

$$\mathbb{R}^{(1)^{-1}}(0) = \begin{bmatrix} R_{\beta \times \beta}^{(1)^{-1}}(0) & 0_{\beta \times \beta} & \cdots & 0_{\beta \times \beta} \\ 0_{\beta \times \beta} & R_{\beta \times \beta}^{(1)^{-1}}(0) & \cdots & 0_{\beta \times \beta} \\ \vdots & \vdots & \ddots & \vdots \\ 0_{\beta \times \beta} & 0_{\beta \times \beta} & \cdots & R_{\beta \times \beta}^{(1)^{-1}}(0) \end{bmatrix}. \tag{3.20}$$

A similar analysis can be derived for the Wiener solution of the second (output) layer. The autocorrelation matrix of the input of the single sub-filter of this layer is

$$R^{(2)}_{M/\beta \times M/\beta} = R^{(2)}_1 = \exp \mathbf{x}^{(2)}_1(n)\mathbf{x}^{(2)^t}_1(n), \quad (3.21)$$

where $\mathbf{x}^{(2)}_1(n) = [x^{(2)}_{11}, x^{(2)}_{11}, \ldots, x^{(2)}_{1\beta}]^t$ and $x^{(2)}_{1i} = \boldsymbol{\theta}^{(1)^t}_i(n)\mathbf{x}^{(1)}_i(n)$. The vector $\mathbf{x}^{(2)}_1(n)$ can be rewritten as

$$\mathbf{x}^{(2)}_1(n) = \begin{bmatrix} \boldsymbol{\theta}^{(1)}_1(n) & \mathbf{0}_{\beta \times 1} & \cdots & \mathbf{0}_{\beta \times 1} \\ \mathbf{0}_{\beta \times 1} & \boldsymbol{\theta}^{(1)}_2(n) & \cdots & \mathbf{0}_{\beta \times 1} \\ \vdots & \vdots & \ddots & \vdots \\ \mathbf{0}_{\beta \times 1} & \mathbf{0}_{\beta \times 1} & \cdots & \boldsymbol{\theta}^{(1)}_{M/\beta}(n) \end{bmatrix}^t \mathbf{x}^{(1)}(n) = [\boldsymbol{\Theta}^{(1)}]^t(n)\mathbf{x}^{(1)}(n). \quad (3.22)$$

The result of the above matrix operation is an $(M \times M/\beta)^t \times (M \times 1) = (M/\beta \times 1)$ vector. Substituting $\mathbf{x}^{(2)}_1(n)$ in (3.21) results in

$$R^{(2)}_{M/\beta \times M/\beta} = \boldsymbol{\Theta}^{(1)^T} \exp \begin{bmatrix} \mathbf{x}^{(1)}_1(n)\mathbf{x}^{(1)^t}_1(n) & \cdots & \mathbf{x}^{(1)}_1(n)\mathbf{x}^{(1)^t}_{M/\beta}(n) \\ \vdots & \ddots & \vdots \\ \mathbf{x}^{(1)}_\beta(n)\mathbf{x}^{(1)^t}_1(n) & \cdots & \mathbf{x}^{(1)}_{M/\beta}(n)\mathbf{x}^{(1)^t}_{M/\beta}(n) \end{bmatrix} \boldsymbol{\Theta}^{(1)}. \quad (3.23)$$

Notice that the matrix in the right-hand side of the previous equation is by definition the $M \times M$ autocorrelation matrix of the input signal of the first layer $R^{(1)}_{M \times M}$. Therefore

$$R^{(2)}_{M/\beta \times M/\beta} = \boldsymbol{\Theta}^{(1)^t} R^{(1)}_{M \times M} \boldsymbol{\Theta}^{(1)}. \quad (3.24)$$

From this equation, it is apparent the the statistical characteristics of the input signal of the second layer depend on the coefficients of the first layer and the characteristics of the input signal $x(n)$. The cross-correlation vector of the input with the desired response of the output layer is

$$\boldsymbol{r}^{(2)}_1 = \exp \boldsymbol{\Theta}^{(1)^t}(n)\mathbf{x}^{(1)}(n)d(n) = \boldsymbol{\Theta}^{(1)^t} \begin{bmatrix} \boldsymbol{r}^{(1)}_1 \\ \vdots \\ \boldsymbol{r}^{(1)}_{M/\beta} \end{bmatrix}. \quad (3.25)$$

Equation (3.17) can be re-written as $\boldsymbol{r}^{(1)}_i = \sum_{j=1}^{M/\beta} R^{(1)}_{\beta \times \beta}((i-j)\beta) \boldsymbol{\theta}^o_j$ where $i = 1, 2, \ldots, M/\beta$. Therefore, (3.25) can be written in matrix form as

$$\begin{bmatrix} \boldsymbol{r}^{(1)}_1 \\ \vdots \\ \boldsymbol{r}^{(1)}_{M/\beta} \end{bmatrix} = R^{(1)}_{M \times M}(0)\boldsymbol{\theta}^o. \quad (3.26)$$

Combining (3.25) and (3.26) results in the following equation for the cross-correlation of the output layer:

$$r_1^{(2)} = \Theta^{(1)^t} R_{M \times M}(0) \Theta^\circ. \tag{3.27}$$

So the Wiener solution for the coefficients of the second layer gives

$$\theta_1^{(2)} = \left[\Theta^{(1)^t} R_{M \times M}^{(1)}(0) \Theta^{(1)} \right]^{-1} \Theta^{(1)^t} R_{M \times M}^{(1)}(0) \Theta^\circ. \tag{3.28}$$

Assuming that there exists an $M \times \beta$ matrix (pseudo-inverse) whose product with the $\beta \times N$ matrix $\Theta^{(1)^t}$ is the unitary $M \times M$ matrix $I_{M \times M}$, and that the autocorrelation matrix of the input signal of the first layer is reversible the above equation can produce

$$\Theta^\circ = \Theta^{(1)} \Theta_1^{(2)}. \tag{3.29}$$

References

1. Balakrishan, J., Sethares, W., Johnson, C.: Approximate channel identification via δ-signed correlation. Int. J. Adapt. Control Signal Proc. **16**, 309–323 (2002)
2. Beaufays, F.: Transform-domain adaptive filters: An analytical approach. IEEE Trans. Signal Proc. **43**, 422–431 (1995)
3. Boukis, C., Mandic, D.: A global gradient descent algorithm for hierarchical FIR adaptive filters. In: International Conference on DSP, pp. 1285–1288 (2002)
4. Boukis, C., Mandic, D., Constantinides, A.: A hierarchical feedforward adaptive filter for system identification. In: IEEE Workshop on Neural Networks for Signal Processing, pp. 269–278 (2002)
5. Boukis, C., Polymenakos, L.: Using hierarchical filters to detect sparseness within unknown channels. In: Knowledge-Based Intelligent Information and Engineering Systems, *Lecture Notes on Artificial Intelligence*, vol. 4253, pp. 1216–1223. Springer (2006)
6. Burdic, W.: Underwater Systems Analysis. Prentice-Hall, Englewood Cliffs, NJ (1984)
7. Farhang-Boroujeny, B.: Adaptive Filters: Theory and Applications. Wiley, New York (1998)
8. Haykin, S.: Neural Networks, A Comprehensive Foundation. Prentice-Hall, Englewood Cliffs (1994)
9. Kalouptsidis, N.: Signal Processing Systems: Theory and Design. Wiley, New York (1997)
10. Kautz, W.: Transient synthesis in the time domain. IRE Trans. Circuit Theory **CT-1**(3), 29–39 (1954)
11. Macleod, M.: Performance of the hierarchical LMS algorithm. IEEE Commun. Lett. **9**(12), 436–437 (2002)
12. Martin, R., Sethares, W., Williamson, R., Johnson, C.: Exploiting sparsity in adaptive filters. IEEE Trans. Signal Proc. **50**(8), 1883–1894 (2002)
13. Nascimento, V.: Analysis of the hierarchical LMS algorithm. IEEE Signal Proc. Lett. **10**(3), 78–81 (2003)

14. Polycarpou, M., Ioannou, P.: Learning and convergence analysis of neural-type structured networks. IEEE Trans. Signal Proc. **3**(1), 39–50 (1992)
15. Principe, J., de Vries, B., de Oliveira, P.: The gamma filter – a new class of adaptive IIR filters with restricted feedback. IEEE Trans. Signal Proc. **41**(2), 649–656 (1993)
16. Stoica, P., Agrawal, M., Ahgren, P.: On the hierarchical least-squares algorithm. IEEE Commun. Lett. **6**(4), 153–155 (2002)
17. Wahlberg, B.: System identification using laguerre models. IEEE Trans. Automat. Contr. **36**(5), 551–562 (1991)
18. Widrow, B., Stearns, S.: Adaptive Signal Processing. Prentice-Hall, Englewood Cliffs (1985)
19. Woo, T.K.: Fast hierarchical least mean square algorithm. IEEE Signal Proc. Lett. **8**(11), 289–291 (2001)
20. Woo, T.K.: HRLS: A more efficient RLS algorithm for adaptive FIR filtering. IEEE Signal Proc. Lett. **5**(3), 81–84 (2001)

4

Acoustic Parameter Extraction From Occupied Rooms Utilizing Blind Source Separation

Yonggang Zhang and Jonathon A. Chambers

Room acoustic parameters such as reverberation time (RT) can be extracted from passively received speech signals by certain 'blind' methods, thereby mitigating the need for good controlled excitation signals or prior information of the room geometry. Observation noise which is inevitable in occupied rooms will, however, degrade such methods greatly. In this chapter, a new noise reducing preprocessing which utilizes blind source separation (BSS) and adaptive noise cancellation (ANC) is proposed to reduce the unknown noise from the passively received reverberant speech signal, so that more accurate room acoustic parameters can be extracted. As a demonstration this noise reducing preprocessing is utilized in combination with a maximum-likelihood estimation (MLE)-based method to estimate the RT of a synthetic noise room. Simulation results show that the proposed new approach can improve the accuracy of the RT estimation in a simulated high noise environment. The potential application of the proposed approach for realistic acoustic environments is also discussed, which motivates the need for further development of more sophisticated frequency domain BSS algorithms.[1]

4.1 Introduction

Room reverberation time (RT) is a very important acoustic parameter for characterizing the quality of an auditory space. The estimation of room RT has been of interest to engineers and acousticians for nearly a century. This parameter is defined as the time taken by a sound to decay 60 dB below its initial level after it has been switched off [14]. RT by this classical definition is referred to as RT60 and denoted as T_{60} in this chapter.

---

[1] Some of the material in this chapter has also been submitted for publication in Neurocomputing, named as 'A combined blind source separation and adaptive noise cancellation scheme with potential application in blind acoustic parameter extraction.'

The reverberation phenomenon is due to multiple reflections of the sound from surfaces within a room. It distorts both the envelope and fine structure of the received sound. Room RT provides a measure of the listening quality of a room; so obtaining an accurate room RT is very important in acoustics. From an application perspective, obtaining accurate room acoustic measures such as room RT is often the first step in applying existing knowledge to engineering practices, diagnosing problems of spaces with poor acoustics and proposing remedial measures. From a scientific research perspective, more realistic and accurate measurements are demanded to enrich the existing knowledge base and correct its imperfections [8].

Many methods have been proposed to estimate RT. Typical methods can be seen in [3, 5, 7, 13, 16]. The first RT estimation method proposed by Sabine is introduced in [7]. It utilizes the geometrical information and absorption characteristics of the room. The methods presented in [5] and [16] extract RT by recording good controlled excitation signals, and measuring the decay rate of the received signal envelope. These traditional methods are not suitable for occupied rooms, where prior information of the room or good controlled excitation signals are normally difficult to obtain. To measure the RT of occupied rooms, an artificial neural network (ANN) method is proposed in [3], and a maximum likelihood estimation (MLE)-based scheme is proposed in [13]. Both of them utilize modern digital signal processing techniques, and can extract RT from occupied rooms by only utilizing passively received speech signals (throughout this chapter RT estimation is assumed to be "in situ", i.e., the excitation signal is assumed to be one generated by someone already within the room, methods based on introducing an external source are not considered). The advantage of these methods is that no prior information of the room or good controlled excitation signals are necessary. However, their performance will be degraded and generally biased by noise, thus they are not suitable for occupied rooms, where high noise is inevitable due to the existence of the audience. In this chapter the term high noise is used to denote signal-to-noise ratios (SNR) of approximately 0 dB and below, and under such conditions both the MLE method and ANN method will generally fail.

To improve the accuracy of RT estimation in high noise occupied rooms, a new approach is proposed in this chapter, which utilizes a combination of blind source separation (BSS) and adaptive noise cancellation (ANC) based upon the least mean square (LMS) algorithm. These adaptive techniques are exploited in a preprocessing stage before the RT estimation. The aim of this study is to improve the accuracy of the existing occupied room RT estimation methods by reducing the unknown noise contained in the received speech signals.

This chapter is organized as follows: the MLE-based RT estimation method and the new noise reducing preprocessing are introduced in Sect. 4.2. A demonstrative study of the proposed RT estimation method is presented in Sect. 4.3. The simulation results are given in Sect. 4.4. A discussion of the proposed approach is provided in Sect. 4.5. Section 4.6 concludes this chapter.

4.2 Blind Estimation of Room RT in Occupied Rooms

In this section, the MLE-based RT estimation method is introduced first. This method is suitable for low noise level conditions. To improve the accuracy of this method in high noise conditions such as occupied rooms, a new noise reducing preprocessing is presented. The new RT estimation method designed for high noise occupied rooms is then set up by combining the noise reducing preprocessing and the MLE-based RT estimation method.

4.2.1 MLE-Based RT Estimation Method

In this method, the RT of an occupied room is extracted from a passively received speech signal [13]. At first, the passively received speech signal is divided into several overlapped segments with the same length. Each segment can be deemed as an observed vector. This observed vector is modelled as an exponentially damped Gaussian random sequence, i.e., it is modelled as an element-by-element product of two vectors, one is a vector with an exponentially damped structure, and the other is composed of independent identical distributed (i.i.d.) Gaussian random samples. Note that the exponentially damped vector also models the envelope of the speech segment. The MLE approach is applied to the observed vector to extract the decay rate of its envelope. The RT can then be easily obtained from the decay rate, according to its definition. An estimation of RT can be extracted from one segment, and a series of RT estimates can be obtained from the whole passively received speech signal. The most likely RT of the room can then be identified from these estimates.

The mathematical formulation for the exponentially damped Gaussian random sequence model is as follows [13]:

$$y(n) = a(n)x(n), \qquad (4.1)$$

where $y(n)$ is the observed reverberant tail signal, $x(n)$ an i.i.d. random sequence with a normal distribution $N(0, \sigma)$ where 0 is the zero mean value and σ is the standard deviation, and $a(n)$ is a time-varying term which formulates the decay envelope of $y(n)$. Denoting the damping rate of the sound envelope by a single decay rate τ, the sequence $a(n)$ can be uniquely determined by

$$a(n) = \exp(-n/\tau). \qquad (4.2)$$

Thus, sequence $a(n)$ can be replaced by a scalar parameter a

$$a(n) = a^n, \qquad (4.3)$$

where

$$a = \exp(-1/\tau). \qquad (4.4)$$

Substituting (4.3) into (4.1), the observed sequence $y(n)$ can be modelled as
$$y(n) = a^n x(n). \quad (4.5)$$
With a set of observed signal samples, the MLE approach can be applied to extract the estimates of both the parameters a and σ. The decay parameter τ can then be obtained from (4.4). According to the definition the RT can be obtained from this decay rate $T_{60} = 6.91\tau$ [13].

Denote the N-dimensional vector of the observed signal samples $y(n)$ by **y**, the likelihood function of **y** from the model described in (4.5) can be formulated as [13]
$$L(\mathbf{y}; a, \sigma) = \left(\frac{1}{2\pi a^{(N-1)}\sigma^2}\right)^{\frac{N}{2}} \times \exp\left(-\frac{\sum_{n=0}^{N-1} a^{-2n}y^2(n)}{2\sigma^2}\right), \quad (4.6)$$
where a and σ are unknown parameters to be estimated from the observation **y**. The log-likelihood function is
$$\ln L(\mathbf{y}; a, \sigma) = -\frac{N(N-1)}{2}\ln(a) - \frac{N}{2}\ln(2\pi\sigma^2) - \frac{1}{2\sigma^2}\sum_{n=0}^{N-1} a^{-2n}y^2(n) \quad (4.7)$$

Thus for a given observation window N and observed signal vector **y**, the log-likelihood function is determined by the parameters a and σ. These two parameters can be estimated using an MLE approach. Differentiating the log-likelihood function in (4.7) with respect to a and σ yields
$$\frac{\partial \ln L(\mathbf{y}; a, \sigma)}{\partial a} = -\frac{N(N-1)}{2a} + \frac{1}{a\sigma^2}\sum_{n=0}^{N-1} na^{-2n}y^2(n) \quad (4.8)$$
and
$$\frac{\partial \ln L(\mathbf{y}; a, \sigma)}{\partial \sigma} = -\frac{N}{\sigma} + \frac{1}{\sigma^3}\sum_{n=0}^{N-1} a^{-2n}y^2(n). \quad (4.9)$$

By setting the partial derivatives of the log-likelihood function (4.8) and (4.9) to zero, the MLE estimates of a and σ can be obtained from the following equations:
$$-\frac{N(N-1)}{2a} + \frac{1}{a\sigma^2}\sum_{n=0}^{N-1} na^{-2n}y^2(n) = 0 \quad (4.10)$$
and
$$\sigma = \sqrt{\frac{1}{N}\sum_{n=0}^{N-1} a^{-2n}y^2(n)} \quad (4.11)$$

The problem of RT estimation from an observed vector now transfers to two equations. Two methods have been proposed in [12] to solve these two equations: one is an online method, and the other a block-based method. Generally, as the online method has a lower computational complexity, it is used in the later simulation.

With a given received noise-free reverberant speech signal, a series of overlapped speech segments can be obtained, from which a series of estimates of RT can be extracted. These estimates can then be used to identify the most likely RT of the room by using an order-statistic filter [13]. A simple and intuitive way to identify the RT from a series of estimations is to choose the first dominant peak of a histogram of the RT estimations. As shown by the authors, it provides reliable RT estimates in noise-free environments with passively received speech signals [13]. However, for high noise conditions, the RT estimates will generally be biased, or even wrong. Next a noise reducing preprocessing is introduced to improve the accuracy of the MLE-based RT estimation method for high noise conditions.

4.2.2 Proposed Noise Reducing Preprocessing

Due to the existence of the audience, passively received speech signals from occupied rooms are normally contaminated by noise, and both the prior knowledge of the noise component and the excitation speech component are unknown. To reduce the noise level contained in the received speech signal, some blind signal processing approaches are necessary. A powerful tool for extracting some unknown noise interference signal from mixtures of speech signals is the convolutive BSS method [11]. Naturally, given two spatially distinct observations, BSS can attempt to separate the mixed signals to yield two independent signals. One of these two signals mainly consists of the excitation speech signal plus residue of the noise and the other signal contains mostly the noise. The estimated noise signal obtained from BSS then serves as a reference signal within an ANC, in which an LMS algorithm is utilized. The output of the ANC is a reverberant speech signal with reduced noise component. In this work, it is assumed that the ANC only locks onto the contaminated noise component. As will be shown by the simulations, the accuracy of RT obtained from the output of the ANC is improved, due to the noise reducing preprocessing. Different stages of this framework are shown in Fig. 4.1.

If RT can be reliably estimated from high noise speech signals, a similar approach can be potentially performed on passively received music signals, and for estimating other acoustic parameters such as the early decay time (EDT) and speech transmission index (STI). With different applications, different BSS, ANC and RT estimation algorithms can be utilized. As a demonstration, a simulated study for the proposed method is given in Sect. 4.3

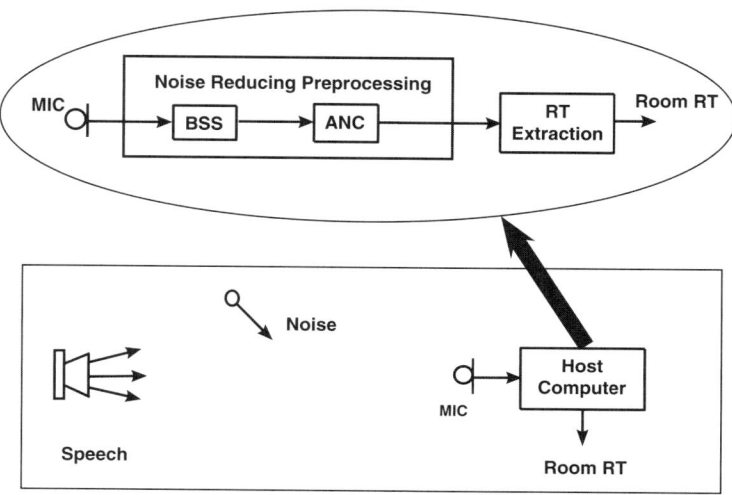

Fig. 4.1. RT estimation from high noise occupied rooms

4.3 A Demonstrative Study

As a demonstration for the proposed approach, a simulated room environment is utilized. The room size is set to be $10 \times 10 \times 5\,\text{m}^3$ and the reflection coefficient is set to be 0.7 in rough correspondence with the actual room. The excitation speech signal and the noise signal are two recorded anechoic 40 s male speech signals with a sampling frequency of 8 kHz, and scaled to have a unit variance over the whole observation. The positions of these two sources are set to be [1 m 3 m 1.5 m] and [3.5 m 2 m 1.5 m]. The positions of the two microphones are set to be [2.45 m 4.5 m 1.5 m] and [2.55 m 4.5 m 1.5 m], respectively. The impulse responses h_{ji} between source i and microphone j are simulated by a simplistic image room model which generates only positive impulse response coefficients [1]. The RT of this room measured by Schroeder's method [16] is 0.27 s. The setup of the simulation can be seen in Fig. 4.2, which is plotted in a two-dimensional way, since all the positions are in the same height. The impulse responses h_{11}, h_{12}, h_{21}, h_{22} are shown in Fig. 4.3. The different stages of the proposed RT estimation method for this simulation are shown in Fig. 4.4.

As shown in Fig. 4.4, the signal $s_1(n)$, which is assumed to be the noise signal in this work, is assumed statistically independent of the excitation speech signal $s_2(n)$. The passively received signals $x_1(n)$ and $x_2(n)$ are modelled as convolutive mixtures of $s_1(n)$ and $s_2(n)$. The room impulse response $h_{ij}(n)$ is the impulse response from source j to microphone i. BSS is used first to obtain the estimated excitation speech signal $\hat{s}_2(n)$ and the estimated noise signal $\hat{s}_1(n)$. The estimated noise signal $\hat{s}_1(n)$ then serves as the reference

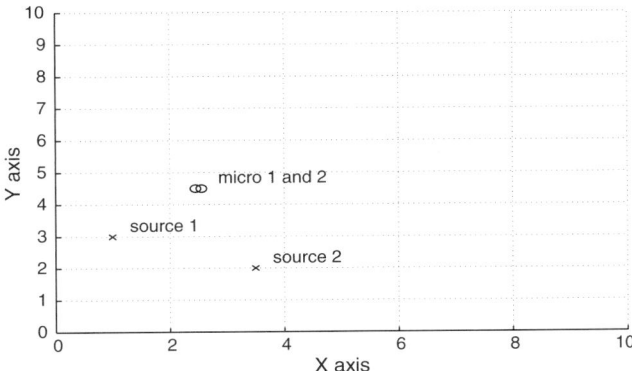

Fig. 4.2. Simulated room

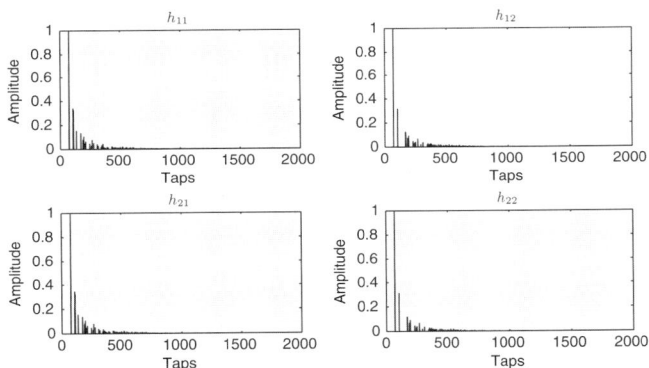

Fig. 4.3. Simulated room impulse responses

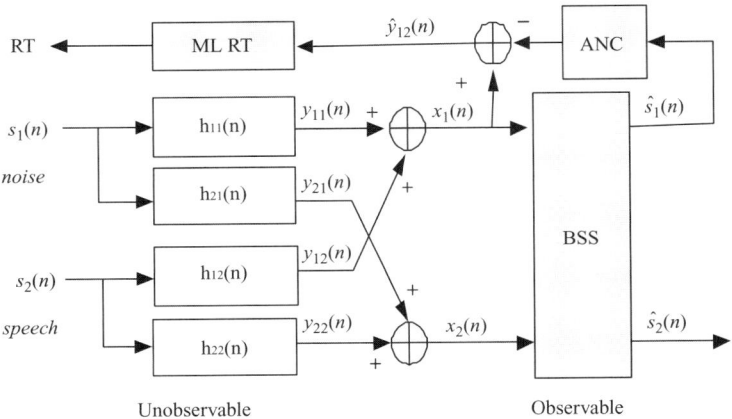

Fig. 4.4. Proposed blind RT estimation framework

signal for the ANC to remove the noise component from $x_1(n)$. The output of the ANC $\hat{y}_{12}(n)$ is an estimation of the noise-free reverberant speech signal $y_{12}(n)$. As compared with $x_1(n)$, it *crucially* retains the reverberant structure of the speech signal and has a low level of noise, therefore, it is more suitable to estimate the RT of the occupied room. As will be shown by the later simulation results, the proposed method can improve the accuracy of the RT estimation in this simulated high noise environment. Note that due to the symmetrical structure of the proposed approach, the signal $s_1(n)$ can also be deemed as an excitation speech signal, the signal $s_2(n)$ can then be deemed as a noise signal, and a similar approach can be performed to extract the RT.

Next, the implementation of the proposed noise reducing preprocessing will be introduced based on this simulation set up.

4.3.1 Blind Source Separation

As shown in Fig. 4.4, the goal of BSS is to extract the estimated noise signal $\hat{s}_1(n)$ from the received mixture signals $x_1(n)$ and $x_2(n)$. If we assume that the room environment is time invariant, the received mixtures $x_1(n)$ and $x_2(n)$ can be modelled as weighted sums of convolutions of the source signals $s_1(n)$ and $s_2(n)$. The equation that describes this convolved mixing process is:

$$x_j(n) = \sum_{i=1}^{2} \sum_{p=0}^{P-1} s_i(n-p) h_{ji}(p) + v_j(n), \quad j = 1, 2, \tag{4.12}$$

where $s_i(n)$ is the source signal from a source i, $x_j(n)$ is the received signal by a microphone j, $h_{ji}(p)$ is the P-point response from source i to microphone j, and $v_j(n)$ is the additive white noise, which in Fig. 4.4 is assumed to be zero. Using a T-point windowed discrete Fourier transformation (DFT), the time domain signal $x_j(n)$ can be converted into the time–frequency domain signal $x_j(\omega, n)$ where ω is a frequency index and n is a time index. For each frequency bin, we have

$$\mathbf{x}(\omega, n) = H(\omega) \mathbf{s}(\omega, n) + \mathbf{v}(\omega, n), \tag{4.13}$$

where $\mathbf{s}(\omega, n) = [s_1(\omega, n), s_2(\omega, n)]^\mathrm{T}$, $\mathbf{x}(\omega, n) = [x_1(\omega, n), x_2(\omega, n)]^\mathrm{T}$ and $\mathbf{v}(\omega, n) = [v_1(\omega, n), v_2(\omega, n)]^\mathrm{T}$ are the time–frequency representations of the source signals, the observed signals and the noise signals, $H(\omega)$ is a 2×2 matrix composed of $h_{ji}(\omega)$, which is the frequency representation for the mixing impulse response $h_{ji}(p)$, $(\cdot)^\mathrm{T}$ denotes vector transpose. The separation can be completed by a 2×2 unmixing matrix $W(\omega)$ of a frequency bin ω

$$\hat{\mathbf{s}}(\omega, n) = W(\omega) \mathbf{x}(\omega, n), \tag{4.14}$$

where $\hat{\mathbf{s}}(\omega,n) = [\hat{s}_1(\omega,n), \hat{s}_2(\omega,n)]^T$ is the time–frequency representation of the estimated source signals and $W(\omega)$ is the frequency representation of the unmixing matrix. $W(\omega)$ is determined so that $\hat{s}_1(\omega,n)$ and $\hat{s}_2(\omega,n)$ become mutually independent.

As a demonstration of the proposed approach, the frequency domain convolutive BSS method which exploits the nonstationary of the observed signals is utilized [11]. Considering the mixing model formulated in (4.12), the autocorrelation matrix of the observed signals at one frequency bin can be approximated by the sample mean

$$\overline{R}_x(\omega,n) = \frac{1}{N}\sum_{i=0}^{N-1} \mathbf{x}(\omega, n+iT)\mathbf{x}^T(\omega, n+iT). \tag{4.15}$$

If the sample averaging size N is large enough, from the independence assumption the estimated autocorrelation matrix can be written as

$$\overline{R}_x(\omega,n) \approx H(\omega)\Lambda_s(\omega,n)H^T(\omega) + \Lambda_v(\omega,n) \tag{4.16}$$

where $\Lambda_s(\omega,n)$ and $\Lambda_v(\omega,n)$ are the time–frequency formulations of the autocorrelation matrices for the source signals and the noise signals, and both are diagonal matrices. Thus the unmixing matrix $W(\omega)$ should satisfy that the estimated autocorrelation matrix of the source signals which is obtained as

$$\hat{\Lambda}_s(\omega,n) = W(\omega)[\overline{R}_x(\omega,n) - \Lambda_v(\omega,n)]W^T(\omega) \tag{4.17}$$

can be a diagonal matrix. Dividing the observed signals into K sections, K estimated autocorrelation matrices can be obtained from (4.15). By utilizing the nonstationary of the observed signals, the unmixing matrix is then updated to simultaneously diagonalize these K autocorrelation matrices, or equivalently, to simultaneously minimize the off-diagonal elements of the K matrices obtained from (4.17). The cost function can then be formulated as

$$J = \sum_{\omega=0}^{T-1}\sum_{k=1}^{K} \|E(\omega,k)\|^2 \tag{4.18}$$

where $\|\cdot\|^2$ denotes the Euclidean norm and

$$\|E(\omega,k)\|^2 = W(\omega)[\overline{R}_x(\omega,k) - \Lambda_v(\omega,k)]W^T(\omega) - \Lambda_s(\omega,k) \tag{4.19}$$

The least squares estimates of the unmixing matrix and the autocorrelation matrices of the source signals and noise signals can be obtained as

$$\hat{W}, \hat{\Lambda}_n, \hat{\Lambda}_s = \arg\min_{W,\Lambda_n,\Lambda_s} J \tag{4.20}$$

The gradients of the cost function are

$$\frac{\partial J}{\partial W(\omega)} = 2\sum_{k=1}^{K} E(\omega,k)W(\omega)[\overline{R}_x(\omega,t) - \Lambda_v(\omega,k)], \quad (4.21)$$

$$\frac{\partial J}{\partial \Lambda_s(\omega,k)} = -\text{diag}[E(\omega,k)], \quad (4.22)$$

$$\frac{\partial J}{\partial \Lambda_n(\omega,k)} = -\text{diag},[W^T(\omega)E(\omega,k)W(\omega)], \quad (4.23)$$

where $diag(\cdot)$ denotes the diagonalization operator which zeros the off-diagonal elements of the matrix. The optimal unmixing matrix $W(\omega)$ and the noise autocorrelation matrix $\Lambda_v(\omega,k)$ can then be obtained by a gradient descent algorithm using the gradients formulated in (4.21) and (4.23), and the autocorrelation matrix of the source signals can be obtained by setting the gradient in (4.22) to zero. With the estimates of $W(\omega)$ the time-frequency domain formulation of the separated signals can be obtained from (4.14), from which the time domain separated signals can then be obtained by using an inverse DFT (IDFT) operation. One advantage of this approach is that it incorporates uncorrelated noise, although in practice the noise may degrade its performance. More details of this approach can be seen in [11].

The performance of BSS can be evaluated by checking the separated noise signal $\hat{s}_1(n)$. According to the above formulation, we have the frequency domain mathematical representation of the separated noise signal:

$$\hat{s}_1(\omega,n) = c_{11}(\omega)s_1(\omega,n) + c_{12}(\omega)s_2(\omega,n), \quad (4.24)$$

where $c_{11}(\omega)$ and $c_{12}(\omega)$ are the frequency-domain representations of the combined system responses:

$$c_{11}(\omega) = h_{11}(\omega)w_{11}(\omega) + h_{21}(\omega)w_{12}(\omega) \quad (4.25)$$

and

$$c_{12}(\omega) = h_{12}(\omega)w_{11}(\omega) + h_{22}(\omega)w_{12}(\omega). \quad (4.26)$$

The performance of BSS can be classified into three possible cases:

1. A perfect performance, which is obtained if the separated noise signal can be approximately deemed as a scaled or delayed version of the original noise signal. In this case, the z-domain representation of the combined system response filter c_{11} can be formulated as $c_{11}(z) = Cz^{-\Delta}$ where C is a scalar and Δ is an integer to denote the delay, and the combined system response c_{12} is close to zero.
2. A good performance, which is obtained if the separated noise signal is approximately a filtered version of the source noise signal. In this case, the filter c_{11} is an unknown filter, and the filter c_{12} is approximately zero.

3. A normal performance, which is obtained if the separated noise signal contains both components of the original noise signal and the excitation speech signal. In this case, both filters c_{11} and c_{12} are two unknown filters.

If the performance of BSS is perfect, the estimation of the structure of the room impulse responses can be obtained from the inverse of the unmixing filters, and the room RT can then be estimated directly from the room impulse responses. However, in real applications, the performance of the most existing BSS algorithms is between case 2 and case 3, and the inverse of the unmixing matrix filters will be seriously biased from the room impulse responses. This is why an extra ANC stage is needed in the proposed framework.

4.3.2 Adaptive Noise Cancellation

After the BSS stage the estimated noise signal $\hat{s}_1(n)$ is obtained, which is highly correlated with the noise signal $s_1(n)$. This signal is then used as a reference signal in the ANC stage to remove the noise component from the received signal $x_1(n)$. Since the target signal $y_{12}(n)$ which is to be recovered is a highly nonstationary speech signal, a modified LMS algorithm namely the sum method [4] is used in this ANC stage. The update of the sum method is as follows:

$$e(n) = x_1(n) - \hat{\mathbf{s}}_1^T(n)\mathbf{w}(n), \tag{4.27}$$

$$\mathbf{w}(n+1) = \mathbf{w}(n) + \frac{\mu e(n)\hat{\mathbf{s}}_1(n)}{L[\hat{\sigma}_e^2(n) + \hat{\sigma}_s^2(n)]}, \tag{4.28}$$

where $e(n)$ is the output error of the adaptive filter, $\hat{\mathbf{s}}_1(n)$ is the input vector with a tap-length of L, $\mathbf{w}(n)$ is the weight vector of the adaptive filter, μ is the step size, $\hat{\sigma}_e^2(n)$ and $\hat{\sigma}_s^2(n)$ are estimations of the temporal error energy and the temporal input energy, which are obtained by first order smoothing filters:

$$\hat{\sigma}_e^2(n) = 0.99\hat{\sigma}_e^2(n-1) + (1-0.99)e^2(n) \tag{4.29}$$

and

$$\hat{\sigma}_s^2(n) = 0.99\hat{\sigma}_s^2(n-1) + (1-0.99)\hat{s}_1^2(n), \tag{4.30}$$

where $\hat{\sigma}_s^2(0) = 0$ and $\hat{\sigma}_e^2(0) = 0$. The choice of 0.99 is related to the window length of such estimates and is approximately $\frac{1}{1-0.99} = 100$. Moreover, the term $1 - 0.99$ ensures unbiased estimates.

The adaptation of the weight vector in (4.28) is based on the sum method in [4]. As explained by the author, the adaptation in (4.28) is adjusted by the input and output error variance automatically, which reduces the influence brought by the fluctuation of the input and the target signals.

If the BSS stage performs well, the output signal of the ANC $\hat{y}_{12}(n)$ should be a good estimation of the noise-free reverberant speech signal $y_{12}(n)$. If we denote the steady-state adaptive filter vector as $\mathbf{w}_s$ and its frequency domain representation as $w_s(\omega)$, the time–frequency domain representation of $\hat{y}_{12}(n)$ can be formulated as follows:

$$\begin{aligned}\hat{y}_{12}(\omega, n) &= x_1(\omega, n) - w_s(\omega)\hat{s}_1(\omega, n) \\ &= g_1(\omega)s_1(\omega, n) + g_2(\omega)s_2(\omega, n)\end{aligned} \quad (4.31)$$

where $g_1(\omega)$ and $g_2(\omega)$ are combined system responses:

$$g_1(\omega) = h_{11}(\omega) - w_s(\omega)c_{11}(\omega) \quad (4.32)$$

and

$$g_2(\omega) = h_{12}(\omega) - w_s(\omega)c_{12}(\omega) \quad (4.33)$$

where $c_{11}(\omega)$ and $c_{12}(\omega)$ are formulated in (4.25) and (4.26). With the three kinds of performances of BSS which are discussed in Sect. 4.3.1, we have three kinds of performance of the ANC, based on the output signal $\hat{y}_{12}(n)$:

1. If the BSS stage has a perfect performance, according to the discussion in Sect. 4.3.1 we have

$$\hat{y}_{12}(\omega, n) = [h_{11}(\omega) - w_s(\omega)Ce^{-j\Delta\omega}]s_1(\omega, n) + h_{12}(\omega)s_2(\omega, n). \quad (4.34)$$

   It is clear to see that if the adaptive filter of the ANC converges to the value of $w_s(\omega) = \frac{h_{11}(\omega)e^{j\Delta\omega}}{C}$, we can obtain a noise-free reverberant speech signal $\hat{y}_{12}(n) = y_{12}(n)$ at the output of the ANC.

2. If the BSS stage has a good performance, we have

$$\hat{y}_{12}(\omega, n) = [h_{11}(\omega) - w_s(\omega)c_{11}(\omega)]s_1(\omega, n) + h_{12}(\omega)s_2(\omega, n). \quad (4.35)$$

   If we want to remove the noise component from $\hat{y}_{12}(n)$, the first term of the right-hand side of (4.35) should be equal to zero, which results in $w_s(\omega) = h_{11}(\omega)/c_{11}(\omega)$. It requires that the combined system response $c_{11}(\omega)$ has an inverse. Thus, if the inverse of the combined system response is not realizable, the ANC cannot remove the noise component completely. According to simulation experience, in most cases the ANC stage can partially remove the noise component from the received mixture signal, and improve the accuracy of the RT estimates.

3. If the BSS stage has a normal performance, the output signal of the ANC will contain both components of the noise signal and the speech signal. The reverberant structure contained in $\hat{y}_{12}$ will be damaged. In practice, if the performance of BSS is poor, the noise contained in the mixture signal can not be removed, and an extra noise component will be introduced.

Next simulation results are presented based on the simulation set up and algorithms provided in this section.

4.4 Simulation Results

In this section, the performance of the proposed approach based on the simulation set up introduced in Sect. 4.3 is examined and compared with that of the original MLE-based method. The parameter setting for the BSS algorithm is as follows: the mixture signals are divided into $K = 5$ sections, so that 5 autocorrelation matrices of the mixture signals at each frequency bin are obtained. The DFT length is set to $T = 2048$. Details of the choice of this parameter can be seen in [2]. The unmixing filter tap-length is set to $Q = 512$, which is much less than T, to reduce the permutation ambiguity [11]. The step size of the update of the frequency domain unmixing matrix is set to unity. The parameter setting for the ANC stage is as follows: the tap-length of the adaptive filter coefficient vector is set to 500, which is chosen according to our simulation experience. The step size μ is set to 0.005. The window width which is used to obtain the observed vector in the MLE-based RT estimation method is set to 1,200, according to the discussion in [13]. All these parameters have been chosen empirically to yield the best performance. For each simulation, the performance of ANC combined with BSS is shown by comparing the combined system responses g_1 and g_2 which are formulated in (4.32) and (4.33) with the room impulse responses h_{11} and h_{12}. According to the motivation of the approach, g_2 should be close to the filter h_{12}, which contains the RT information, and g_1 should contain less energy as compared with h_{11}, so that the noise contained in $\hat{y}_{12}(n)$ is reduced as compared with the mixture signal $x_1(n)$.

The output signal of the ANC stage $\hat{y}_{12}(n)$ will then be used to extract the RT by using the MLE-based method. The RT results extracted from $\hat{y}_{12}(n)$ and $x_1(n)$ will be compared with the RT results extracted from the noise-free reverberant speech signal $y_{12}(n)$, to show the advantages of the proposed approach. The histogram of the RT results extracted from $y_{12}(n)$ and $x_1(n)$ can be seen in Figs. 4.5a, b. It is easy to see from these two figures that RT can be easily identified from Fig. 4.5a, which is obtained by using the noise-free reverberant speech signal $y_{12}(n)$: the peak of the RT estimation results appears at 0.3 s, and it is close to the real RT 0.27 s. There are many peaks in Fig. 4.5b which are obtained from the mixture signal $x_1(n)$ due to the high-level noise; thus, RT is difficult to be identified.

In the first simulation, we assume that the BSS stage has a perfect performance, and the separated signal is equal to the original signal, i.e., $\hat{s}_1 = s_1$. In this case, the combined system response g_2 is equal to h_{12}. To show the performance of ANC combined with BSS, we plot both combined system responses g_1 and g_2 in Fig. 4.6. It can be clearly seen that the combined system response g_1 is close to zero, which indicates that the output signal $\hat{y}_{12}(n)$ is very close to the noise-free reverberant speech signal $y_{12}(n)$, according to (4.31).

Then the MLE method is used to extract the RT. The RT from signal $\hat{y}_{12}$ is shown in Fig. 4.5c. It is clear to see in Fig. 4.5 that the histogram of the RT estimations obtained with the extracted signal $\hat{y}_{12}(n)$ is very similar to that

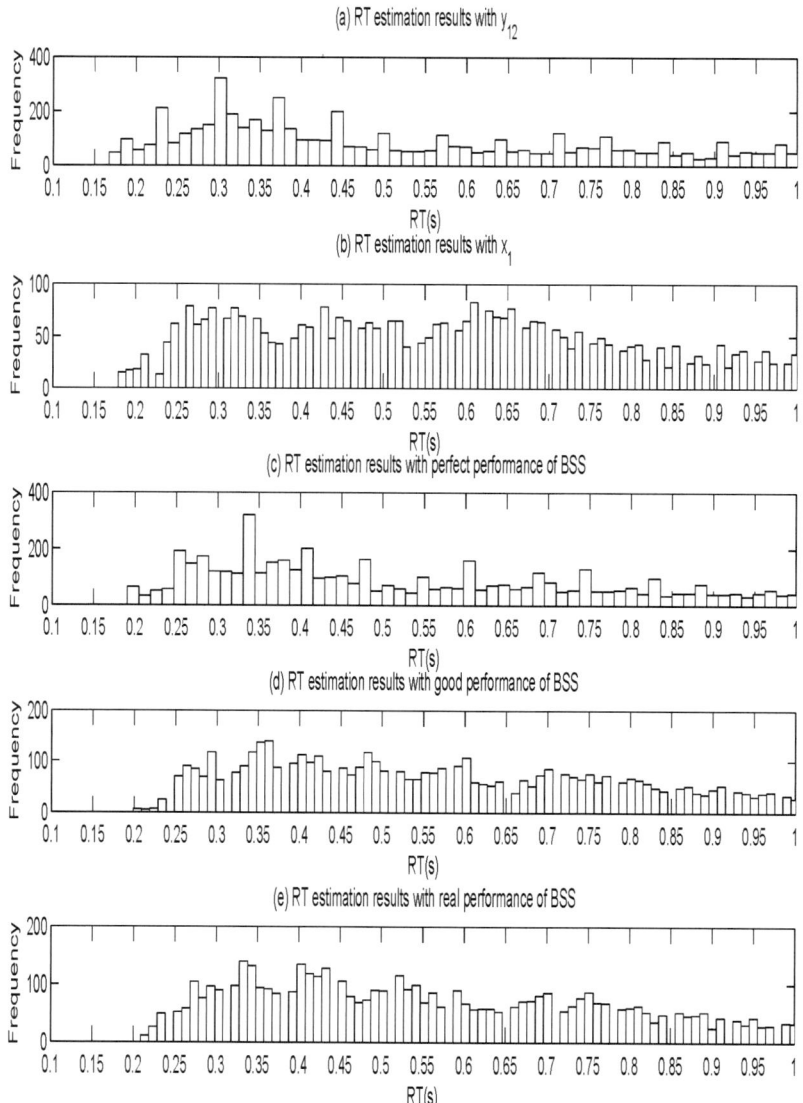

Fig. 4.5. The histogram of the RT estimation results with different signals

obtained by using the noise-free signal $y_{12}(n)$. An obvious peak of the RT estimation results appears at 0.34 s. Note that both the estimates in Fig. 4.5a, c are larger than the real RT due to the lack of sharp transients in the clean speech [13]. The RT results in Fig. 4.5c are slightly larger than the results in Fig. 4.5a due to the noise interference.

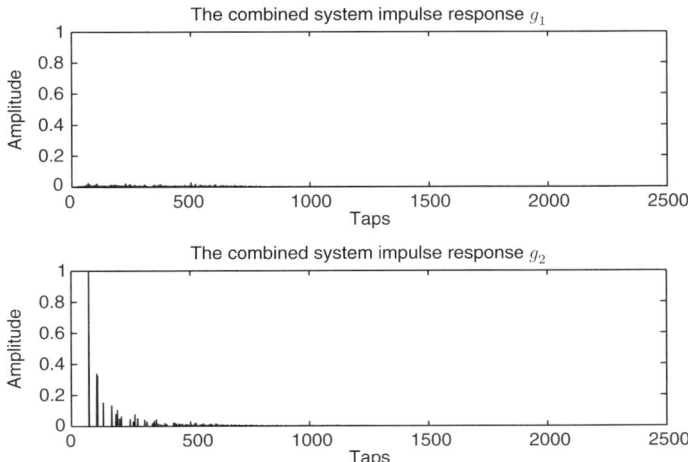

Fig. 4.6. Combined system responses with a perfect performance of BSS

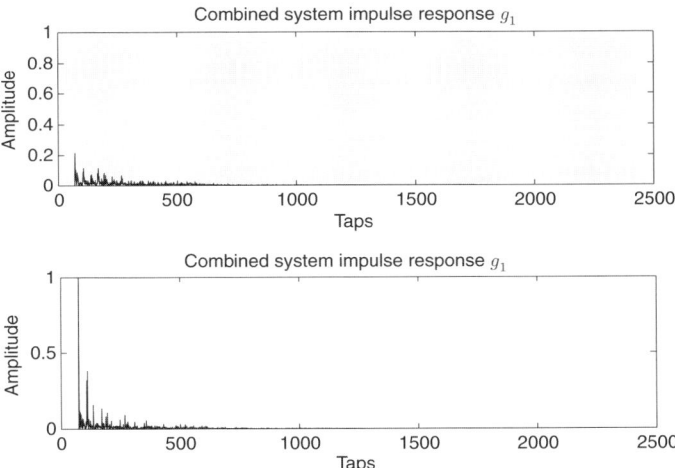

Fig. 4.7. Combined system responses with a good performance of BSS

In the second simulation, all the settings are the same as those of the first simulation, except that the reference signal is replaced with a filtered version of the noise signal. In this case, the frequency domain representation of the reference signal of the ANC can be formulated as

$$\hat{s}_1(\omega, n) = c_{11}(\omega) s_1(\omega, n), \tag{4.36}$$

where $c_{11}(\omega)$ is formulated in (4.25). The combined system responses g_1 and g_2 are shown in Fig. 4.7. From Fig. 4.7 we can see that the combined system response g_2 is very close to h_{12}, which contains the information of the RT. The

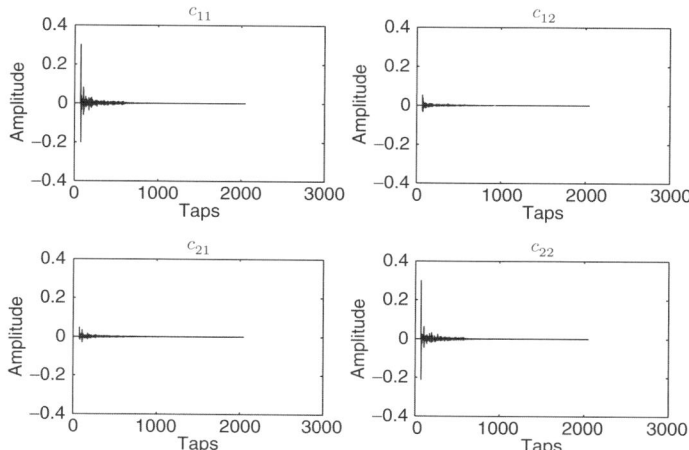

Fig. 4.8. Combined system responses c_{11}, c_{12}, c_{21}, and c_{22} of BSS

combined system response g_1, which contains the noise component, has less energy as compared with the filter h_{11}. Thus as compared with the mixture signal $x_1(n)$, the output signal $\hat{y}_{12}(n)$ has a much lower level noise component.

Similar to that of the previous simulation, the MLE method is then used to extract the RT from $\hat{y}_{12}(n)$. The histogram of the RT estimations are shown in Fig. 4.5d. The peak of the RT estimations appears at 0.36 s, and this peak is much clearer as compared with that in Fig. 4.5b, which indicates that the result obtained from $\hat{y}_{12}(n)$ is better than the result obtained from $x_1(n)$.

In the last simulation, the reference input of the ANC stage is the real output of the BSS stage. To show the performance of the BSS, we plot the combined system responses c_{11}, c_{12}, c_{21}, c_{22} in Fig. 4.8. It is clear to see from Fig. 4.8 that the separated signal $\hat{s}_1(n)$ mainly comes from the original noise signal $s_1(n)$, and the separated signal $\hat{s}_2(n)$ mainly comes from the original speech signal $s_2(n)$, thus the separated signal $\hat{s}_1(n)$ can be approximately deemed as a filtered version of the original noise signal $s_1(n)$, and is highly correlated with $s_1(n)$.

The separated signal $\hat{s}_1(n)$ serves as a reference signal within the ANC, to remove the noise component contained in the mixture signal $x_1(n)$. The MLE-based method is then used to extract the RT estimations from the output signal $\hat{y}_{12}(n)$. The combined system responses g_1 and g_2 are shown in Fig. 4.9, and the histogram of the RT estimations are shown in Fig. 4.5e. From Fig. 4.9 we can see, although the combined filter g_1 contains more energy as compared with that in Fig. 4.7, it is still much smaller as compared with h_{11}, which indicates the noise level contained in $\hat{y}_{12}(n)$ is still much less than that of $x_1(n)$. The peak of the RT estimation results in Fig. 4.5e appears at 0.33 s, and it is also clearer as compared with Fig. 4.5b.

4 Blind Extraction of Acoustic Parameters   71

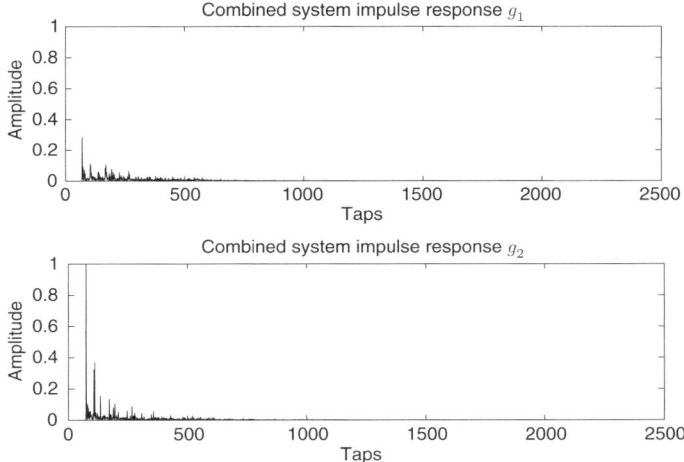

Fig. 4.9. Combined system responses with a real performance of BSS

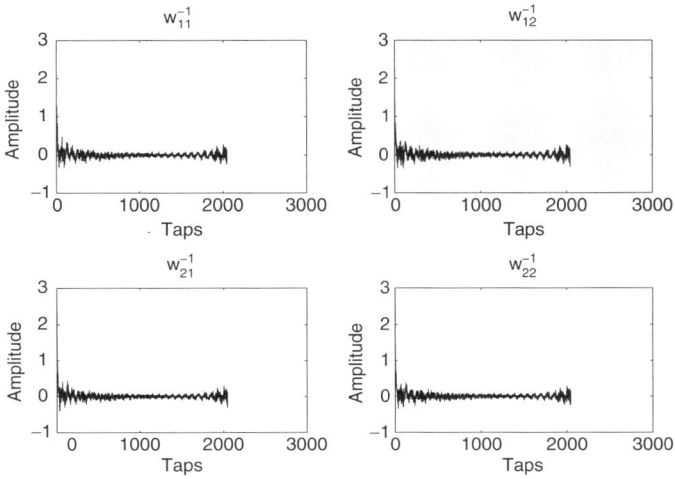

Fig. 4.10. Inverse of the unmixing filters

Furthermore, to show the limitation of the performance of the BSS stage, the inverse of the unmixing filters, w_{11}^{-1}, w_{12}^{-1}, w_{21}^{-1} and w_{22}^{-1} are also plotted in Fig. 4.10. Ideally, they should be estimations of the original mixing filters. However, it is evident in Fig. 4.10 that these filters are seriously biased from the original mixing filters plotted in Fig. 4.3. The RT measured by Schroeder's method [16] from these filters is -1.99 s, which is obviously wrong, since the RT can only be positive values. The negative result of RT is because the structure of the filters is no longer exponential decay. The simulation results plotted in Fig. 4.10 confirm the discussion of the performance of the BSS stage in Sect. 4.3.1, and explain why another ANC stage is necessary.

From the above simulations, we can conclude that utilizing BSS combined with ANC can reduce the noise component from the mixture signal while retaining the reverberant structure, and a more accurate RT result can be obtained. Although the performance of the proposed approach highly depends on the performance of the BSS stage, as shown by our simulations above, nonetheless, the accuracy of the RT estimation is improved, and reliable RT can be extracted by using this method within a simulated highly noisy room, something that has not previously been possible.

4.5 Discussion

It is clear to see in the above simulations that the key stage of the proposed approach is the BSS stage. Although, as shown in the simulation in Sect. 4.4, BSS is successfully used in a simple simulated room impulse response model, the application of BSS and ANC in acoustic parameter extraction is still limited in practice, mainly because of the performance of BSS for a real environment. Generally, there are three fundamental limitations of the BSS stage:

1. The permutation problem in the frequency domain BSS algorithms. Since the blind estimated unmixing matrix at one frequency bin can at best be obtained up to a scale and permutation, at each frequency bin the separated signal can be recovered at an arbitrary output channel. Consequently, the recovered source signal is not necessarily a consistent estimate of the real source over all frequencies. So it is necessary to solve the permutation problem and align the unmixing matrix to an appropriate order, so that the source signals can be recovered correctly. Many methods have been proposed to solve the permutation problem. In [11] the authors suggest to solve the permutation problem by imposing a smoothness constraint on the unmixing filters, i.e., exploiting unmixing matrix spectral continuity. The authors in [9] solve the permutation problem by exploiting the similarity of the signal envelope structure at different frequency bins from the same source signal. Geometrical constraints are used in [6, 10] to solve the permutation problem. A combined approach which utilizes both the similarity of the signal envelope structure and geometrical constraints is proposed in [15]. It has been shown that the combined approach is more robust and precise as compared with other methods, and can solve the permutation problem quite well. We have to point out that solving the permutation problem can improve the performance of frequency domain BSS algorithms only when the mixture signals have been well separated. Thus the separation process at each frequency bin is still the key problem for BSS.

2. The choice of the FFT frame size T. This parameter provides a trade-off between maintaining the independence assumption which is related with the sample number at each frequency bin, and covering the whole reverberation in frequency domain BSS [2]. On one hand, it is constrained that $T > P$ where P is the tap-length of the room impulse response, more strictly, $T > 2P$, so that a linear convolution can be approximated by a circular convolution. On the other hand, if T is too large, the sample number at each frequency bin may be too small, and the independence assumption will collapse. Thus, it is important to choose a proper value of the parameter T, as explained in [2].
3. As discussed in [2], the frequency domain BSS system can be understood as two sets of adaptive beamformers (ABFs). As shown in the simulation in [2], although BSS can remove the reverberant jammer sound to some extent, it mainly removes the sound from the jammer direction, i.e., the unmixing matrix $W(\omega)$ mainly removes the direct sound of the jammer signal, and the other reverberant components which arrive from different directions cannot be separated completely. For the room impulse response model used in this chapter, the energy contained in the reverberant tails is quite small as compared with that of the direct signal, and a good performance of BSS is obtained. For real room impulses, however, the performance of current frequency domain BSS algorithms degrade.

Due to the fundamental limitations of current frequency domain BSS algorithms, more research is needed to make the proposed approach work in more realistic acoustic environments.

4.6 Conclusion

In this chapter, a new framework has been proposed for blind RT estimation in a high noise environment by utilizing BSS and ANC. Detailed discussions and simulations are also provided for the performance of the proposed approach, and its potential application in practice. As can be seen in the simulations, the noise is removed greatly by the proposed framework from the received speech signal and the performance of this framework is good in a simulated high noise room environment. Due to the motivation of the framework, BSS and ANC can be potentially used together in many acoustic parameter estimation methods as a preprocessing. This framework provides a new way to overcome the noise disturbance in RT estimation. However, to make the proposed approach work in real acoustic parameter extraction, more research is needed, especially on the convolutive BSS algorithms. Future work will focus on the theoretical analysis of this blind RT estimation framework and the improvement of its stages, especially the improvement of convolutive BSS in long reverberation environments.

References

1. Allen, J.B., Berkley, D.A.: Image method for efficiently simulating small-room acoustics. J. Acoust. Soc. Am. **65**, 943–950 (1979)
2. Araki, S., Mukai, R., Makino, S., Nishikawa, T., Saruwatari, H.: The fundamental limitation of frequency domain blind source separation for convolutive mixtures of speech. IEEE Trans. Speech Audio Proces. **11**(2), 109–116 (2003)
3. Cox, T.J., Li, F., Darlington, P.: Extracting room reverberation time from speech using artificial neural networks. J. Audio Eng. Soc. **49**, 219–230 (2001)
4. Greenberg, J.E.: Modified LMS algorithm for speech processing with an adaptive noise canceller. IEEE Trans. Signal Proces. **6**(4), 338–351 (1998)
5. ISO 3382: Acoustics-measurement of the reverberation time of rooms with reference to other acoustical parameters. International Organization for Standardization (1997)
6. Knaak, M., Araki, S., Makino, S.: Geometrically constrained independent component analysis. IEEE Trans. Speech Audio Proces. **15**(2), 715–726 (2007)
7. Kuttruff, H.: Room Acoustics 4th ed. Spon, London (2000)
8. Li, F.F.: Extracting room acoustic parameters from received speech signals using artificial neural networks. Ph.D. thesis, Salford University (2002)
9. Murata, N., Ikeda, S., Ziehe, A.: An approach to blind source separation based on temporal structure of speech. Technical Report BSIS Technical Reports No.98-2, RIKEN Brain Science Institute (1998)
10. Parra, L., Alvino, C.V.: Geometric source separation: merging convolutive source separation with geometric beamforming. IEEE Trans. Speech Audio Proces. **10**(6), 352–362 (2002)
11. Parra, L., Spence, C.: Convolutive blind source separation of nonstationary sources. IEEE Trans. Speech Audio Proces. **8**(3), 320–327 (2000)
12. Ratnam, R., Jones, D.L., Jr. O'Brien, W.D.: Fast algorithms for blind estimation of reverberation time. IEEE Signal Proces. Lett. **11**(6), 537–540 (2004)
13. Ratnam, R., Jones, D.L., Wheeler, B.C., Jr. O'Brien, W.D., Lansing, C.R., Feng, A.S.: Blind estimation of reverberation time. J. Acoust. Soc. Am. **114**(5), 2877–2892 (2003)
14. Sabine, W.C.: Collected papers on acoustics. Harvard U.P. (1922)
15. Sawada, H., Mukai, R., Araki, S., Makino, S.: A robust and precise method for solving the permutation problem of frequency-domain blind source separation. IEEE Trans. Speech Audio Proces. **12**(5), 530–538 (2001)
16. Schroeder, M.R.: New method for measuring reverberation time. J. Acoust. Soc. Am. **37**, 409–412 (1965)

Part II

Signal Processing for Source Localization

5

Sensor Network Localization Using Least Squares Kernel Regression

Anthony Kuh and Chaopin Zhu

This chapter considers the sensor network localization problem using signal strength. Signal strength information is stored in a kernel matrix. Least squares kernel regression methods are then used to get an estimate of the location of unknown sensors. Locations are represented as complex numbers with the estimate function consisting of a linear weighted sum of kernel entries. The regression estimates have similar performance as previous localization methods using kernel classification methods, but at reduced complexity. Simulations are conducted to test the performance of the least squares kernel regression algorithm. We also consider the cases where sensors are mobile and on-line kernel regression learning algorithms are formulated to track moving sensors. Finally, we discuss some physical constraints on the sensor networks (i.e., communication and power constraints). To deal with these constraints, we proposed using distributed learning algorithms to cut down on communications between sensors. An ensemble of learners each solve a kernel regression algorithm and then communicate among each other to reach a solution. The communication costs are lowered using distributed learning algorithms and through simulations we show that the performance is comparable to the centralized kernel regression solution.

5.1 Introduction

Information gathering is relying more on distributed communication and distributed networking systems. Ad hoc sensor networks are being deployed in a variety of applications from environmental sensing to security and intrusion detection to medical monitoring [1]. These sensor networks are becoming increasingly more complex with sensors responsible for different tasks. We will consider a sensor network consisting of two different types of sensors: base sensors where the locations are known (the base sensors could have GPS) and simple sensors called motes where the locations of the motes are unknown. Here we consider the problem of estimating the location of the motes also referred to as the sensor localization problem. Given base node locations and signal strength information (between sensor pairs) we estimate location of motes. The problem is solved by storing the signal strength information in kernel matrices and then solving a complex least squares subspace problem. The chapter considers three different cases: centralized kernel regression, online kernel regression (used when sensors are mobile to track motes), and distributed kernel regression (used when sensors are subject to communication and power constraints). Simulations are conducted to assess the performance of the different algorithms.

Before information can be gathered for many sensor networks applications, we need to develop localization algorithms to estimate the location of the motes. Location information is indispensable for ad hoc network applications such as gathering military information, disaster relief, and location-based routing schemes [31]. When signal strength is available a common method to perform sensor localization is using ranging information as discussed in [5, 15]. Range-based methods commonly use a two-step approach when performing localization: first signal distance is estimated between pairs of devices from signal strength and then a localization algorithm is used based on these estimated distances.

Due to the rich scattering in the real world, the received signal strength is quite noisy. This makes it more difficult for range-based methods such as [5, 15] to get accurate readings of the distance between different devices. Statistical methods such as maximum likelihood estimation (MLE) [23], EM algorithm [28], and Bayesian networks [6] were used to alleviate the scattering effect. These methods usually have high computing complexity.

We use an alternative approach based on using machine learning methods first established in [22] where signal strength information is stored in a kernel matrix. The kernel matrix is a nonnegative definite matrix that contains information about signal strength. The kernel matrix can then be used to solve classification and regression problems as discussed in [21, 33]. A popular learning algorithm using kernels that is commonly used is the support vector machine (SVM) [21, 33]. The SVM has been used in many applications ranging from character recognition to bioinformatics to recovery of communication signals to text classification. The SVM and kernel methods in general have achieved popularity as they are based on principles of structural risk minimization, reproducing kernel Hilbert space (RKHS), and solving quadratic programming programs [8]. Many of the methods discussed in the earlier paragraph are based on solving nonlinear optimization problems that are not necessarily convex.

The learning algorithm discussed here uses signal strength information between base stations as training data. The locations of these sensors are known. Unknown parameters of the learning algorithm are trained using the kernel matrix formed from signal strength information between base stations. These parameters are then used to estimate the location of motes. The motes can be viewed as testing data.

In [22, 34] a classification problem is solved to determine whether a sensor lies in a given region or not. In [22] the classification problem is solved using the SVM (with hinge loss function) [21, 33]. In [34] the classification problem is solved using the least squares SVM (LS-SVM) (with quadratic loss function with equality constraints) [19, 32]. Fine localization is achieved by performing the classification problem several times with different overlapping regions. A mote's location is estimated from the average of the centroids of each region it belongs to.

This method gives reasonably accurate estimates of the locations of motes, but is computationally expensive as the classification problem must be solved for each region considered. This chapter estimates the location of motes using complex least squares kernel regression. Sensors are located on a two-dimensional grid with their location represented by a complex number. The first algorithm we consider is centralized kernel regression where a base station or a central processor serves as a fusion center that receives signal strength information between all base stations, forms the kernel matrix, and learns the parameters of the system using a least squares kernel regression algorithm. This is then extended to an on-line recursive least squares kernel regression algorithm to track motes that are mobile.

We then examine methods to reduce energy consumption for each base sensor by reducing communication costs. Communication costs can be reduced if we employ a distributed learning algorithm. In recent years there has been growing interest in distributed learning as this is often the only feasible way to handle data generated from multiple sources such as distributed sensor networks and to satisfy physical and communication constraints imposed on such networks. For wireless sensor networks there is a recent survey article on current research in distributed learning, [25]. A distributed kernel regression algorithm was formulated in [13] for sensor networks as the algorithm considered is based on Gaussian elimination and a message passing algorithm. Distributed learning is also considered for sensor networks in [24] using alternate projections. We discuss some simple distributed kernel regression algorithms that reduce communication costs. These algorithm have similarities to ensemble learning methods such as bagging [3] and boosting [12].

The chapter is organized as follows. In Sect. 5.2 we discuss the ad hoc sensor network model used and how the sensor localization problem can be viewed as an inverse problem. Signal strength information from each base node is stored in an information vector. Then kernels are constructed from the information vectors. Section 5.3 discusses previous work by [22, 34] where kernel classification algorithms were used. Section 5.4 presents the complex least squares subspace kernel regression algorithm, how support vectors are chosen, and on-line versions of the algorithm. Section 5.5 solves the localization algorithm using three kernel regression methods: centralized kernel regression algorithm, on-line kernel regression algorithm to track mobile motes, and a distributed kernel regression algorithm to reduce power consumption and save on communication costs. Section 5.6 discusses simulations of the algorithms for sensors deployed on a two-dimensional grid. Finally, Sect. 5.7 summarizes the chapter and discusses further directions for this research.

5.2 Sensor Network Model

Here we assume an ad hoc network of size N is deployed in a connected two-dimension geographical area $\mathcal{T}$. Sensor (also called node) i is denoted by n_i and the set of all nodes is denoted by $\mathcal{N} = \{n_i, 1 \leq i \leq N\}$. The location of node n_i is denoted by $p_i \in \mathcal{T}$. Let the nodes n_i, $1 \leq i \leq M < N$ be *base nodes* and the nodes n_i, $M+1 \leq i \leq N$ be *motes*. Base node locations, p_i, $1 \leq i \leq M$ are known and these sensors have more computational power than motes. We do not have the location of the motes, p_i, $M+1 \leq i \leq N$, but will estimate these locations from base node locations and signal strength information.

In most scenarios, wireless radio is used in ad hoc networks to transfer data because of the availability of technology and devices. Here we assume that each node is capable of transmitting signal strength to each of the other nodes. Signal strength between two sensors is a function of distance between the sensors and there are also mitigating effects such as interference, fading, scattering, and additive noise that can adversely affect signal strength [29]. In this chapter, we assume a simple signal strength model that was used also in [22, 34]:

$$s(p_i, p_j) = \exp\left\{-\frac{\|p_i - p_j\|^2}{\Sigma} + V\right\}, \tag{5.1}$$

where V is a zero mean Gaussian random variable with standard deviation τ.

The sensor localization problem can be viewed as an inverse problem as shown in Fig. 5.1. As mentioned above, the signal strength information

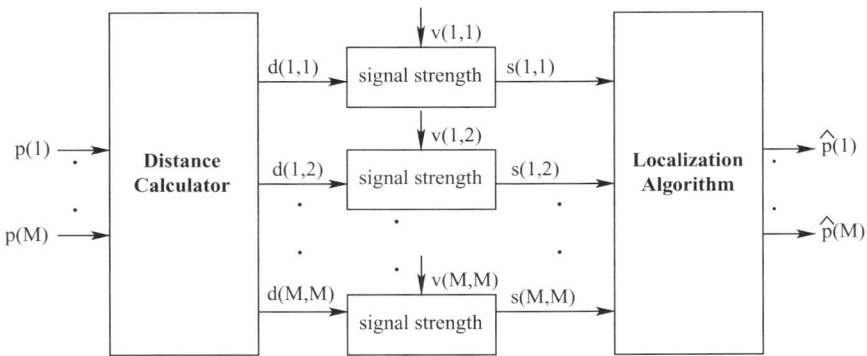

Fig. 5.1. Sensor localization based on signal strength can be viewed as an inverse problem

between two nodes depends on the distance between the two nodes, additive noise, V, and possibly other mitigating effects (e.g., interference, fading, and scattering). The localization problem takes signal strength as inputs and makes an estimate of the location for the ith node by estimating $\hat{p}_i$.

We use an approach first discussed in [22] where signal strength information is stored in a kernel matrix. This method avoids the problem of estimating distances or solving nonconvex optimization algorithms. The procedure consists of first forming an information vector $x(i)$ that contains the signal strength information from a set of base nodes to the sensor n_i. Then feature vectors $z(i) = \phi(x(i))$ are formed from the information vector. Kernels are then just inner products of the feature vectors $K(i,j) = \phi(x(i))^{\mathrm{T}}\phi(x(j))$. The construction of the feature vectors and the kernels is usually done in one step and can be done without determining the feature vectors $\phi(x(i))$. The next step is using the kernel matrix in a learning algorithm to train the parameters: support vector weights $\alpha(i)$ and threshold value b.

Once these parameters are learned, these parameter values are sent to appropriate base nodes. The parameter values, signal strength information between base nodes and desired mote, and kernel function are used to estimate the location of the mote. For distributed kernel regression algorithms there may also be additional queries of base stations by motes and also possibly message passing algorithms.

5.3 Localization Using Classification Methods

In [22, 34] the localization is performed in two stages. A classification problem is solved first to determine whether a sensor lies in a given region $\mathcal{A}$ (i.e., the region could be a circular or elliptical disc) or not. This is referred to as coarse localization. The classification problem uses signal strength training data from base nodes (whose location is known) to first form information vectors $x(i)$ and then construct the kernel matrix K as discussed in Sect. 5.2. The learning algorithm then takes the kernel matrix, K, the information vectors, $x(i)$ and the desired binary outputs $y(i)$ to learn the parameters $\alpha(i)$ and threshold value b given by the following equation:

$$f(x) = \mathrm{sgn}\left(\sum_{i=1}^{l} \alpha(i)y(i)K(x,x(i)) + b\right) \tag{5.2}$$

where $f(x)$ is the decision function to determine whether base node $i \in \mathcal{A}$ or not. In [22] the classification problem is solved using the SVM (with hinge loss function) [21, 33]. In [34] the classification problem is solved by using a quadratic loss function using the least squares SVM, [32]. This involves solving a system of linear equations. Sparsity is achieved with this method by using an intelligent selection method to selectively choose a subset of the training vectors as support vectors.

The second step performs fine localization by repeating the classification problem several times with different overlapping regions (each region has the same shape, but is translated to a different location). When using a quadratic cost function, each classification problem involves solving a set of linear equations. The data matrix is unchanged for all classification problems as only the outputs $y(i)$ change. Solving all classification problems, therefore, involves inverting the one data matrix, A when the data matrix is nonsingular or finding the psuedo-inverse if the data matrix is singular. Each classification problem is then solved by multiplying A^{-1} ($A^{\dagger}$) by the desired binary output vector to find $\alpha(i)$ and b. A mote's location is then estimated from the average of the centroids of each region it belongs to.

Simulations were performed for this classification localization algorithm on both synthetic data [22, 34] and a real deployed sensor network [22]. The key performance measure was mean localization error on the motes and results showed that this error decreased monotonically as the number of base nodes increased. A key concern about this method is the high complexity of the algorithm due to the large number of classification problems that need to be learned. This led us to consider a complex kernel regression algorithm.

5.4 Least Squares Subspace Kernel Regression Algorithm

This section discusses the least squares kernel subspace algorithm presented in [19]. Similar methods are also discussed in [7, 9, 10, 18, 27, 30]. Here we consider kernel methods where the output and weight vectors are complex-valued.

5.4.1 Least Squares Kernel Subspace Algorithm

We are given m training samples or observations drawn from input space $\mathcal{X} \in \mathcal{R}^l$ and output space $\mathcal{Y} \in \mathcal{C}$. Assume each observation $(x(i), y(i)), 1 \leq i \leq m$ is independent. Rewrite it compactly as $(\mathbf{x}, \mathbf{y})$ where $\mathbf{x} = [x(1)|\cdots|x(m)]$ and $\mathbf{y} = [y(1), \cdots, y(m)]^{\mathrm{T}}$. The inputs are transformed via kernel functions $\phi(x)$ that map from input space $\mathcal{X}$ to feature space $\mathcal{R}^d$. Let $\Phi(\mathbf{x}) = [\phi(x(1))|\cdots|\phi(x(m)))]$. In testing stage, when input x is presented, the output/estimate is given by

$$\hat{y}(x) = \phi(x)^{\mathrm{T}} w + b, \qquad (5.3)$$

where $w \in \mathcal{C}^d$ is the weight vector and b is a complex-valued scalar threshold.

The weight vector can be expressed as a linear combination of feature vectors, i.e., $w = \Phi(\mathbf{x})\alpha$ where α is a complex-valued vector. Any training sample $x(i)$ associated with a nonzero $\alpha(i)$ (an element of α) is called a support vector. For the standard SVM [33] only a fraction of the training samples are support vectors as an ϵ – insensitive cost function is used to remove training inputs as support vectors close to the zero error solution. For the standard least squares SVM [32], the weight vector depends on all training feature vectors. An external procedure needs to be established to reduce the number of training samples. Methods to intelligently choosing training examples can be found in [19]. Assume that a subset of training samples, $\mathbf{x}_S$ is chosen which is a matrix containing $m_S < m$ columns from $\mathbf{x}$. The corresponding feature vectors are denoted by $\Phi(\mathbf{x}_S)$.

Given the constraints on training samples the least squares kernel regression problem reduces to the following quadratic programming problem with equality constraints:

$$\min J(w,b) = \min_{w,b} \frac{1}{2}||w||^2 + \frac{\gamma}{2}||e||^2 \qquad (5.4)$$

subject to

$$e = \mathbf{y} - \Phi^{\mathrm{T}}(\mathbf{x})w - \mathbf{1}b \qquad (5.5)$$

and

$$w = \Phi(\mathbf{x}_S)\alpha, \qquad (5.6)$$

where now α is a complex-valued m_s vector weighting the training feature vectors and $\mathbf{1}$ is an m vector of 1s. Define $K_{SS} = \Phi^{\mathrm{T}}(\mathbf{x}_S)\Phi(\mathbf{x}_S)$ and $K_S = \Phi^{\mathrm{T}}(\mathbf{x}_S)\Phi(\mathbf{x})$ and substitute (5.6) into (5.4) and (5.5). We have

$$\min Q(\alpha,b) = \min \frac{1}{2}\alpha^{\mathrm{H}} K_{SS}\alpha + \frac{\gamma}{2}||\mathbf{y} - K_S^{\mathrm{T}}\alpha - \mathbf{1}b||^2. \qquad (5.7)$$

where superscript 'H' is the hermitian operation. This problem is solved by finding the solution to the following set of linear equations:

$$\begin{bmatrix} m & \mathbf{1}^{\mathrm{T}} K_S^{\mathrm{T}} \\ K_S \mathbf{1} & K_{SS}/\gamma + K_S K_S^{\mathrm{T}} \end{bmatrix} \begin{bmatrix} b \\ \alpha \end{bmatrix} = \begin{bmatrix} \mathbf{1}^{\mathrm{T}}\mathbf{y} \\ K_S \mathbf{y} \end{bmatrix}. \qquad (5.8)$$

Assume $A = K_{SS}/\gamma + K_S K_S^{\mathrm{T}}$ is invertible. By elimination we then get that

$$b = \frac{\mathbf{1}^{\mathrm{T}}\mathbf{y} - \mathbf{1}^{\mathrm{T}} K_S^{\mathrm{T}} A^{-1} K_S \mathbf{y}}{m - \mathbf{1}^{\mathrm{T}} K_S^{\mathrm{T}} A^{-1} K_S \mathbf{1}} \qquad (5.9)$$

and

$$\alpha = A^{-1} K_S (\mathbf{y} - \mathbf{1}b). \qquad (5.10)$$

Substituting (5.6) into (5.3), we get that

$$\hat{y}(x) = K(x, \mathbf{x}_S)\alpha + b, \qquad (5.11)$$

where $K(x, \mathbf{x}_S) = \phi(x)^{\mathrm{T}}\Phi(\mathbf{x}_S)$.

How are the support vectors chosen? A number of approaches have been tried. One approach is to randomly choose the support vectors where as other approaches choose support vectors based on an information criteria. In [19] support vectors are added if they can reduce the overall training error significantly. In simulations conducted in [19], the subspace methods perform quite well in a number of applications from function approximation in additive noise to tests on the UC Irvine machine learning repository [2] to simulations we later show for the sensor network localization problem. The subspace methods works better when support vectors are chosen intelligently rather than randomly. The subspace methods have advantages that the matrices that are inverted are of dimension m_S with performance comparable to full-sized systems of dimension m. Other information criteria could be based on information about the application. For the sensor network localization problem clustering algorithms could be used as the information vectors $x(i)$ contain signal strength information that decay as distances between sensors increase.

5.4.2 Recursive Kernel Subspace Least Squares Algorithm

When dealing with kernel matrices that change with time updates it becomes more efficient to use on-line algorithms. For an on-line algorithm, we have an old estimate and based on new training data we modify the update. Here we use a windowed recursive least square algorithm where that can be described as follows:

1. Train parameters on initial set of data using batch or on-line methods.
2. Get new training data and add to information set.
3. If new data satisfies specified criteria add as a support vector.
4. Selectively prune support vectors.
5. Selectively prune information data.
6. If there is more data go to 2.

When adding and pruning support vectors the overall solution will change and the key is the inversion of the information data matrix shown in (5.8) Rather than invert the matrix from scratch every time, there is an addition or deletion of support vectors we can use the following block inversion formula that reduces computation:

$$\begin{bmatrix} F & G \\ E & H \end{bmatrix}^{-1} = \begin{bmatrix} F^{-1} & 0 \\ 0 & 0 \end{bmatrix} + UW^{-1}V, \tag{5.12}$$

where $U = \begin{bmatrix} F^{-1}G \\ -1 \end{bmatrix}$, $W = H - EF^{-1}G$, and $V = \begin{bmatrix} EF^{-1} & -1 \end{bmatrix}$. This on-line updating is a key to the recursive least squares algorithm developed here and those used in adaptive filtering [14].

For the applications that we consider, we will use a fixed window size where if we decide to add a support vector from training data we must also delete a support vector. Simple criteria are used to add and delete support vectors according to the discussion in Sect. 5.4.1. This can be implemented as follows: evaluate the kernel vector between the newest data point and all other training data; compare with the training error vector e. We normalize each of the vectors to have magnitude 1 and compute the inner product between the two vectors which is the same as computing the cosine of the angle between the two vectors. If the magnitude of the inner products is above a specified threshold value, then the new training data is added as a support vector. This criteria is also used in [18]. During the deletion process, we delete the support vector that makes the least contribution to the weight vector.

5.5 Localization Using Kernel Regression Algorithms

This section discusses three kernel regression algorithms using the least squares kernel subspace algorithm presented in Sect. 5.4. The differences between the algorithms depend on where computations are done. We will consider two scenarios.

In the first situation, we have a sensor fusion center where learning takes place. The sensor fusion center can be a separate processor or even one of the base sensors. During learning, other base sensors send their signal strength information to the fusion center and parameters $\alpha(i)$ and b are learned. During the testing phase the localization of motes is performed by the mote sending its signal strength information to the fusion center. The motes location is estimated by the fusion center using (5.11).

In the second case, we have an ad hoc network where computations are done distributively. These can take place at several of the base sensors. We call base sensors where computations are done distributed computation centers (DCCs). Each DCC gets its respective training data information and employs a kernel regression algorithm to learn its parameters. The DCCs then communicate with each other, possibly using message passing algorithms to improve their solution. After training is completed a mote will send signal strength information to local DCCs and each local DCC will give an estimate of the mote's location. By weighted averaging we get the estimate of the mote's location.

5.5.1 Centralized Kernel Regression

These methods assume that there is a base node or a central processor that acts as a fusion center to collect all the signal strength readings between base nodes and performs the training using the algorithm of Sect. 5.3. This method works well if the number of base nodes is not too large and base nodes can easily communicate signal strength information to the fusion center.

The first step is to choose an information vector from the signal strength readings of the base nodes. For each base node n_i, $1 \leq i \leq M$ there is an information vector of length M where the jth component describes the signal strength at time t, $s(p_j, p_i)(t)$ between sensors n_j and n_i. The time index is dropped for signal strengths between base nodes unless the base nodes are mobile. The information vector is denoted and described by

$$x(i) = \begin{bmatrix} s(p_1, p_i) \\ s(p_2, p_i) \\ \vdots \\ s(p_m, p_i) \end{bmatrix}.$$

The M information vectors correspond to $M = m$ training examples. The labels for each training example is the location of the base node. We then use m_S of these information vectors as support vectors. We can use a selection criteria discussed in [19] or an unsupervised algorithm such as the k-means algorithm (based on locations) to choose support vectors. The kernel least squares regression algorithm from Sect. 5.3 is then applied to get the parameters α and b.

To estimate the location of mote n_j, $M + 1 \leq j \leq N$ we then form an information vector from signal strengths to get

$$x(j, t) = \begin{bmatrix} s(p_1, p_j)(t) \\ s(p_2, p_j)(t) \\ \vdots \\ s(p_m, p_j)(t) \end{bmatrix}.$$

This information vector is sent to the fusion center and the estimate of location is found using (5.11).

When the number of base nodes is very large the number of computations can be reduced by introducing neighborhood functions. As a simple example, the components of information vectors are set to zero if the associated signal strength are not larger than a specified threshold value s_0.

5.5.2 Kernel Regression for Mobile Sensors

Here we consider when sensors are mobile. Let us consider where base nodes are stationary during a period of length T seconds. Motes are mobile and can move during this T seconds interval. Assume that mote positions are estimated every $T_m \ll T$ seconds. If mote information vectors can be sent to the fusion center every T_m seconds, then the centralized kernel regression algorithm is used to track the motes. Note that training is only done outside the T seconds period as base node locations do not move in this time period and therefore parameters α and b remain unchanged. If signal strength is unavailable every T_m seconds. estimates can be made of the mote's location based on previous estimates of the mote's location. The previous estimates can be stored at the fusion center.

Outside this T-second interval we can consider the case where some base nodes might change. The fusion center should receive a signal from the base node that its position has changed by a significant amount. We will then have to relearn the parameters α and b as the information vector for the mobile base sensor will have changed resulting in a changed kernel matrix. Here we can use recursive on-line learning to relearn the parameters α and b. The information vector associated with the new location of the base node will be added and the information vector associated with the old location of the base node will be pruned. Recursive on-line least squares algorithms discussed in Sect. 5.3 can then be used.

5.5.3 Distributed Kernel Regression

In sensor networks a key concern is conserving energy. Energy can be conserved by reducing communication costs. When a fusion center is used to gather information from sensors each sensor must transmit with enough power so that the fusion center can receive information about signal strength. The centralized algorithms in Sect. 5.5.2 require that each base node receive M signal strength information from all other base nodes (including itself). Then each base node must send an M vector of signal strength information to the fusion center. This requires many communications between base nodes and may not be practical if the number of base nodes is large.

To save communication costs and reduce energy consumption distributed kernel regression is proposed. In distributed kernel regression, there are more than one distributed computation centers (DCC)s with each DCC performing a kernel regression or updating of parameter estimates. The DCCs then have the option of communicating with each other to improve their solution. This can be done via message passing algorithms discussed in [17]. Nonparametric distributed kernel regression has been discussed in a survey article in [25]. Some of the base sensors can serve as DCCs and the communication between DCC and base sensors can be via a graph describing the ad hoc sensor network. Research using DCCs and message passing algorithms include [13] which discusses kernel regression using a distributed Gaussian elimination and a junction tree algorithm. In [24] kernel regression is performed via an on-line updating algorithm using alternating projections. In [26] a distributed incremental subgradient algorithm is proposed where estimates are passed from base node to base node.

Here we will focus on simple distributed algorithms where each DCC performs a kernel regression followed by simple weighted averaging operations to estimate the mote location. The training and testing phases are described in detail in rest of this section and were presented in [35].

Training Base Nodes

The distributed algorithm consists of two steps. To reduce communication costs no message passing is performed. A key to the algorithm is partitioning the base nodes into different regions with each region having a DCC that performs a kernel regression and makes an estimate of a sensor based on parameters α and threshold value b. The location estimate will usually be more accurate in the region where base nodes are located, then outside the region.

The base nodes are partitioned into r different sets. Set $R(i)$, $1 \leq i \leq r$ will also be called region $R(i)$ which consists of base nodes $n_{j_1}, \ldots, n_{j_{r(i)}}$ where $r(i) \geq 1$ is the number of base nodes in the ith region. Since estimation algorithm is based on signal strength which is dependent on distance, base nodes within a region should for the most part be closer in distance than when compared to base nodes outside the region. Different algorithms can be used to partition the base nodes into the different regions with a goal of having a computationally efficient algorithm that can take advantage of parallel processing such as a distributed k-means algorithm discussed in [16]. Base nodes in each region will also have a tag identifying region.

After partitioning into regions, each DCC in the region will use the subspace kernel regression algorithm to come up with a set of parameters α that are associated with base nodes that are support vectors and threshold value b. Let $S(i)$ denote the set of all nodes that will be used in the regression algorithm for the ith region with $R(i) \subset S(i)$. Each node in $S(i)$ will have signal strength with at least one node of $R(i)$ that is greater than some specified threshold value s_0. For region i information vector for sensor i' will be given by

$$x(i')[i] = (s(j, i'), n_j \in S(i))^{\mathrm{T}}$$

and the estimate for the location of sensor i' is given by

$$\hat{p}(i')[i] = \sum_{n_j \in S(i)} \alpha_j[i] K(x(i')[i], x(j)[i]) + b[i]. \tag{5.13}$$

When the subspace algorithm is used a subset of base nodes in $S(i)$ will be support vectors and only those nodes will have nonzero α values. Each region will have its own estimate of location so there will be a total of r different estimates for location of sensor i'. Region i will have an estimate with parameters given by $\alpha[i]$ and $b[i]$ with each base node in i storing these set of parameters.

The algorithm's most critical parameter is r the number of regions. If $r = 1$, then this corresponds to centralized base node regression where one subspace kernel regression is performed on the m base nodes. If $r = M$, then every base node performs a subspace kernel regression algorithm and this is similar to the k nearest neighbors algorithm. Good values for r depend on the number of base nodes M, their physical deployment, and signal strength. Let $m_{S(i)}$ denote the number of support vectors for region i and $m(i)$ be the cardinality of $S(i)$. Note that $m_{S(i)}$ will depend on the signal strength of nodes in $S(i)$ and the desired accuracy we would like. Finding $\alpha[i]$ and $b[i]$ for least squares kernel regression involves matrix multiplication and inversion operations. This will take $\mathcal{O}(m_{S(i)}^3 + m_{S(i)}^2 m(i))$ operations.

5 Sensor Network Localization Using Least Squares Kernel Regression

For most ad hoc sensor networks, the sensors will be deployed over a wide geographical range with only a few base node sensors having strong signal strength with a given mote. This will result in only a few of the regions having significant nonzero information vectors for a given mote. In general, only a few regions will be part of the location estimate of a given mote.

Mote Localization

Each mote will store information about regions that are in close proximity to the mote. The set of these regions is denoted by $N(j)$ where j refers to mote n_j. For each region $i \in N(j)$ the associated DCC will use $a[i]$ and $b[i]$ to compute its estimate of the location of mote n_j using the information vector for mote n_j. Each DCC in $N(j)$ will then send the estimate of location to the mote n_j. The mote will then compute a location estimate that is the weighted average of the location estimate of each DCC in $N(j)$. The weighting will depend on the number of base nodes in i that are close to mote n_j and their signal strengths. The location estimate is given by

$$\hat{p}(j) = \sum_{i \in N(j)} w(i)\hat{p}(j)[i], \qquad (5.14)$$

where the weights $w(i)$ are positive numbers that sum to 1. Weights are given by

$$w(i) = \frac{\sum_{j \in R(i)} g(s(j,i))}{\sum_{j=1}^{m} g(s(j,i))} \qquad (5.15)$$

with possible values for weighting functions $g()$ including $g(z) = z$ and $g(z) = \mathbf{1}(z > s_0)$ with $\mathbf{1}()$ denoting an indicator function.

If the motes are mobile, then the regions that are in close proximity to the mote will change. The mote will have to update the regions that are in close proximity to the mote and get parameters and information vectors as needed.

The computational and communication costs to estimate a mote's location will depend on the number of regions and signal strengths. Again, the key parameter is r, the number of regions chosen. The number of regions should be chosen to balance the error rate, computational, and communication costs.

5.6 Simulations

Several simulations were conducted to test the performance of the kernel regression localization algorithms. Sensors were placed on a 10×10 grid. There were $M = k^2$ base sensors placed on a grid and then the locations were perturbed by additive Gaussian noise. There were 400 motes placed randomly on the grid. Signal strength had additive noise with deviation $\tau = 0.2$. Here we show three sets of simulations: using centralized kernel regression for stationary sensors, using centralized kernel regression for mobile motes, and using distributed kernel regression for stationary sensors. We used Gaussian kernels described by

$$K(x(i), x(j)) = \exp(||x(i) - x(j)||^2/(2\sigma^2)).$$

5.6.1 Stationary Motes

Centralized kernel regression was used with the least squares kernel subspace algorithm. Here we chose $m_s = 3k$ support vectors. The algorithm worked well with the regularization parameter set at $\gamma = 60$ and the width of the Gaussian kernel function at $\sigma = 2.7$. Figure 5.2 shows how the mean localization error varies as the number of base sensors increases and Σ is varied. Simulations were conducted 100 times for each setting with average curves and standard deviations shown.

We define mean localization error as the average of the absolute localization error over all tests of each test mote. The plots show that the mean localization estimation error decreases monotonically as the number of base sensors increases. The results are similar to the results for fine localization shown in [22, 34]. When $\Sigma = 3, 5$ signal strength decreases rapidly as distance and the error rate remains roughly constant at a little below 0.5 when more than 100 base sensors are used.

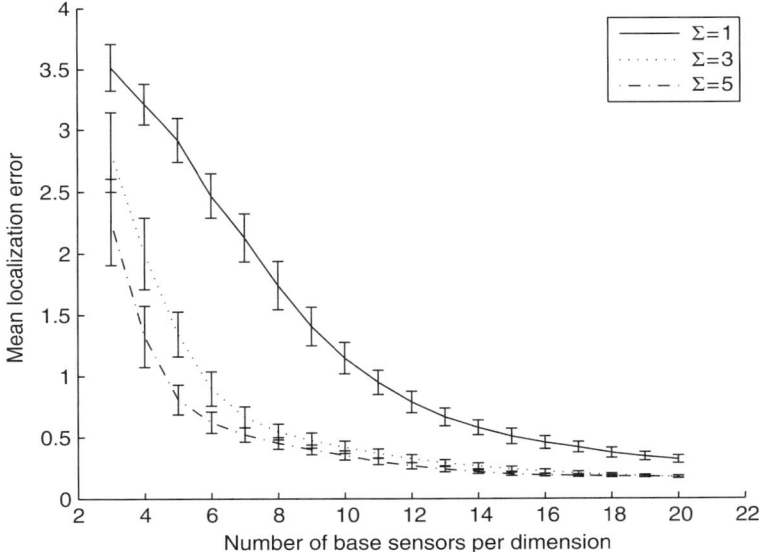

Fig. 5.2. Centralized regression with Gaussian kernels, $3k$ SV

5.6.2 Mobile Motes

We then conducted an experiment where motes moved. The *random waypoint* model [4] is used to simulate the movement of all motes. Any mote n_j $M+1 \leq j \leq N$ chooses a random destination uniformly in the 10×10 region and moves to this destination at a random speed uniformly distributed in the interval from 0 to $V_{rmmax} = 1$. When it arrives at a destination, it stays at this location for T_{pause} time units. After T_{pause} time units has passed, it chooses a random destination with a random speed and continues moving.

We study the relationship between performance of these methods and the number of base nodes per dimension k and the mobility by testing 100 motes every 1 time unit in a 400-time unit window with different T_{pause} values. For these simulations we again use Gaussian kernels.

As expected, the mean localization error for mobile motes is very similar to the stationary case. As long as the fusion center has access to the signal strength information vector when mote location is estimated, the average absolute error will be the same as the stationary case. Here we again let $\gamma = 60$ and $\sigma = 2.7$. We again use $m_s = 3k$ support vectors.

Similar performance is obtained using the full kernel regression method by setting $\gamma = 20$ and $\sigma = 2.9$. We also find similar performance when using simple threshold neighborhood functions. Using neighborhood functions will likely give more robust results when obtaining signal strengths from noisy environments.

In Fig. 5.3, we show one sample simulation of a mote moving and the kernel subspace regression algorithm used to estimate the location of the mote. The

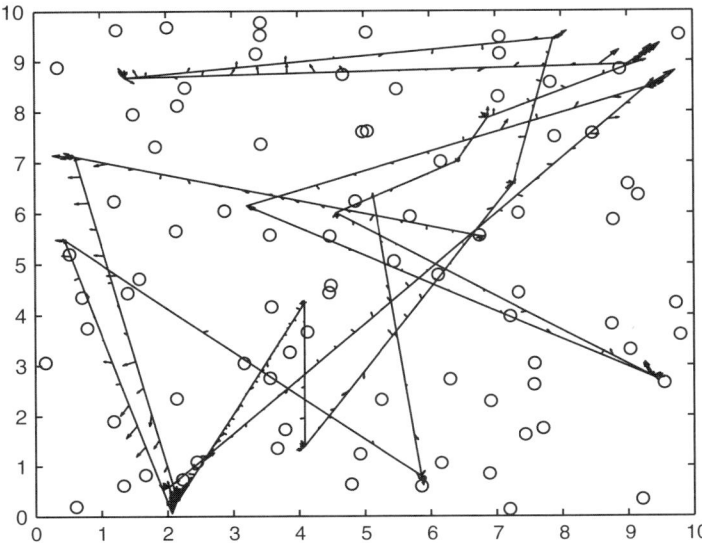

Fig. 5.3. Visualization of localization error with 100 base nodes

mobile mote movement is described by the solid line. Each arrow denotes an estimate with the tip of the arrow denoting the estimate of position and the head of the arrow denoting the actual position. Each circle represents one of the 100 base nodes.

5.6.3 Distributed Algorithm

We then implemented distributed kernel regression using the method described in Sect. 5.5.3. Nodes were partitioned into regions with a DCC performing a kernel regression for each region. We again used Gaussian kernels with the same parameters as the stationary case. Figure 5.4 considers distributed learning with k regions and seven support vectors used for each region. The base nodes were partitioned using the k-means algorithm. The error performance is very close to the centralized base node performance and actually give slightly better results when Σ is small. Computationally, the distributed algorithm requires slightly more computations than the centralized method, but large savings can be realized if the base nodes perform computations in parallel. The distributed algorithm saves communication costs as motes need to get signal strength information only from base nodes that are close to motes.

Figure 5.5 considers an extreme version of distributed learning where every base node is a DCC. Here we use three support vectors for each kernel regression performed. The performance for this case is slightly worse than the

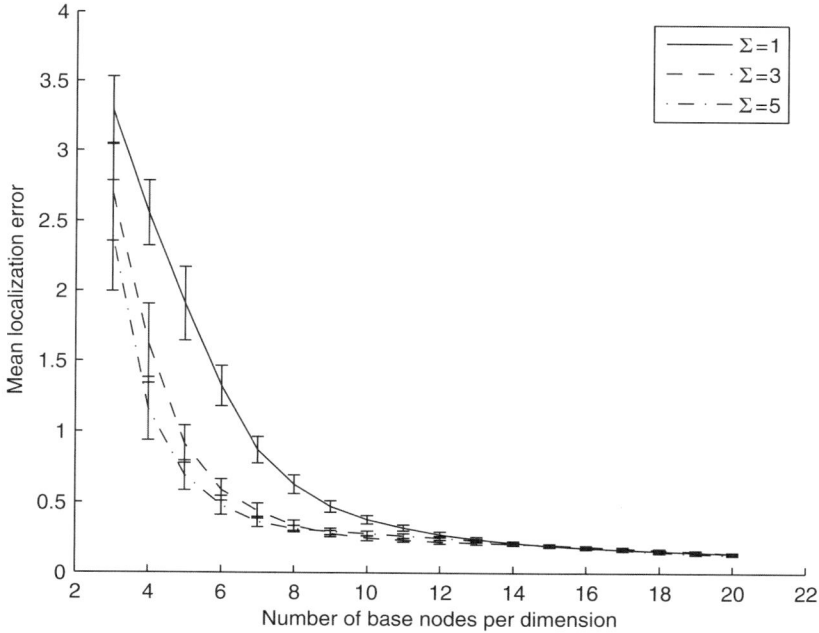

Fig. 5.4. Distributed regression with Gaussian kernels, $r = k$, 7 SV

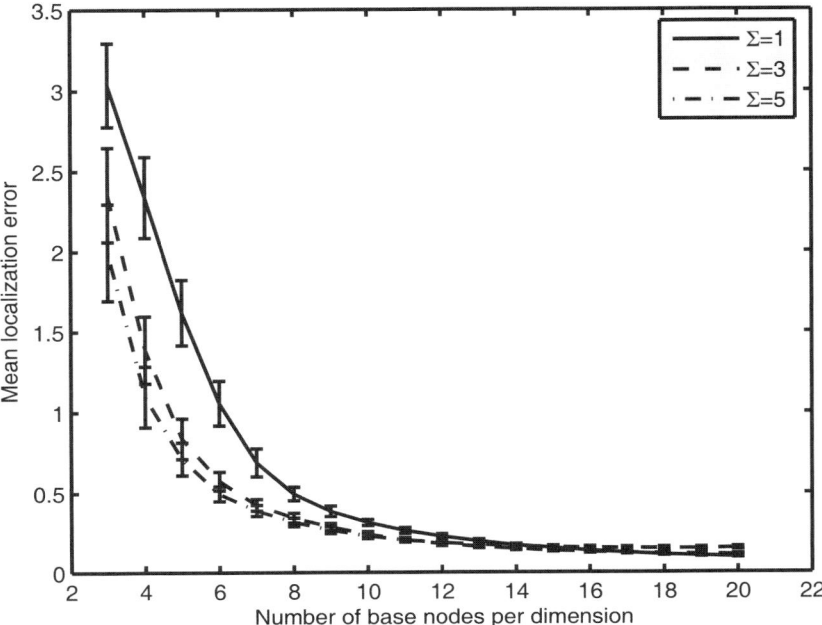

Fig. 5.5. Distributed regression with Gaussian kernels, $r = k^2$, 3 SV

previous distributed learning case and the centralized case when the number of base nodes is small. However, as the number of base nodes increase, the performance of all three algorithms in terms of mean absolute localization error is comparable.

5.7 Summary and Further Directions

This chapter shows that sensor localization can be successfully performed using a least squares kernel subspace regression algorithm. This algorithm has good performance and has computational savings over kernel classification algorithms. The algorithms can also be modified to track mobile sensors. To reduce energy consumption, we developed a distributed algorithm where a group of distributed computation centers (DCC)s each perform a kernel regression. Location estimates can then be made by a mote by taking weighted averages of DCCs localization estimates. The distributed algorithms cut down on communications and therefore reduce energy consumption. There are several further directions for this work.

We considered a simple model for how signal strength varies in the sensor network. This was to demonstrate the viability of the different least squares kernel regression localization algorithms. For these models both centralized and distributed kernel algorithms give similar good performance. We chose Gaussian kernels for our experiments, but other kernels would give similar performance. Further work would consider sensor networks in more complex environments with impairments including scattering, fading, and interference. For more complex environments the choice of kernels could be more critical. We would also like to examine the performance of the algorithms in real environments such as [22].

The kernel regression algorithm performs well for localization estimation, but we can use these same algorithms to gather other information from sensor networks. In [13, 24] distributed kernel regression algorithms are used to approximate sensor information such as temperature information. Distributed kernel learning algorithms are proposed to learn this information. We would also like to compare different distributed learning algorithms. Here we discussed simple learning algorithms without message passing between DCCs. How does this compare to distributed algorithms that employ message passing.

The distributed learning algorithms that we propose have similarities to other ensemble learning methods. For harder estimation problems we may need to employ message passing algorithms to improve learning algorithm performance. It would be fruitful to employ machine learning methods such as bagging and boosting to the kernel learning algorithms. A DCC could implement an ensemble of learning algorithms with newer learning algorithms conducting regression algorithms emphasizing training examples where base node localization is poor. Algorithms similar to Adaboost [11] or distributed boosting [20] could also be implemented.

In [22] an analysis was conducted discussing how performance is related to parameters of the kernel classification problem. This can be extended to kernel regression. We can also explore the performance of the various distributed regression algorithms. In [24] convergence is shown for a distributed algorithms with message passing using alternating projections. For the simple distributed algorithms that we proposed, we can explore bounds for the mean absolute localization error.

References

1. Akyildiz, I., Su, W., Sankarasubramaniam, Y., Cayirci, E.: A survey on sensor networks. IEEE Communications Magazine, 102–114 (2002)
2. Blake, C., Merz, C.: UCI repository of machine learning databases. Department of Information and Computer Science, UC Irvine, Irvine, CA (1998)
3. Breiman, L.: Bagging predictors. Machine Learning **26**(2), 1579–1619 (1996)
4. Broch, J., Maltz, D., Johnson, D., Hu, Y.C., Jetcheva, J.: A performance comparison of multi-hop wireless ad hoc network routing protocols. In: Proceedings of the Fourth Annual ACM/IEEE International Conference on Mobile Computing and Networking (MobiCom'98). Dallas, TX (1998)
5. Bulusu, N., Heidemann, J., Estrin, D.: GPS-less low cost outdoor localization for very small devices. Technical Report 00-0729, Computer Science Department, University of Southern California (2000)

6. Castro, P., Chiu, P., Kremenek, T., Muntz, R.: A probabilistic room location service for wireless networked environments. In: ACM Ubicomp 2001. Atlanta, GA (2001)
7. Cawley, G., Talbot, N.: A greedy training algorithm for sparse least-squares support vector machines. In: Proceedings of the International Conference on Artificial Neural Networks ICANN 2002, pp. 681–686. Madrid, Spain (2002)
8. Cristianini, N., Shawe-Taylor, J.: An Introduction to Support Vector Machines. Cambridge University Press, Cambridge, UK (2000)
9. Csato, L., Opper, M.: Sparse on-line gaussian processes. Neural Computation **14**, 641–668 (2002)
10. Engel, Y., Mannor, S., Meir, R.: The kernel recursive least-squares algorithm. IEEE Transactions on Signal Processing **52**(8), 2275–2285 (2004)
11. Freund, Y., Schapire, R.: Experiments with a new boosting algorithm. In: Machine Learning: Proceedings of the Thirteenth International Conference, pp. 148–156 (1996)
12. Freund, Y., Schapire, R.: A decision-theoretic generalization of on-line learning and an application to boosting. Journal of Computer System and Sciences **55**(1), 119–139 (1997)
13. Guestrin, C., Bodik, P., Thibaux, R., Paskin, M., Madden, S.: Distributed regression; an efficient framework for modeling sensor network data. In: Information Processing in Sensor Networks 2004. Berkeley, CA (2004)
14. Haykin, S.: Adaptive Filter Theory, 4th edn. Prentice-Hall, Englewood Cliffs, NJ (2003)
15. Hightower, J., Borriello., G.: Real-time error in location modeling for ubiquitous computing. In: Location, Modeling for Ubiquitous Computing, Ubicomp 2001 Workshop Proceedings, pp. 21–27 (2001)
16. Jagannathan, G., Wright, R.: Privacy-preserving distributed k-means clustering over arbitrarily partitioned data. In: Proceedings of the 11th ACM SIGKDD International Conference on Knowledge Discovery and Data Mining (KDD) (2005)
17. Jordan, M.: Learning in Graphical Models. MIT Press, Cambridge, MA (1999)
18. de Kruif, B.: Function approximation for learning control, a key sample based approach. Ph.D. thesis, University of Twente, Netherland (2004)
19. Kuh, A.: Intelligent recursive kernel subspace estimation algorithms. In: The 39th Annual Conference of Information Sciences and Systems (CISS 2005), pp. 216–221. Baltimore, MD (2005)
20. Lazarevic, A., Obradovic, D.: The distributed boosting algorithm. In: KDD '01, Proceedings of the seventh ACM KDD conference on Knowledge Discovery and Data Mining. San Francisco, CA (2001)
21. Muller, K., Mika, S., Ratsch, G., Tsuda, K., Scholkopf, B.: An introduction to kernel-based learning algorithms. IEEE Transactions on Neural Networks **12**(2), 181–202 (2001)
22. Nguyen, X., Jordan, M., Sinopoli, B.: A kernel-based learning approach to ad hoc sensor network localization. ACM Transactions on Sensor Networks **1**(1), 134–152 (2005)
23. Patwari, N., Hero, A., Perkins, M., Correat, N., O'Dea, R.: relative location estimation in wireless sensor networks. IEEE Transaction on Signal Processing **51**(8), 2137–2148 (2003)

24. Predd, J., Kulkarni, S., Poor, H.: Distributed regression in sensor networks: Training distributively with alternating projections. In: Proceedings of the SPIE Conference and Advanced Signal Processing Algorithms Architectures, and Implementations XV. San Diego, CA (2005)
25. Predd, J., Kulkarni, S., Poor, V.: Distributed learning in wireless sensor networks. IEEE Signal Processing Magazine **23**(4), 56–69 (2006)
26. Rabbat, M., Nowak, R.: Quantized incremental algorithms for distributed optimization. IEEE Journal on Selected Areas in Communications **23**(4), 798–808 (2006)
27. Rifkin, R.: Learning with kernels: Support vector machines, regularization, optimization and beyond. Ph.D. thesis, MIT (2002)
28. Roos, T., Myllymaki, P., Tirri, H.: A statistical modeling approach to location estimation. IEEE Transactions on Mobile Computing **1**(1), 59–69 (2002)
29. Seidel, S., Rappaport, T.: 914 MHz path loss prediction models for indoor wireless communications in multifloored buildings. IEEE Transactions on Antennas and Propagation **40**(2), 207–217 (1992)
30. Smola, A., Schölkopf, B.: Sparse greedy matrix approximation for machine learning. In: Proceedings of the 17th International Conference on Machine Learning, pp. 911–918. Morgan Kaufmann, USA (2000)
31. Stojmenovic, I.: Position-based routing in ad hoc networks. IEEE Communications Magazine **40**(7), 128–134 (2002)
32. Suykens, J., Gestel, T.V., Brabanter, J.D., Moor, B.D., Vandewalle, J.: Least Squares Support Vector Machines. World Scientific, Singapore (2002)
33. Vapnik, V.: Statistical Learning Theory. Wiley, New York City, NY (1998)
34. Zhu, C., Kuh, A.: Sensor network localization using pattern recognition and least squares kernel methods. In: Proceedings of 2005 Hawaii, IEICE and SITA Joint Conference on Information Theory, (HISC 2005). Honolulu, HI (2005)
35. Zhu, C., Kuh, A.: Ad hoc sensor network localization using distributed kern regression algorithms. In: 2007 International Conference on Acoustics, Speech, and Signal Processing, vol. 2, pp. 497–500. Honolulu, HI (2007)

6
Adaptive Localization in Wireless Networks

Henning Lenz, Bruno Betoni Parodi, Hui Wang, Andrei Szabo,
Joachim Bamberger, Dragan Obradovic, Joachim Horn,
and Uwe D. Hanebeck

Indoor positioning approaches based on communication systems typically use the received signal strength (RSS) as measurements. To work properly, such a system often requires many calibration points before its start. Based on theoretical-propagation models (RF planning) and on self-organizing maps (SOM) an adaptive approach for Simultaneous Localization and Learning (SLL) has been developed. The algorithm extracts out of online measurements the true model of the RF propagation. Applying SLL, a self-calibrating RSS-based positioning system with high accuracies can be realized without the need of cost intensive calibration measurements during system installation or maintenance.

The main aspects of SLL are addressed as well as convergence and statistical properties. Results for real-world DECT and WLAN setups are given, showing that the localization starts with a basic performance slightly better than Cell-ID, finally reaching the accuracy of pattern matching using calibration points.

6.1 Introduction

Localization is almost a synonym for the ubiquitous global positioning system (GPS). Within the car navigation systems GPS can successfully be applied [17]. Unfortunately GPS does not achieve the same accuracy in indoor or campus environments as in outdoor due to signal attenuation and multi-path propagation. The indoor localization systems can be classified in systems using dedicated sensors and those that use existing infrastructure, as a communication system. Many systems like those which make use of infrared beacons, e.g., the active badge location system [27], ultrasound time of arrival, e.g., the active bat system [10] and the cricket location support system [20], and received signal strength (RSS), e.g., the LANDMARC [16] with RFIDs fall in the first category. In the second category, there are systems based on GSM networks that combine the RSS with the time advance of mobiles terminals [12], use smart antennas to measure angles [14], or based on WLAN/DECT networks that measure the RSS [1, 2, 22, 25, 26] or the propagation time [9]. There are still many other systems which use even contact sensors and images retrieved from cameras. Those systems are most suitable for robotics application.

For localization systems based on communication systems, the computation of location out of measured features can be done using theoretical-propagation models. The advantage of such solution is the low complexity due to the absence of pre-calibration. The disadvantage of such systems is that its accuracy is generally low, being more accurate only for special cases, e.g., measuring propagation time in line of sight (LOS) conditions [18, 19]. Unfortunately, this highly limits the solution applicability. Another approach is the measurement of RSS and use pattern matching on pre-calibration measurements. The achievable accuracy is suitable for localization of humans and objects, but the pre-calibration and maintenance costs are high.

In this chapter an unsupervised learning algorithm is developed. The simultaneous localization and learning (SLL) avoids the requirement for manually obtained reference measurements using an initial propagation model with few parameters, which can be adapted by a few measurements like the mutual measurements of the access points. Linear propagation models and more involved dominant path models incorporating map information are applied for the initialization. Thus, a feature map is obtained with the predicted RSS measurements at the map grid points. After the initialization the operating phase starts, which performs two tasks: localization and learning. Localization is done by pattern matching using the feature map. In addition, the RSS measurements are collected and used batch wise for learning. The learning is a refinement of the feature map, so that it corresponds better to the reality, hence reducing the localization error. This new developed method uses a modified Kohonen's SOM learning algorithm. A closed form formulation for the algorithm, as well as algebraic and statistical conditions that need to be satisfied are given, deriving some convergence properties.

In Sect. 6.2 an overview of signal propagation models typically used for localization is given. Section 6.3 describes localization algorithms using the propagation models. The SLL approach is first presented in Sect. 6.4 and tested in different real-world scenarios in Sect. 6.5.

6.2 RF Propagation Modelling

For RSS-based indoor positioning system, it is important to know the RSS distribution accurately. Typically, two kinds of approaches are used: model based and pattern matching. The model-based approach assumes that the propagation of radio signals follows a parameterized channel model. If the parameters are known, the RSS distribution can be predicted. In contrast, a pattern matching approach does not assume any prior information of radio

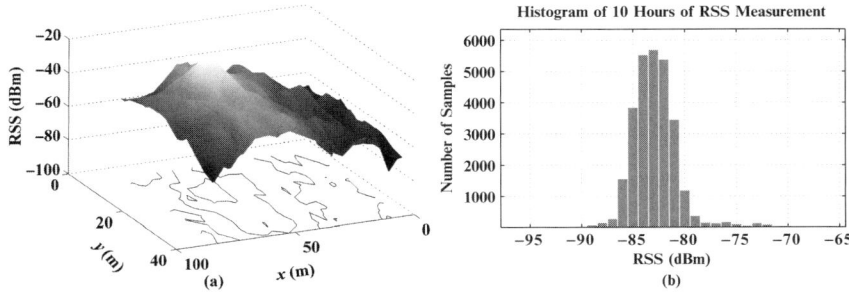

Fig. 6.1. (a) Spatial distribution of received power and (b) temporal distribution of received power at a fixed point

propagation. It measures the received power at various sample points to construct the radio distribution of the whole area. In the following section, these two categories of approaches are discussed.

6.2.1 Characteristics of the Indoor Propagation Channel

The propagation of radio frequency (RF) signals can be modelled at two scales. At a large scale the RSS decreases with the propagation distance. At a small scale, obstacles in the propagation path (due to walls, windows, cabinets and people), shadowing and multi-path effects cause a strongly varying RSS on small distances. Due to the complexity of indoor environments, this effect is unpredictable and typically modelled as a random variable.

Figure 6.1 shows an example of the spatial variation of the RSS in a real scenario. In subplot (a) it can be observed that the RSS has a decreasing trend, when the transmitter–receiver distance increases. This tendency is superimposed with different distortions effects due to shadowing and fading. In subplot (b) the time variation for the RSS measurement at a fixed position is shown.

6.2.2 Parametric Channel Models

The following models use a parameterized description of the radio channel. The parameters can be fit to a set of measurements so that the model is more accurate.

Linear Model

The theoretical propagation in free space follows a linear decay of the received power with distance on a logarithmic scale, [21]:

$$p_{\rm r} = p_0 - 10\gamma \log\left(\frac{d}{d_0}\right), \tag{6.1}$$

where p_r is the received power in dBm as a function of distance d and p_0 is a constant term representing the received power in dBm at distance d_0. The model parameter γ is in free space equal with 2. In indoor environments, γ typically has a value between 2 and 6.

There are two ways to obtain the parameters of the LM: One way is to use standard values found in literature. However, since radio properties in indoor scenarios vary a lot, this approach may cause a large model error. Another way is to use the values from several calibration points to tune the model. With m calibration points, the linear model can be defined as a system with m equations and two variables, the parameters p_0 and γ, while d_0 is fixed. These parameters can be estimated using least squares optimization [24].

Piecewise Liner Model

Another common model is the piecewise linear model (PLM) [8]. A PLM is fitted to calibration measurements with N segments, with slopes given by $\{\gamma_1, \ldots, \gamma_N\}$, specified by $N-1$ breakpoints $\{d_1, \ldots, d_{N-1}\}$. Different methods can be used to determine the number and location of breakpoints [8]. Once these are fixed, the slopes corresponding to each segment can be obtained by linear regression, similarly to the least square estimation used for the LM. An example of LM and PLM is shown in Fig. 6.2.

6.2.3 Geo Map-Based Models

The following models use also a parameterized description of the radio channel. Additionally, information about the environment geometry, i.e., a map, and physical properties are used with the measurements to fit the model parameters.

Multi-wall Model

In indoor environments, walls are major obstacles and they cause large attenuation of the RSS. In Fig. 6.3, the received signals in an indoor environment

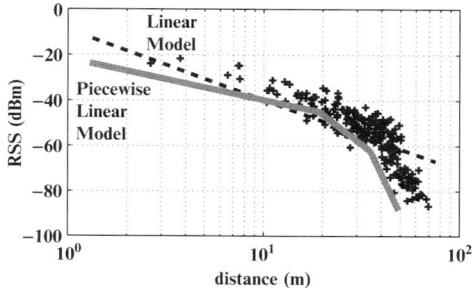

Fig. 6.2. Example of linear model and piecewise linear model with three segments

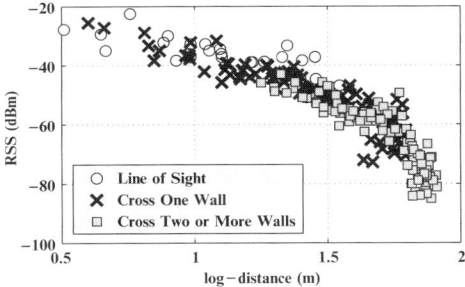

Fig. 6.3. Groups of calibration points with different number of walls

are plotted. Additionally, the number of walls between the transmitter and the receiver is shown. From the plot, we can see that at the same distance, the lower received power typically refers to a high number of walls between the transmitter and the receiver. To compensate the influence of walls, a new channel model based on the building layout is needed.

One such model is named multi-wall model (MWM) or Motley–Keenan model [21], described by the following equation:

$$p_\mathrm{r} = p_0 - 10 \cdot \gamma \cdot \log\left(\frac{d}{d_0}\right) - \begin{cases} \sum \mathrm{WAF} & \text{if } l < w \\ C & \text{if } l \geq w \end{cases}, \quad (6.2)$$

where γ and p_0 have the same meaning as in the LM. WAF is a wall attenuation factor, which represents the partition value of the walls encountered by a single ray between the transmitter and the receiver. l is the number of walls encountered. The equation means that if the number of encountered walls is less than w, the power loss by the walls can be computed as the sum of each WAF. And if the number of the encountered walls is more than w, the maximum path loss takes the value C.

The key parameter in the multi-wall model is the WAF. This parameter can either be set to a standard value or be estimated from the measurements at calibration points. A lot of experiments and simulations are made to get the value of WAF [6, 8, 15, 23]. These reported values differ due to the signal frequency, the type, the material, the shape and the position of the walls. Typically, some standard values, which average the results of several different experimental environments, are used in the modeling. A table with standard *WAF* values at 2 GHz can be found at [21].

If there are calibration data available, it is possible to obtain the mean value of WAF from the real measurements by the regression estimation. Assuming uniform WAF for each wall, (6.2) can be expressed as:

$$p_\mathrm{r} = p_0 - 10 \cdot \gamma \cdot \log\left(\frac{d}{d_0}\right) - \begin{cases} l \cdot \mathrm{WAF} & \text{if } l < w, \\ w \cdot \mathrm{WAF} & \text{if } l \geq w. \end{cases} \quad (6.3)$$

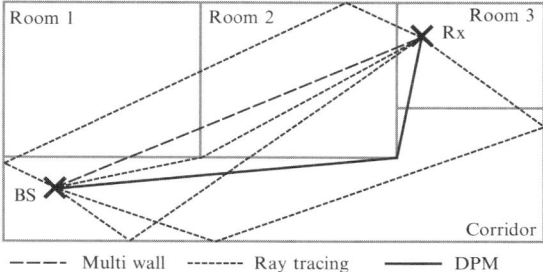

Fig. 6.4. Possible paths between a transmitter and a receiver

Using m calibration points, a system with m equations and three variables, the parameters p_0, γ and WAF, can be resolved also by least square estimation, similarly to LM.

Dominant Path Model

The MWM only takes the direct ray between the transmitter and the receiver into consideration. This will lead to an underestimation in many cases. Ray tracing algorithms take into consideration every possible path between transmitter and receiver, requiring a high computation effort. In Fig. 6.4, the BS emits many possible rays to the receiver Rx. The direct path penetrates three walls and some paths bounce, reflect and refract until reaching Rx, each of them with single contribution to the RSS that will be added by ray tracing. One path particularly has the strongest contribution among all others, presenting the optimal combination of distance, number of walls crossed and other attenuation factors. This is the dominant path.

The DPM [28] aims to find the dominant path, which could be a direct path as well, and use this single contribution to estimate the RSS. This overcomes the underestimation of only using the direct path and is considerably faster than ray tracing. References [24, 28] give the algorithm to find the dominant path, which will not be described here.

6.2.4 Non-Parametric Models

Another radio map generation approach is based only on recording of calibration data. Considering the propagation profile as a continuous function, then the calibration procedure is actually the sampling of this function at the positions specified for the radio map. The task can be understood as a regression problem, i.e., to estimate a continuous function from a finite number of observations.

The simplest way is to use a sampled radio map instead of a continuous one. This approach takes only the calibration points in consideration, neglecting other locations in the map. Therefore, estimated locations using matching algorithms are also restricted to these calibration points or the interpolation of them. Such radio map sampling is used in the pattern matching localization.

6.3 Localization Solution

Since the RSS vectors can be modelled as random variables, the statistical theory can be used to solve the matching problem. Being $\boldsymbol{x}$ the location of a mobile terminal, and $\boldsymbol{p} = [p_1, \ldots, p_N]^T$ the vector with RSS from N BSs, where $\{p_1, \ldots, p_N\}$ are the respective RSS values from BS_1 to BS_N. Then the probability of being at the true location $\boldsymbol{x}$ given the measurement $\boldsymbol{p}$ is expressed as $\Pr(\boldsymbol{x}|\boldsymbol{p})$, which can be written using the Bayesian rule as:

$$\Pr(\boldsymbol{x}|\boldsymbol{p}) = \frac{\Pr(\boldsymbol{p}|\boldsymbol{x}) \cdot \Pr(\boldsymbol{x})}{\int \Pr(\boldsymbol{p}|\boldsymbol{x}) \cdot \Pr(\boldsymbol{x}) \, \mathrm{d}\boldsymbol{x}}. \tag{6.4}$$

Usually the prior probability $\Pr(\boldsymbol{x})$ is set as uniformly distributed, assuming that all positions are equally probable, hence independent of $\boldsymbol{x}$. Then (6.4) can be simplified as:

$$\Pr(\boldsymbol{x}|\boldsymbol{p}) = \frac{\Pr(\boldsymbol{p}|\boldsymbol{x})}{\int \Pr(\boldsymbol{p}|\boldsymbol{x}) \, \mathrm{d}\boldsymbol{x}}. \tag{6.5}$$

The conditional probability $\Pr(\boldsymbol{p}|\boldsymbol{x})$ is determined from the data stored in the radio map. Using $\Pr(\boldsymbol{x}|\boldsymbol{p})$ a location estimate of the mobile terminal can be obtained using Bayesian inference.

There are two common methods for solving Bayesian estimation problems. The first one uses the minimum mean variance Bayesian estimator or minimum mean square error (MMSE) estimator. This method gives an unbiased estimator with minimum mean square error. Considering $\boldsymbol{x}$ as the true location and $\hat{\boldsymbol{x}}$ as its estimate, the mean square error is written as $\mathrm{Cost} = \int \|\boldsymbol{x} - \hat{\boldsymbol{x}}\|_2^2 \cdot \Pr(\boldsymbol{x}|\boldsymbol{p}) \, \mathrm{d}\boldsymbol{x}$. Making $\partial \mathrm{Cost}/\partial \boldsymbol{x} = 0$, the best estimation for $\boldsymbol{x}$ is given by

$$\hat{\boldsymbol{x}} = \mathrm{E}[\boldsymbol{x}|\boldsymbol{p}] = \int \boldsymbol{x} \cdot \Pr(\boldsymbol{x}|\boldsymbol{p}) \, \mathrm{d}\boldsymbol{x}, \tag{6.6}$$

where $\mathrm{E}[\boldsymbol{x}|\boldsymbol{p}]$ is the expected value for $\boldsymbol{x}$ given $\boldsymbol{p}$.

Another method uses the Maximum A Posteriori (MAP) estimator:

$$\hat{\boldsymbol{x}} = \arg\max_{x} \Pr(\boldsymbol{x}|\boldsymbol{p}) \tag{6.7}$$

From (6.5) and taking into account that $\int \Pr(\boldsymbol{p}|\boldsymbol{x}) \, \mathrm{d}\boldsymbol{x}$ is a normalizing constant, it follows that (6.7) can also be written as:

$$\hat{\boldsymbol{x}} = \arg\max_{x} \Pr(\boldsymbol{p}|\boldsymbol{x}), \tag{6.8}$$

which is also known as the maximum likelihood (ML) method.

In practice, an expression for $\Pr(\boldsymbol{p}|\boldsymbol{x})$ is hard to model, so several simplifying assumptions are made: First, the measurements from several BSs are assumed independent so that $\Pr(\boldsymbol{p}|\boldsymbol{x})$ can be factorized. Further on, it is assumed that the measurements follow a Gaussian distribution. Then:

$$\Pr(\boldsymbol{p}|\boldsymbol{x}) = \prod_{n=1}^{N} \frac{1}{\sigma_n \sqrt{2\pi}} \exp\left(-\frac{(p_n - \bar{p}_n)^2}{2\sigma_n^2}\right), \quad (6.9)$$

where $\bar{p}_n$ is the mean RSS from BS_n at position $\boldsymbol{x}$, and σ_n is the standard deviation at this position.

With these assumptions, the maximum likelihood solution becomes:

$$\hat{\boldsymbol{x}} = \arg\min_{x} \sum_{n=1}^{N} \left(\ln \sigma_n + \frac{(p_n - \bar{p}_n)^2}{2\sigma_n^2}\right). \quad (6.10)$$

If the standard deviation σ_n for every position is constant, then (6.10) reduces to:

$$\hat{\boldsymbol{x}} = \arg\min_{x} \sum_{n=1}^{N} (p_n - \bar{p}_n)^2. \quad (6.11)$$

This equation describes the method known as nearest neighbor (NN), which compares the distances in signal space between an input vector (here as $\boldsymbol{p}$) and each stored data (here the recorded radio map calibration data). The nearest stored data to the input, i.e., with smallest distance in signal space, is chosen as best match and the position of this data is returned as the estimation for $\boldsymbol{x}$. The advantage of this algorithm is that it does not require any statistical prior knowledge of the calibration data since only the mean is recorded. However, due to this simplicity, the accuracy is degraded if the hypothesis of constant σ does not correspond to the measurements.

6.4 Simultaneous Localization and Learning

To work with a reasonable accuracy, pattern matching often requires many measurements as a way to build a detailed feature map. Thus, the calibration effort prior to system start (offline phase) is usually very time consuming and expensive. The collection of measurements is labelled with the true location where they were taken until the samples are significant enough to represent the desired feature.

For this reason, the research aiming at the so-called calibration free systems has risen rapidly in the last few years. In [5] a first version of a new algorithm was presented to reduce the calibration effort significantly: the Simultaneous Localization and Learning (SLL). The SLL is a modified version of the Kohonen Self Organizing Map (SOM) [13] that can straightforwardly be used for RSS-based localization in environments with already available infrastructure as WLAN or DECT.

6.4.1 Kohonen SOM

SOMs are a special class of neural networks, which are based on competitive learning. In an SOM, the neurons are placed at the nodes of a lattice that is typically one- or two-dimensional. The neurons become selectively adapted to various input patterns in the course of a competitive learning process. The locations of the adapted, i.e., winning neurons become ordered with respect to each other in such a way that a meaningful coordinate system for different input feature is created over the lattice [13]. An SOM is therefore characterized by the formation of a topological map of input patterns in which the spatial locations of the neurons are indicative of intrinsic statistical features contained in the input patterns [11].

The principal goal of Kohonen's SOM is to transform an incoming signal pattern of arbitrary dimension into a one- or two-dimensional (1D or 2D) discrete map, and to perform this transformation adaptively in a topologically ordered fashion. This is achieved by iteratively performing three steps in addition to the initialization: competition, cooperation and adaptation. During the initialization, the synaptic weights in the neural network are set, typically using a random number generator. In the competitive step the winning neuron i with the weight vector $w_i = [w_{i1}, \ldots, w_{in}]$ in the n dimensional input space shows the smallest cost with respect to a given input feature vector $\xi = [\xi_1, \ldots, \xi_n]^T$, that is, $i = \arg\min\{|\xi - w_j|\}$, with the index j going through all neurons in the lattice. The winning neuron will be the center for the adaptation process.

The cooperation determines which neurons will be adapted together with the winning neuron i. A neighbourhood function $h_{ij}(k)$, dependent on the discrete time step k, is used to find the neuron j close to the winner and to weigh it accordingly with the distance to the winner in the lattice. The amount of adaptation decreases monotonically with distance from the center neuron, i.e., the winner.

A typical choice for the neighbourhood function at 1D problems is the constant function, set to 1 for the winner and for an equal number of neighbours, forward and backward (usually just 2 neighbours are taken). For 2D or 3D maps the Gaussian function is usually chosen, so that:

$$h_{ij}(k) = \eta(k) \exp\left(-\left(\frac{d_{ij}}{2\sigma(k)}\right)^2\right), \tag{6.12}$$

where $\eta(k)$ is the learning rate and $\sigma(k)$ is the effective width of the topological neighbourhood, both dependent on k. d_{ij} is the distance from neuron j to neuron i at the center. The adaptation law, given by

$$w_j(k+1) = w_j(k) + h_{ij}(k)\bigl(\xi(k) - w_j(k)\bigr), \tag{6.13}$$

ensures that the response of the winning neuron to the subsequent application of a similar input pattern is enhanced [11].

The adaptive process consists of two phases: the self-organizing or ordering phase and the convergence phase. In the ordering phase the topological ordering of the weight vectors takes place. During this phase the learning rate and the neighbourhood area should decrease. The neighbourhood area goes from complete coverage to a few neurons or even to the winning neuron itself. In the convergence phase the fine tuning of the feature map takes place to provide an accurate statistical quantification of the input space. The learning rate should stay constant or it could decay exponentially [11].

The Kohonen algorithm is surprisingly resistant to a complete mathematical study (cf. [7]). The only thorough analyses could be achieved for 1D case in a linear network. For higher dimensions, the results are only partial.

6.4.2 Main Algorithm

The SLL is an iterative algorithm that can be understood as the following scenario describes: a measurement is taken and then used to locate a user, based on a coarse feature map. The found location is then used as center for a learning step, where the neighbourhood surrounding this center is adapted towards the measurement. These operations continue repeatedly at each new measurement, improving the feature map.

The modelling used for the SLL originally assumed that the feature map contained the mean value of RSS of each BS as a function of position. This feature map was then termed radio map [5]. However, the SLL is by no means constrained to this feature only. Other features with spatial gradients, like propagation times or angles of arrival, could be used as well.

The model $\boldsymbol{p}_k(\boldsymbol{x})$ describes the RSS propagation through space at the discrete time k. The dimension of $\boldsymbol{p}_k(\boldsymbol{x})$ defines the number of BSs considered for the localization scenario. $\boldsymbol{x}$ defines a position in some fixed reference frame. $\boldsymbol{p}_0(\boldsymbol{x})$ represents the initial model at the initial time $k = 0$. The measurement $\boldsymbol{p}_\mathrm{M}$ is associated to $\boldsymbol{x}_\mathrm{M}$, a (not necessarily known) measurement position. $\boldsymbol{p}_{\mathrm{M},k}$ is the measurement taken at the discrete time k.

Starting from this model, the SLL is defined by the following feedback or update law (implicitly described in [5]):

$$\boldsymbol{p}_{k+1}(\boldsymbol{x}) = \boldsymbol{p}_k(\boldsymbol{x}) + f_{\mathrm{c},k+1}\big(\boldsymbol{p}_{\mathrm{M},k+1} - \boldsymbol{p}_k(\boldsymbol{x})\big), \tag{6.14}$$

where $f_{\mathrm{c},k} = f(\boldsymbol{x}_{\mathrm{c},k}, \boldsymbol{x}, \kappa, \phi)$ is a function of the centering position $\boldsymbol{x}_{\mathrm{c},k}$ at time k, of $\boldsymbol{x}$, and of the SLL control variables κ and ϕ. $f_{\mathrm{c},k}$ spatially bounds and weights the update based on the difference between the actual measurement $\boldsymbol{p}_{\mathrm{M},k+1}$ and the present model $\boldsymbol{p}_k(\boldsymbol{x})$.

$f_{\mathrm{c},k}$ can have different forms, like a polynomial or Gauss distribution. Its important characteristics are that it is symmetric around $\boldsymbol{x}_{\mathrm{c},k}$ and has its magnitude bounded by the interval $[0; \kappa]$, with $\kappa < 1$. If the distance from $\boldsymbol{x}$ to $\boldsymbol{x}_{\mathrm{c},k}$ is greater than ϕ then $f_{\mathrm{c},k} = 0$. The function $f_{\mathrm{c},k}$ reaches its maximum at $\boldsymbol{x}_{\mathrm{c},k}$ with value κ, and falls to smaller values until the boundary defined

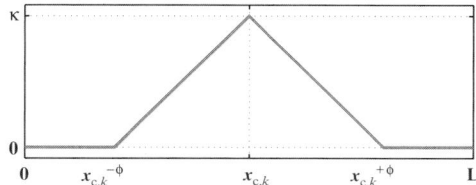

Fig. 6.5. Weighting function $f_{c,k}$

by ϕ. The location $\boldsymbol{x}_{c,k} = \boldsymbol{x}_{c,k}(\boldsymbol{p}_{M,k})$ corresponds to the measurement $\boldsymbol{p}_{M,k}$. The determination of $\boldsymbol{x}_{c,k}$ depends on the localization technique chosen (see Sect. 6.3). Figure 6.5 shows a qualitative 1D example for the function $f_{c,k}$ and the meaning of the function arguments.

6.4.3 Comparison Between SOM and SLL

The resemblance between (6.13) and (6.14) is noteworthy; in fact, the update law for SLL is the same as that for SOM. However, the SOMs exist over a discrete lattice with a finite number of neurons, as the SLL can work in a continuous space (in which case the concept for a single neuron vanishes, since there are infinite neurons). The neighbourhood function h_{ij} becomes the weighting function $f_{c,k}$ and the input vector ξ becomes the measurement vector $\boldsymbol{p}_M$.

The SLL starts with a coarse initial model, which at least presents some plausible physical property (for example, a radio map where the maximum RSS is placed at the BS position and decays with increasing distance). This ensures that the initial model can be used for localization queries, even if it is not very accurate. Proceeding with the analogy to SOM, the initial model represents an already ordered state, as the feature map is never initialized with random values, therefore, only the cooperation and convergence phases are of interest for the SLL.

The localization has also an important role for the SLL. Since it determines the location where the update will be made, if the localization delivers a $\boldsymbol{x}_c$ very far from the (usually) unknown measurement position, the SLL will fail to improve the initial model.

In [3, 4] some algebraic properties as well as the statistical conditions for a successful use of SLL were presented. The analysis was then focused on the 1D problem. Now some generalizations for more dimensions are presented.

6.4.4 Convergence Properties of SLL

The recursive SLL formula given in (6.14) can be written in closed form:

$$\boldsymbol{p}_k(\boldsymbol{x}) = \boldsymbol{p}_0(\boldsymbol{x}) + \sum_{j=1}^{k} \left(f_{c,j}(\boldsymbol{p}_{M,j} - \boldsymbol{p}_0(\boldsymbol{x})) \prod_{i=j+1}^{k} (1 - f_{c,i}) \right) \quad (6.15)$$

with
$$\prod_{i=a}^{b}(\cdot) = 1 \ \forall a > b, \tag{6.16}$$

as it has been proved in [3]. Defining the utility functions:

$$F_k = \sum_{j=1}^{k}\left(f_{c,j}\prod_{i=j+1}^{k}(1-f_{c,i})\right), \tag{6.17}$$

and

$$\mathbf{P}_{\mathrm{M},k} = \sum_{j=1}^{k}\left(f_{c,j}\cdot\boldsymbol{p}_{\mathrm{M},j}\prod_{i=j+1}^{k}(1-f_{c,i})\right). \tag{6.18}$$

Then (6.15) can be compactly written as:

$$\boldsymbol{p}_k(\boldsymbol{x}) = \boldsymbol{p}_0(\boldsymbol{x}) - \boldsymbol{p}_0(\boldsymbol{x})F_k + \mathbf{P}_{\mathrm{M},k}. \tag{6.19}$$

Limit Value

In [3, 4] it has been proved that $\lim_{k\to\infty}F_k = 1$ inside a closed interval for the 1D case. This result can be extended to higher dimensions, as long as the weighting function $f_{c,k}$ has the properties explained in the last section and that the series $\boldsymbol{x}_{c,k}$, where the $f_{c,k}$s are centered, cover repeatedly the entire closed interval.

$\mathbf{P}_{\mathrm{M},k}$, in contrast to F_k, cannot reach a steady state. Each measurement $\boldsymbol{p}_{\mathrm{M},j}$, as it appears in (6.18), pushes $\mathbf{P}_{\mathrm{M},k}$ towards $\boldsymbol{p}_{\mathrm{M},j}$ inside the hypersphere centered in $\boldsymbol{x}_{c,j}$ and with radius ϕ. Since the measurements vary through space, $\mathbf{P}_{\mathrm{M},k}$ continuously changes.

In this way, when $k\to\infty$, $\boldsymbol{p}_k(\boldsymbol{x})$, as given in (6.19) tends to:

$$\lim_{k\to\infty}\boldsymbol{p}_k(\boldsymbol{x}) = \boldsymbol{p}_0(\boldsymbol{x}) - \boldsymbol{p}_0(\boldsymbol{x}) + \lim_{k\to\infty}\mathbf{P}_{\mathrm{M},k} = \lim_{k\to\infty}\mathbf{P}_{\mathrm{M},k}. \tag{6.20}$$

This shows an important result of SLL: Eventually, the initial model $\boldsymbol{p}_0(\boldsymbol{x})$ will be replaced entirely by $\mathbf{P}_{\mathrm{M},k}$, a term that depends on the measurements and on the location estimation. Since $\boldsymbol{p}_0(\boldsymbol{x})$ disappears with increasing iterations, there is no need to make the initial model extremely precise. It suffices to start with a coarse and hence relatively simple model, e.g. linear model. The requirement for feature map initialization is a reasonable location estimation. Another effect of SLL is that old measurements have a smaller contribution to $\mathbf{P}_{\mathrm{M},k}$ than newer ones. This can clearly be seen in (6.18), where the $\boldsymbol{p}_{\mathrm{M},j}$s are multiplied by products of $(1 - f_{c,i}) \leq 1\ \forall i$. The older the measurements are, the bigger is the number of terms in the product, which will tend to zero. The consequence is that the model is always updated, as long as new measurements are considered and as the control parameters are non-zero.

Measurement Noise

Assuming that each measurement $\boldsymbol{p}_\text{M}$ is corrupted by stationary Gaussian noise $\boldsymbol{\zeta}(\boldsymbol{x})$ with mean $\boldsymbol{\mu}(\boldsymbol{x})$ and variance $\boldsymbol{\sigma}^2(\boldsymbol{x})$, it is desirable to know the remaining effect of this noise after some iterations of the SLL.

Returning to (6.14), the recursive equation regarding the noise $\boldsymbol{\zeta}_{k+1} = \boldsymbol{\zeta}(\boldsymbol{x}_{\text{M},k+1})$ at the new measurement position $\boldsymbol{x}_{\text{M},k+1}$ becomes:

$$\boldsymbol{p}_{k+1}(\boldsymbol{x}) = \boldsymbol{p}_k(\boldsymbol{x}) + f_{\text{c},k+1}\big(\boldsymbol{p}_{\text{M},k+1} + \boldsymbol{\zeta}_{k+1} - \boldsymbol{p}_k(\boldsymbol{x})\big), \qquad (6.21)$$

which, similarly to (6.15), leads to the closed form:

$$\boldsymbol{p}_k(\boldsymbol{x}) = \boldsymbol{p}_0(\boldsymbol{x}) + \sum_{j=1}^{k}\left(f_{\text{c},j}\cdot(\boldsymbol{p}_{\text{M},j} + \boldsymbol{\zeta}_j - \boldsymbol{p}_0(\boldsymbol{x}))\prod_{i=j+1}^{k}(1 - f_{\text{c},i})\right) \qquad (6.22)$$

The noise term can be separated from (6.22) defining the utility function

$$\mathbf{Z}(\boldsymbol{x}, \boldsymbol{x}_{\text{M},1:k}, \boldsymbol{x}_{\text{c},1:k}) = \mathbf{Z}_k = \sum_{j=1}^{k}\left(f_{\text{c},j}\boldsymbol{\zeta}_j \prod_{i=j+1}^{k}(1 - f_{\text{c},i})\right), \qquad (6.23)$$

such that the following short form is attained using (6.17) and (6.18):

$$\boldsymbol{p}_k(\boldsymbol{x}) = \boldsymbol{p}_0(\boldsymbol{x}) - \boldsymbol{p}_0(\boldsymbol{x})\mathrm{F}_k + \mathbf{P}_{\text{M},k} + \mathbf{Z}_k, \qquad (6.24)$$

which corresponds to (6.19) with the extra term $\mathbf{Z}_k$ modelling the influence of the measurement noise. It is important to note that $\mathbf{Z}_k$ depends not only on the considered location $\boldsymbol{x}$, but also on the sequence of true measurement locations, defined by $\boldsymbol{x}_{\text{M},1:k} = \{\boldsymbol{x}_{\text{M},1}, \ldots, \boldsymbol{x}_{\text{M},k}\}$ as well as on the sequence of estimated locations, defined by $\boldsymbol{x}_{\text{c},1:k} = \{\boldsymbol{x}_{\text{c},1}, \ldots, \boldsymbol{x}_{\text{c},k}\}$, where the weighting functions are centered.

The similarity between $\mathbf{P}_{\text{M},k}$ in (6.18) with $\mathbf{Z}_k$ in (6.23) is notable. They differ only in the scalar term introduced with new iterations: i.e., $\boldsymbol{p}_{\text{M},j}$ and $\boldsymbol{\zeta}_j$, respectively.

$\mathbf{Z}_k$ cannot reach a steady state for the same reason as $\mathbf{P}_{\text{M},k}$. However, departing from the assumption that each $\boldsymbol{\zeta}_j$ is an independent Gaussian random variable, it is possible to calculate expectations of mean and variance of $\mathbf{Z}_k$ based on the mean $\boldsymbol{\mu}_j$ and variance $\boldsymbol{\sigma}_j^2$ of each $\boldsymbol{\zeta}_j$.

For one particular fixed point $\boldsymbol{x}_\text{f}$, (6.23) shows that $\mathbf{Z}(\boldsymbol{x}_\text{f}, \boldsymbol{x}_{\text{M},1:k}, \boldsymbol{x}_{\text{c},1:k}) = \mathbf{Z}_k(\boldsymbol{x}_\text{f})$ is formed as a weighted sum of random variables. And therefore, the following properties for linear operations on independent random variables can be used, provided that a and b are scalars:

$\text{mean}\{a + b\zeta_i\} = a + b\mu_i$	$\text{mean}\{\zeta_i + \zeta_j\} = \mu_i + \mu_j$
$\text{var}\{a + b\zeta_i\} = b^2\sigma_i^2$	$\text{var}\{\zeta_i + \zeta_j\} = \sigma_i^2 + \sigma_j^2$

In this way, using the recursive formulation for (6.23):

$$\mathbf{Z}_{k+1}(\boldsymbol{x_f}) = \mathbf{Z}_k(\boldsymbol{x_f}) \cdot (1 - f_{c,k+1}) + f_{c,k+1}\boldsymbol{\zeta}_{k+1}, \qquad (6.25)$$

setting $mean\{\mathbf{Z}_k(\boldsymbol{x_f})\} = \mathbf{M}_\mathbf{Z}(\boldsymbol{x_f}, \boldsymbol{x}_{\mathbf{M},1:k}, \boldsymbol{x}_{\mathbf{c},1:k}) = \mathbf{M}_{\mathbf{Z},k}(\boldsymbol{x_f})$ and $\text{var}\{\mathbf{Z}_k(\boldsymbol{x_f})\} = \mathbf{S}_\mathbf{Z}^2(\boldsymbol{x_f}, \boldsymbol{x}_{\mathbf{M},1:k}, \boldsymbol{x}_{\mathbf{c},1:k}) = \mathbf{S}_{\mathbf{Z},k}^2(\boldsymbol{x_f})$, and using the properties above listed, it is possible to express the mean of $\mathbf{Z}_k(\boldsymbol{x_f})$ recursively as:

$$\mathbf{M}_{\mathbf{Z},k+1}(\boldsymbol{x_f}) = \mathbf{M}_{\mathbf{Z},k}(\boldsymbol{x_f})\bigl(1 - f_{c,k+1}(\boldsymbol{x_f})\bigr) + f_{c,k+1}(\boldsymbol{x_f})\boldsymbol{\mu}(\boldsymbol{x}_{\mathbf{M},k+1}), \qquad (6.26)$$

and similarly for its variance as:

$$\mathbf{S}_{\mathbf{Z},k+1}^2(\boldsymbol{x_f}) = \mathbf{S}_{\mathbf{Z},k}^2(\boldsymbol{x_f})\bigl(1 - f_{c,k+1}(\boldsymbol{x_f})\bigr)^2 + f_{c,k+1}^2(\boldsymbol{x_f})\boldsymbol{\sigma}^2(\boldsymbol{x}_{\mathbf{M},k+1}). \qquad (6.27)$$

Comparing (6.26) with the recursive formulation for F_k in [3], it follows that (6.26) has also the form of an exponential filter with the variable parameter $f_{c,k+1}(\boldsymbol{x_f})$ and with the variable input $\boldsymbol{\mu}(\boldsymbol{x}_{\mathbf{M},k+1})$. $\lim_{k\to\infty} \mathbf{M}_{\mathbf{Z},k}(\boldsymbol{x_f}) = \boldsymbol{\mu}$ holds for constant $\boldsymbol{\mu}(\boldsymbol{x}_{\mathbf{M},k}) = \boldsymbol{\mu} \; \forall k$, constant measurement location $\boldsymbol{x_\mathbf{M}}$, and constant estimated location $\boldsymbol{x_\mathbf{c}}$.

Notwithstanding the similarity between (6.26) and (6.27), the latter cannot be treated as an exponential filter due to its quadratic terms. Even if $\boldsymbol{\sigma}^2$ is constant in all space, $\mathbf{S}_{\mathbf{Z},k}^2(\boldsymbol{x})$ will vary according to the sequence of $\boldsymbol{x_c}$s. However, $\mathbf{S}_{\mathbf{Z},k}^2(\boldsymbol{x})$ is upper-bounded by a maximum value. This maximum can be estimated considering a constant update center, i.e., $\boldsymbol{x}_{\mathbf{c},k} = \boldsymbol{x_c}$ for all k and assuming space-invariant and therefore also time-constant noise, i.e., $\boldsymbol{\sigma}^2(\boldsymbol{x}) = \boldsymbol{\sigma}^2$ and $\boldsymbol{\mu}(\boldsymbol{x}) = \boldsymbol{\mu}$.

Since $\boldsymbol{x}_{\mathbf{c},k}$ is constant in time, so is $f_{c,k} = f_c$ too for all k (assuming that neither κ nor ϕ vary with time). The recursive equation for $\mathbf{S}_{\mathbf{Z},k+1}^2$ can be written as:

$$\mathbf{S}_{\mathbf{Z},k+1}^2 = \mathbf{S}_{\mathbf{Z},k}^2(1 - f_c)^2 + f_c^2 \cdot \boldsymbol{\sigma}^2. \qquad (6.28)$$

Assuming a steady state, i.e., $\mathbf{S}_{\mathbf{Z},k+1}^2 = \mathbf{S}_{\mathbf{Z},k}^2 = \mathbf{S}_{\mathbf{Z},steady}^2$,

$$\mathbf{S}_{\mathbf{Z},steady}^2 = \frac{f_c^2 \cdot \boldsymbol{\sigma}^2}{1 - (1 - f_c)^2} = \frac{f_c \cdot \boldsymbol{\sigma}^2}{2 - f_c}, \qquad (6.29)$$

holds. In particular at the position $\boldsymbol{x} = \boldsymbol{x_f} = \boldsymbol{x_c}$:

$$\mathbf{S}_{\mathbf{Z},steady}^2(\boldsymbol{x_c}) = \frac{\kappa \cdot \boldsymbol{\sigma}^2}{2 - \kappa}, \qquad (6.30)$$

which is the maximum for this function.

Considering that $\kappa \in [0; 1]$, and that the upper bound is given by (6.30), it is easy to verify that $\mathbf{S}_{\mathbf{Z},steady}^2 \leq \boldsymbol{\sigma}^2$. This indicates that the variance of $\mathbf{Z}_k$ at one particular position will be at most $\boldsymbol{\sigma}^2$, and that only if $\kappa = 1$.

The important result is the noise reduction property of SLL: by exponential filtering and spatial weighting due to f_c, the variance of the learned radio map is reduced. At one particular position $\boldsymbol{x_f}$, this noise averaging is achieved not only using the single measurements at $\boldsymbol{x_f}$, but also using the noisy measurement of neighbouring positions (e.g. the measurement sequence $\boldsymbol{x}_{\mathbf{M},1:k}$).

Limit Area for Perfect Localization

The concept of limit area appears if only one dimension in space is considered and with perfect localization. The extension of this result in more dimensions would result in limit hyper volumes and will not be treated here.

Perfect localization implies that, for a given measurement $p_{M,k}$, the associated position $x_{c,k}$ corresponds exactly to the real measurement position $x_{M,k}$, i.e., $x_{c,k} = x_{M,k}$. The measurements follow a propagation law g:

$$p_M(x_M) = g(x_M, p_{\text{out}}, \gamma), \qquad (6.31)$$

where p_{out} is the output power of a BS and γ is an attenuation factor. The signal propagation is assumed to be monotonic, i.e., $\partial g/\partial x$ is always negative. Without loss of generality, the BS is assumed to be placed on the left side of the radio map such that the measurement at $x_f - \phi$ is:

$$p_M(x_f - \phi) = p_{M+} = g(x_f - \phi, p_{\text{out}}, \gamma), \qquad (6.32)$$

being x_f a fixed position and the measurement at $x_f + \phi > x_f - \phi$ is:

$$p_M(x_f + \phi) = p_{M-} = g(x_f + \phi, p_{\text{out}}, \gamma) < p_{M+}, \qquad (6.33)$$

A perfect initialization means that the starting radio map at the instant $k = 0$ has exactly the same propagation profile as the measurements. Hence, the start model at some position x_f is:

$$p_0(x_f) = g(x_f, p_{\text{out}}, \gamma). \qquad (6.34)$$

As the weighting function f_c has a bounded support, the measurement positions x_M which can change the radio map at the considered position x_f belong to the interval $[x_f - \phi; x_f + \phi]$. Hence, there are two measurement cases to consider:

If x_M lies outside of the support of the weighting function f, i.e., $x_M \notin (x_f - \phi; x_f + \phi)$, then $f_{c,k+1}(x_f) = 0$ and the update law in (6.14) results in $p_{k+1}(x_f) = p_k(x_f)$. In this case, at x_f no update is made.

If $x_M \in (x_f - \phi; x_f + \phi)$ then $f_{c,k+1}(x_f) \in (0; \kappa]$ and the update will cover x_f. Accordingly, the update law reduces the difference between the radio map at x_f and the current measurement: $|p_{k+1}(x_f) - p_{M,k+1}| \leq |p_k(x_f) - p_{M,k+1}|$. Due to perfect initialization, for $k = 0$ the model will stay unchanged at x_f only if the measurement equals the radio map, i.e., $p_M = p_0(x_f)$ and $x_M = x_f$. In this case no improvement can be accomplished since the radio map is already perfect. For all other measurement values in the interval $[p_{M+}; p_{M-}]$ the model will be disturbed at x_f towards the actual measurement. Since $p_{M,k+1} \in [p_{M+}; p_{M-}]$ and considering the assumed monotony, $p_{k+1}(x_f)$ is also limited by this interval. The maximal positive disturbance at x_f is given by p_{M+} and the maximal negative disturbance is given by p_{M-}. p_{M+} or p_{M-} can be obtained, if and only if measurements are repetitively taken either at $x_f - \phi$ or at $x_f + \phi$, respectively.

Since there is no assumption on the placement of x_f, this result can be generalized to the whole radio map interval $[0; L]$. Thus a upper and lower limit curve can be defined for the whole interval $[0; L]$, where $p_k(x)$ is not expected to cross. The radio map is restricted to a limit area, which is defined by moving the true propagation model by ϕ to the left and to the right:

$$g(x + \phi) \leq p_k(x) \leq g(x - \phi). \tag{6.35}$$

If the measurements are not taken repeatedly at the same position, i.e., any position inside the interval $[0; L]$ could be taken, then the radio map $p_k(x)$ still stays inside the defined limit area.

Relaxing the initialization constraint, i.e., allowing an arbitrary wrong initial model, it can be shown that with perfect localization and sufficient iterations the model can be brought in finite time to the limit area.

The starting model can, e.g., be defined as:

$$p_0(x) = g(x, p_{\text{out}} + \Delta p_{\text{out}}, \gamma + \Delta \gamma), \tag{6.36}$$

where Δp_{out} and $\Delta \gamma$ are offsets on the output power and on the attenuation factor, respectively. Considering x_f, there are again two measurement cases to be looked at. Case 1 behaves as with perfect initialization: if x_M lies outside of the support of the weighting function no update is made.

Considering the measurement case 2, the inequality $|p_{k+1}(x_f) - p_{M,k+1}| \leq |p_k(x_f) - p_{M,k+1}|$ still holds, but now $p_0(x_f)$ is arbitrary wrong and a finite number of iterations is required to bring the radio map at x_f into the limit area, defined before. To reach a radio map that is inside the limit area, all positions on the radio map must be sufficiently updated and therefore the measurements must be taken to cover the complete area.

Assuming that the initial radio map lies completely outside the limit area and that the considered fixed position x_f is given by x_M, i.e., $x_f = x_M \; \forall k$, so that $p_M(x_f)$ is constant, then applying (6.36) on (6.19) at x_f gives:

$$p_k(x_f) = g(x_f, p_{\text{out}} + \Delta p_{\text{out}}, \gamma + \Delta \gamma)(1 - F_k) + p_M(x_f) F_k. \tag{6.37}$$

Due to the constant measurement positions and the exponential filter form of F_k [3], the following relation holds:

$$F_k(x_f) = 1 - (1 - \kappa)^k. \tag{6.38}$$

Hence, the influence of the initialization will decrease according to $(1-\kappa)^k$ and the measurement term will reach $p_M(x_f)$ according to $1-(1-\kappa)^k$. In finite time $p_k(x_f)$ will reach the bounded region defined by p_{M+} and p_{M-}.

Assuming a linear propagation model, making $p_M(x_f) = p_{\text{out}} - \gamma x_f$, making $g(x_f, p_{\text{out}} + \Delta p_{\text{out}}, \gamma + \Delta \gamma) = p_{\text{out}} + \Delta p_{\text{out}} - (\gamma + \Delta \gamma) x_f$ and setting $p_k(x_f)$ at the bounded region as $p_{\text{out}} - \gamma x_{\text{f}} + \gamma \phi$, then substituting these terms in (6.37) an inequality can be written:

$$(1 - \kappa)^k \leq \gamma \phi / (\Delta \gamma x_{\text{f}} - \Delta p_{\text{out}}), \tag{6.39}$$

which can be used to determine the required convergence time k.

The analytical investigations show that both the SLL parameters should be small: ϕ should be small to enable a tight limit area; κ should be small to reduce noise effects. Otherwise the larger the two parameters are, the faster (large κ) and wider (large ϕ) the radio map learning is. A good trade-off between accuracy and speed could be achieved by starting with larger parameters reducing them over time.

6.4.5 Statistical Conditions for SLL

In this section the influence of real localization on the performance of SLL is investigated. After definition of real localization, statistical conditions for convergence towards the analytical bound are identified using simulations. The statistical conditions are confirmed by real-world experiments.

Real Localization

When the localization is not perfect, then the estimated position x_{c}, where the update function f_{c} is centered, is not guaranteed anymore to be the measurement position x_{M}.

Now $x_{\text{c},k}$ is estimated using the measurement $p_{\text{M},k}$ and an NN search (cf. Sect. 6.3) on $p_{k-1}(x)$, looking for the best $x = x_{\text{c},k}$ for whom $p_{k-1}(x_{\text{c},k})$ is the closest match for $p_{\text{M},k}$. For the simulations the NN algorithm is run on the quantized 1D space comprised in the interval $[0; L]$.

Simulations

In [5] first results with SLL applied to real-world data have been shown. In [3, 4] an analysis was performed to reveal the statistical conditions that need to be satisfied to reliably obtain a good result. For that, some experiments were defined to determine those statistical conditions:

The experiments performed differ with respect to the distribution and the order of measurements. From the analytical considerations, it is known that both the SLL parameters, κ and ϕ, should be small at the end of learning time to achieve a high accuracy and larger at the beginning to achieve fast learning. To avoid a superposition of the statistical effects here investigated with effects possibly introduced by time-varying parameters, κ and ϕ are set to fixed values and kept constant over time.

The radio map is defined on the interval $[0; 15]$ meters. The distance Δx between consecutive positions on the discrete radio map is 0.1 m. The learning parameter κ is set to 0.5. The measurement positions x_{M} are uniformly

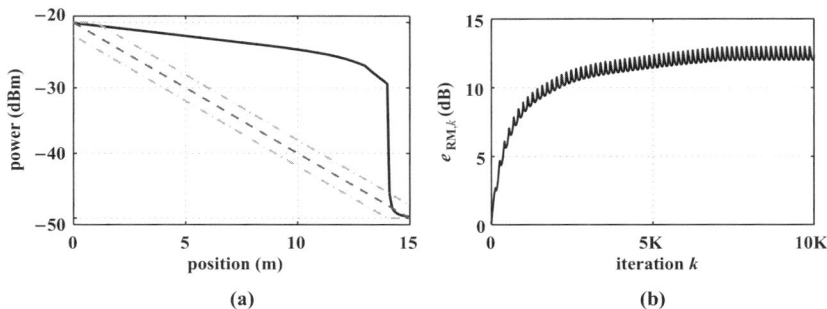

Fig. 6.6. Radio map learning for sequentially ordered measurements. The distance between two succeeding measurement positions is smaller than ϕ

distributed over space. The plots show the radio map at the initialization (dashed lines) and at the end of simulation (bold lines), the measurements (thin line or circles) and the limit area (dash dotted lines), with power as a function of position. The plots also show the radio map error, defined as the RMS error between the measurements (as labelled data and taken at once) and the actual radio map at some instant k:

$$e_{\text{RM},k} = \sqrt{\left(p_k(x) - p_\text{M}(x)\right)^\text{T}\left(p_k(x) - p_\text{M}(x)\right)/m} \qquad (6.40)$$

with m as the number of reference positions on the radio map.

Figure 6.6a, b depicts the experiment 1. The distance Δx_M between two succeeding measurement positions is 0.1 m, which results in 150 measurements. ϕ is set to 1 m, which is bigger than Δx_M. The initial radio map, as well as the measurements are given by the linear equation $p_0(x) = p_\text{out} - \gamma x$, with $p_\text{out} = -20$ dBm, $\gamma = 2$ dBm m^{-1} (since the initial radio map and measurements are coincident, only the dashed line is plotted).

The measurements are sequentially ordered, taken at increasing coordinate values, i.e., from 0 to L. The final radio map is given by the solid line after the ordered sequence of measurements is cyclically used 70 times, which results in 10,570 iterations. The black line shows the final radio map. Noteworthy is that the slope of the final radio map depends directly on Δx_M.

$e_{\text{RM},k}$ departs from 0, as the initialization is perfect and increases until a steady state is reached. The small oscillations visible in the radio map error are caused by the periodicity of the measurement locations. An equivalent and complementary experiment, with the sequence of positions going from L to 0 was shown in [3]. e_RM for that case was exactly the same and the same properties were verified.

An important feature of SLL, called edge effect, is shown in this figure. Due to the sequence of narrowing close positions, almost the entire radio map lay outside the theoretical limit area. However, because of the limitation of the localization space between 0 and L together with the monotonicity of the

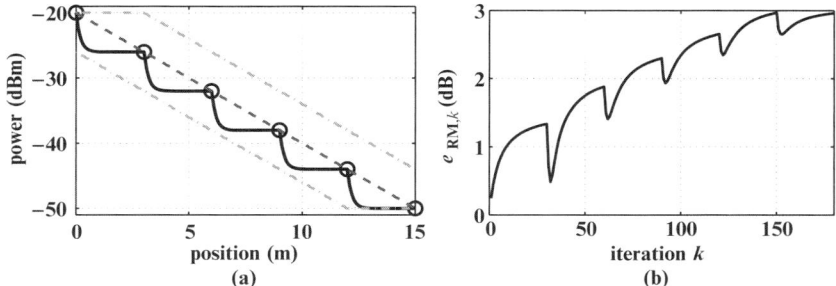

Fig. 6.7. Radio map learning for sequentially ordered measurements. The distance between two succeeding measurement positions is equal to ϕ

propagation function, the strongest RSS will be learned at $x = 0$ and the weakest RSS will be learned at $x = L$, even for real localization. If the localization tries to locate a measurement outside the given interval, the estimated position is set to the corresponding edge.

Figure 6.7a,b shows the experiment 2, which has a slightly different setup as the first one. Here, the distance between measurement positions Δx_M has the same value as ϕ, set to 3 m, i.e., $x_\mathrm{M} = \{0, 3, 6, 9, 12, 15\}$ m, marked with circles on the plot. The initial radio map and the measurements follow the same linear propagation as in the last simulation, for which reason once again only the dashed line is shown. Each measurement is taken repeatedly at the same position 30 times before going to the next position, which gives 180 iterations.

The final radio map is given by stair steps, with a spacing of ϕ and achieved with the measurement positions going from 0 to L. This result can be explained: At the beginning the radio map is perfect and real localization delivers the exact true position. After 30 updates at this same position a step is formed, with wideness defined by ϕ. On the next measurement position the real localization still delivers the exact position, since the radio map has not been changed at this position by the last update series. A new step is formed at this position, and one half of the previous step is entirely replaced by this new one, the other half remains.

$e_{\mathrm{RM},k}$ again starts from 0 and rises, although the radio map remains inside the limit area this time. This flattening effect, which forms the stair steps, is a direct result of (6.20), with the replacement of the initial model by a term dependent on the measurements p_MS.

In [3] another experiment shows that the uniform distribution of measurement positions is a requirement for SLL to work well. In this example, a logarithmic propagation profile is learnt as a linear profile because the distribution was uniform in respect to the measurements rather to their positions.

Yet in [3], a discontinuity imposed by a wall is smoothly learnt, showing the versatility of the SLL.

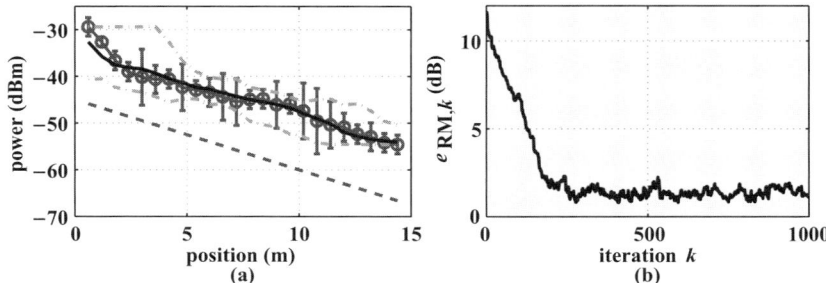

Fig. 6.8. Radio map learning for real-world data

1D Real-World Experiment

A real-world experiment with data collected at an LOS scenario has been performed. Real-world experiments are different from the previously described simulations in the sense that real-world signal propagation is more complex and not necessarily linear but close to logarithmic as can be seen in Fig. 6.8. The radio map has been initialized with a linear propagation profile, and with a large offset, being far from the true measurements such that the improvement by the SLL can be clearly seen.

Due to the findings in the section "Simulations", SLL has been applied using a uniform distribution of measurement positions in combination with a random order. Δx_M is set to 0.6 m. Each of the 24 positions was measured 30 times, so that a representative collection of noisy measurements for each position was achieved, being then randomly selected during 1,000 iterations.

The final radio map (bold black line) shows that the real-world propagation, which is close to logarithmic, is learned. The noise (see error bars) is significantly reduced by SLL such that the final error is within the analytical boundary, given by $\phi = 3$ m.

Hence, the found statistical conditions, that is, the uniform distribution in space and random ordering of measurements, are verified by the real-world experiment.

6.5 Results on 2D Real-World Scenarios

Part of the proofs so far has only been shown for 1D examples, and for the same reason the counterpart proofs for SOM: they are unfeasible for more dimensions, although there always been practical examples in 2D or 3D that do converge.

The SLL was tested in real-world scenarios with successful results and the results here verify the validity of this algorithm beyond theoretical set of containments.

Since SLL is independent of the radio technology used, examples with DECT and WLAN are shown here. The usual pattern matching with NN is compared with SLL using the same validation data and for different initial models.

The localization error e_x is calculated as the mean localization error among all validation positions at one given instant, i.e., the distance between the location of all validation positions and their correspondent located positions with the actual radio map $\boldsymbol{p}_k(\boldsymbol{x})$.

The weighting function $f_{c,k}$ used for the 2D SLL has the form:

$$f_{c,k}(\boldsymbol{x}) = \begin{cases} \kappa \cdot \left(1 - \dfrac{d_{c,k}(\boldsymbol{x})}{\phi}\right), & \text{if } d_{c,k}(\boldsymbol{x}) \leq \phi, \\ 0, & \text{if } d_{c,k}(\boldsymbol{x}) > \phi \end{cases} \quad (6.41)$$

with $d_{c,k}(\boldsymbol{x}) = \sqrt{(\boldsymbol{x} - \boldsymbol{x}_{c,k})^{\mathrm{T}}(\boldsymbol{x} - \boldsymbol{x}_{c,k})}$ and which correspond to Fig. 6.5, only that here $f_{c,k}$ has the shape of a cone.

The first test environment is an office plant with nine DECT BSs installed (marked as squares), as depicted in Fig. 6.9a. There are 223 training positions (crosses) which are also used as validation points. The second test environment is also an office plant with 14 WLAN BSs installed (squares), as seen in Fig. 6.9b. There are 114 training positions (crosses) used also as validation points.

Considering the DECT office plant, the 223 training positions were randomly selected, with repetitions allowed, during 5,000 iterations and their measurements used as unlabeled input data for the SLL. The initial models considered were the LM and the DPM, as described in Sect. 6.2.

Figure 6.10a shows the evolution of SLL for 5,000 iterations, displaying e_x and the SLL parameters κ and ϕ. From this plot it can be stated that the linear model is simpler than the DPM as the initial error is bigger for the first. In fact, the DPM is more complex than the LM, since environment layout information is built in the model. Nevertheless, the SLL steps improve both models, as it

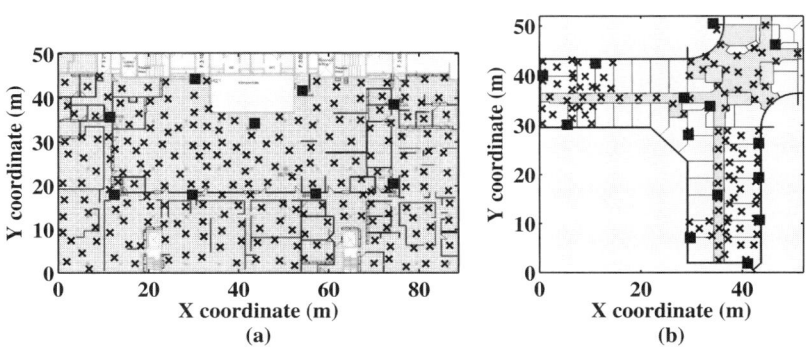

Fig. 6.9. Office plant with DECT (**a**) and WLAN (**b**) BSs

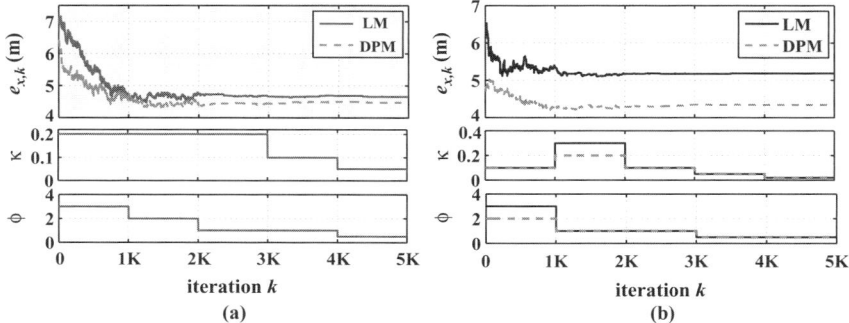

Fig. 6.10. Experiments at the DECT (**a**) and WLAN (**b**) office plant

can be seen with the fall of the localization error. A steady state is achieved very near the theoretical lower bound for accuracy using common pattern matching, which for this scenario is calculated as $\sqrt{A_{\text{DECT}}/223} = 4.5\,\text{m}$, where A_{DECT} is the area of the considered office facility.

For the WLAN office plant, the same proceeding as for the DECT set up was used: 114 training positions were randomly selected, with repetitions allowed, and their measurements were used as unlabeled input data during 5,000 iterations. The initial models were again the LM and the DPM.

Figure 6.10b shows the results of SLL for this setup. Here, the initial error difference between the linear model and the DPM is even bigger. This can be explained by the fact that the walls at this office are covered with metallic foils that attenuates highly the signals. Accordingly, the information about the walls already gives a very good initial model, leading to a low localization error prior to the start of learning (4.8 m). Further learning by the SLL brings improvement, as it can be seen with the fall of the localization error, but not much. A steady state is again achieved near the theoretical accuracy lower bound using common pattern matching, calculated as $\sqrt{A_{\text{WLAN}}/114} = 3.3\,\text{m}$, where A_{WLAN} is the area of the WLAN office facility.

6.6 Conclusions

Indoor positioning based on communication systems typically use the RSS as measurements. For proper operation such a system often requires many calibration points before its start. Applying SLL a self-calibrating RSS-based positioning system can be realized.

The algebraic and statistical conditions required to perform SLL have been explored. Important properties of SLL are the replacement of the initial radio map by the measurements, the reduction of noise by exponential filtering of different measurements, and the existence of a boundary limit defined by the adaptation width ϕ and by the profile of the signal propagation. ϕ and the learning rate κ should be kept as small as possible to achieve a high accuracy. A good trade-off between accuracy and speed can be achieved by starting with larger parameters and reducing them over time.

The statistical conditions impose the use of a uniform distribution of measurement positions over a limited interval and in combination with random ordering. Such implementation can be easily achieved by starting the system and performing SLL in batch, i.e., measurements are collected until the space is sufficiently covered, then the measurement collection is randomly ordered and SLL is performed to self-calibrate the RSS-based positioning system.

The initial model must be physically plausible and its complexity reflects directly the starting accuracy. Nevertheless, SLL iterations will improve the initial model and finally reach the accuracy boundary imposed by the measurement position density. The advantage of SLL is significant: in contrast to existing solutions no manual calibrations are required. The approach is self-calibrating thereby realizing an RSS-based localization system with truly low costs for installation and maintenance.

References

1. Bahl, P., Padmanabhan, V.N.: RADAR: An in-building RF-based user location and tracking system. In: IEEE InfoCom'00, pp. 775–784. Tel Aviv, Israel (2000)
2. Battiti, R., Brunato, M., Villani, A.: Statistical learning theory for location fingerprinting in wireless LANs. Technical Report, Universitá di Trento (2002)
3. Betoni Parodi, B., Lenz, H., Szabo, A., Bamberger, J., Horn, J.: Algebraic and statistical conditions for the use of SLL. In: ECC'07. Kos, Greece (2007)
4. Betoni Parodi, B., Lenz, H., Szabo, A., Bamberger, J., Horn, J.: SLL: Statistical conditions and algebraic properties. In: IEEE 4th WPNC'07, pp. 113–120. Hannover, Germany (2007)
5. Betoni Parodi, B., Lenz, H., Szabo, A., Wang, H., Horn, J., Bamberger, J., Obradovic, D.: Initialization and online-learning of RSS maps for indoor/campus localization. In: PLANS 2006, pp. 164–172. San Diego, CA, USA (2006)
6. Correia, L.M.: Wireless Flexible Personalized Communications. Wiley, New York, NY, USA (2001)
7. Cottrell, M., Fort, J.C., Pagés, G.: Two or three things that we know about the Kohonen algorithm. In: ESANN'94, pp. 235–244. Brussels, Belgium (1994)
8. Goldsmith, A.: Wireless Communications. Cambridge University Press (2005)
9. Günther, A., Hoene, C.: Measuring round trip times to determine the distance between WLAN nodes. In: Networking 2005. Waterloo, Canada (2005)
10. Harter, A., Hopper, A., Steggles, P., Ward, A., Webster, P.: The anatomy of a context-aware application. In: MobiCom, pp. 59–68. Seattle, WA, USA (1999)
11. Haykin, S.: Neural Networks: A Comprehensive Foundation. Prentice-Hall, Upper Saddle River, NJ, USA (1998)
12. Horn, J., Hanebeck, U.D., Riegel, K., Heesche, K., Hauptmann, W.: Nonlinear set-theoretic position estimation of cellular phones. In: ECC'03. Cambridge, UK (2003)
13. Kohonen, T.: The self-organizing map. Proceedings of the IEEE **78**, 1464–1480 (1990)

14. Liberti, J.C., Rappaport, T.S.: Smart Antennas for Wireless Communications: IS-95 and Third Generation CDMA Applications. Prentice-Hall, Upper Saddle River, NJ, USA (1999)
15. Lott, M., Forkel, I.: A multi-wall-and-floor model for indoor radio propagation. In: IEEE 53rd VTS'01 Spring, Berlin Heidelberg, New York, vol. 1, pp. 464–468. Rhodes, Greece (2001)
16. Ni, L.M., Liu, Y., Lau, Y.C., Patil, A.P.: LANDMARC: Indoor location sensing using active RFID. Wireless Networks **10**, 701–710 (2004)
17. Obradovic, D., Lenz, H., Schupfner, M.: Sensor fusion in Siemens car navigation system. In: 14th IEEE Machine Learning for Signal Processing, pp. 655–664. São Luiz, MA, Brazil (2004)
18. Oppermann, I., Karlsson, A., Linderbäck, H.: Novel phase based, cross-correlation position estimation technique. In: IEEE ISSSTA'04, pp. 340–345 (2004)
19. Pahlavan, K., Li, X., Makela, J.P.: Indoor geolocation science and technology. IEEE Communications Magazine **40**(2), 112–118 (2002)
20. Priyantha, N.B., Chakraborty, A., Balakrishnan, H.: The Cricket location-support system. In: 6th MobiCom, pp. 32–43. Boston, MA, USA (2000)
21. Rappaport, T.S.: Wireless communications principles and practice, 2nd edn. Prentice-Hall, Englewood Cliffs, NJ (2002)
22. Roos, T., Myllymäki, P., Tirri, H., Misikangas, P., Sievänen, J.: A probabilistic approach to WLAN user location estimation. International Journal of Wireless Information Networks **9**(3), 155–164 (2002)
23. Tila, F., Shepherd, P.R., Pennock, S.R.: 2 GHz propagation and diversity evaluation for in-building communications up to 4 MHz using high altitude platforms (HAP). In: IEEE 54th VTS'01 Fall, vol. 1, pp. 121–125 (2001)
24. Wang, H.: Fusion of information sources for indoor positioning with field strength measurements. Master's thesis, TU-München (2005)
25. Wang, H., Lenz, H., Szabo, A., Bamberger, J., Hanebeck, U.D.: WLAN-based pedestrian tracking using particle filters and low-cost MEMS sensors. In: IEEE 4th WPNC'07, pp. 1–7. Hannover, Germany (2007)
26. Wang, H., Lenz, H., Szabo, A., Hanebeck, U.D., Bamberger, J.: Fusion of barometric sensors, WLAN signals and building information for 3-D indoor/campus localization. In: IEEE MFI'06, pp. 426–432. Heidelberg, Germany (2006)
27. Want, R., Hopper, A., Falcão, V., Gibbons, J.: The Active Badge location system. ACM Transactions on Informatic Systems **10**(1), 91–102 (1992)
28. Wölfle, G., Wahl, R., Wertz, P., Wildbolz, P., Landstorfer, F.: Dominant path prediction model for indoor scenarios. In: GeMIC'05, pp. 176–179. Ulm, Germany (2005)

7

Signal Processing Methods for Doppler Radar Heart Rate Monitoring

Anders Høst-Madsen, Nicolas Petrochilos, Olga Boric-Lubecke,
Victor M. Lubecke, Byung-Kwon Park, and Qin Zhou

A practical means for unobtrusive and ubiquitous detection and monitoring of heart and respiration activity from a distance could be a powerful tool for health care, emergency, and surveillance applications, yet remains a largely unrealized goal. Without the need for contact or subject preparation (special clothing, attachments, etc.), this could better extend health monitoring to the chronically ill in routine life, allow wellness monitoring for a large population without known predisposition for risk or harm, and provide alarm and data in emergencies. Such technology could also be used to detect lost or hidden subjects, to help assess emotional state, and to compliment more cumbersome measurements as pre-screening. Doppler radar remote sensing of vital signs has shown promise to this end, with proof of concept demonstrated for various applications. Unfortunately, this principle has not been developed to the level of practical application, mainly due to a lack of an effective way to isolate desired target motion from interference. However, by leveraging recent advances in signal processing and wireless communications technologies, this technique has the potential to transcend mere novelty and make a profound impact on health and welfare in society.

7.1 Introduction

Practical non-contact detection and monitoring of human cardiopulmonary activity could be a powerful tool for health care, emergency, military, and security applications. Doppler radar remote sensing of heart and respiration activity has shown promise toward this end, with proof of concept demonstrated for various applications [3, 12, 14]. By estimating the associated Doppler shift in a radio signal reflected by the body, cardiopulmonary-related movement can be discerned without physical contact. Through its non-invasive nature, this approach is well suited to applications where it is important to minimize disruption of the subject's activity, particularly where prolonged monitoring is needed. A robust Doppler radar system would be well suited to collection of long-term heartbeat interval data for heart rate variability (HRV) diagnosis and prognosis [8]. Additional benefits of microwave Doppler radar include the versatile ability to function at a distance through clothing, walls, or debris. This allows the system to be applied both in medical health care scenarios, where some degree of subject cooperation can be assumed, and in emergency response or security applications, where subjects do not or cannot cooperate. Alternative techniques for long-term medical monitoring typically require direct contact (ECG and Holter monitors, piezoelectric sensors), while minimally invasive techniques tend to require very accurate control or placement (laser Doppler vibrometer), which might not always be possible or desirable. The use of infrared (IR) body heat sensors in search and rescue operations is limited due to poor IR propagation properties through walls, rubble, and weather. While Doppler radar offers distinct advantages, this approach has not been developed to the level of practical application, mainly because there has been no genuine effective way to isolate desired target motion from other motion and other targets. Fortunately, through the application of recent advances in wireless communications and signal processing technologies, Doppler cardiopulmonary radar now has the potential to transcend mere novelty and make a significant impact on health care and national security.

The use of Doppler radar was demonstrated for detection of respiratory rate in 1975 [10], and heart rate in 1979 [11], using commercially available waveguide X-band Doppler transceivers. Our recent work to implement this concept by leveraging telecommunications technology, includes the detection of heart and respiration signals with existing wireless terminals [14, 15], implementation of dedicated low-cost microwave Doppler radars [5–7], and development of related software for automated rate detection [13]. Doppler sensing with communications signals in the 800–2400 MHz range has been demonstrated with very promising results for both detection of surface and internal heart and respiration motion [14]. Higher frequency signals like those used for motion-controlled doors and traffic lights, in the 10 GHz range, also work well for detection of cardiopulmonary motion at the chest surface, even through clothing [1]. While reliable heart and respiration rate extraction can be performed for relatively still and isolated subjects [6], it is a major challenge to obtain useful data in the presence of random motion of the human target, radar, peripheral human subjects, and other moving objects. Many contact (such as ECG, EEG) and non-contact medical measurements (such as fMRI) also suffer from motion artifacts due to random motion of the subject during the measurements. Various DSP techniques are used to extract useful data from such measurements [19]. The problem of background noise has been a barrier to bringing Doppler vital signs sensing into practical, everyday applications. We propose to explore promising new solutions to this problem, taking advantage of recent developments in wireless communications. These have the potential to not only improve the robustness of Doppler radar sensing to practical levels, but to also make possible the gathering of additional

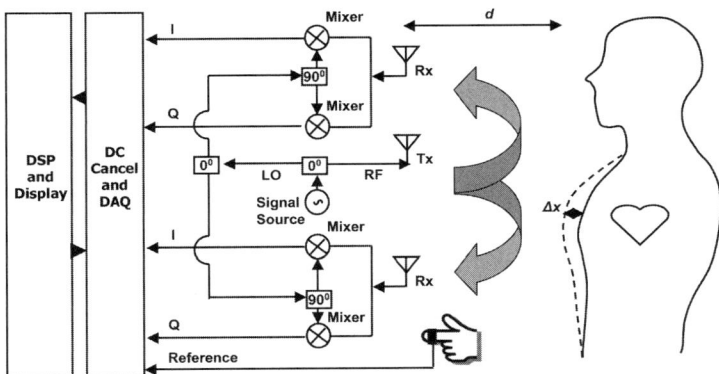

Fig. 7.1. A Doppler radar system with one transmit and two receive antennas. Each receive antenna is connected to a quadrature receiver chain to provide two orthonormal baseband signals (I and Q). DC components are eliminated via a dc canceller. Demodulated output results are compared with a wired finger pulse sensor reference

information, such as determining the number of subjects in a particular environment. The application of multiple input multiple output (MIMO) system techniques to provide robust Doppler radar heart signal detection as well as detection and count of multiple subjects will be discussed.

Multiple antennas can be used to detect multiple copies of the same signal with different phase information, with coherent combining used to provide a greatly improved estimate of desired Doppler motion. Figure 7.1 shows the block diagram of experimental set-up with one transmit and two receive antennas. Each receive antenna is connected to a quadrature receiver chain to provide two orthonormal baseband signals (I and Q). DC components are eliminated via a dc canceller. Demodulated output results are compared with a wired finger pulse sensor reference. When more than one target is in view, multiple transmitters and receivers providing multiple signal copies could be used to distinguish between the different sources of Doppler motion, and isolate the desired signal. We will first discuss demodulation methods for recovery of phase information in Doppler radar system, followed by signal processing methods for heart signal detection and estimation, and finally separation of multiple heartbeat signals.

7.2 Signal Model

In this section, we will discuss modeling of the signal. We assume a continuous wave (CW) radar system transmitting a single tone signal at frequency ω. The transmitted signal is

$$s(t) = \cos(\omega t + \phi(t)), \tag{7.1}$$

where $\phi(t)$ is phase noise in the oscillator.

This signal is reflected from a target at a nominal distance d, with a time-varying displacement given by $x(t)$. Suppose at first that the signal from the subject arrives from a single path. The received signal at the kth antenna is then

$$r_k(t) = A_k \cos\left(\omega t - \frac{4\pi}{\lambda}(d + x(t) - k\tau) + \phi\left(t - \frac{2d}{c} - k\tau\right) + \varphi_0\right) + w_k(t)$$

$$\tau = d_A \sin\alpha \tag{7.2}$$

Here d_A is the inter-antenna spacing, α the angle of arrival, $w_k(t)$ the noise at the antenna, and A_k is the received amplitude at the kth antenna, which depends on the antenna pattern, e.g., if the antennas are omnidirectional, A_k is independent of k, and φ_0 some initial phase offset. To arrive at this, we have used some key properties of $x(t)$. First, $x(t)$ is slowly varying, so that $x(t - d/c - k\tau) \approx x(t)$. Second, $x(t)$ is small compared with d so that $\phi(t-(2d-x(t))/c-k\tau) \approx \phi(t-2d/c-k\tau)$. At the receiver, $r_k(t)$ is multiplied by $\cos(\omega t + \phi(t))$ and the phase-shifted signal $\sin(\omega t + \phi(t))$ and then lowpass filtered resulting in the complex signal

$$r_k(t) = A_k \exp\left(\mathrm{j}\left(\frac{4\pi}{\lambda}(x(t) - k\tau) + \phi\left(t - \frac{2d}{c} - k\tau\right) - \phi(t) + \varphi\right)\right) + w_k(t)$$

$$= C_k\, s_k(\tau)\, \exp\left(\mathrm{j}\Delta\phi_k(t)\right)\, \exp\left(\mathrm{j}\frac{4\pi}{\lambda}x(t)\right) + w_k(t) \tag{7.3}$$

$$C_k = A_k \mathrm{e}^{\mathrm{j}\varphi}$$

$$s_k(\tau) = \mathrm{e}^{-\mathrm{j}\frac{4\pi}{\lambda}k\tau}$$

$$\Delta\phi_k(t) = \phi\left(t - \frac{2d}{c} - k\tau\right) - \phi(t)$$

$$\varphi = \frac{4\pi}{\lambda}d + \varphi_0$$

If $x(t)$ is small compared to λ and $\Delta\phi_k(t)$ is small, we can use a first order (i.e., linear) approximation for the complex exponential, so that

$$r_k(t) \approx C_k s_k(\tau) + C_k s_k(\tau)\mathrm{j}\frac{4\pi}{\lambda}x(t) + C_k s_k(\tau)\mathrm{j}\Delta\phi_k(t) + w_k(t) \tag{7.4}$$

So far we have assumed that the RF signal is only reflected by a single object, the subject of interest, and that only the vital sign signal is returned. However, there are multiple other reflections that must be taken into account. The signal is reflected by various stationary objects (e.g., walls) (and potentially moving objects, e.g., rustling leaves, although in many applications this is less likely). These reflections results in a dc-offset, as well as some additional noise. Thus, in most cases the dc of the signal contains no information and can be removed. This turns out to be of practical value as well. The signal due to the heartbeat is very small, and if the dc is not removed this gives dynamic range problems with the quantizer. Additionally, the signal reflected off the body of the subject is also reflected by stationary objects, resulting in multipath. In that case the received signal is (after dc-removal)

$$r_k(t) = \exp\left(j\frac{4\pi}{\lambda}x(t)\right) \sum_{i=1}^{M} C_{k,i} s_k(\tau_i) \exp\left(j\Delta\phi_{k,i}(t)\right) + \tilde{w}_k(t) \quad (7.5)$$

where M is the number of multipath components. Each signal has its own signal strength $C_{k,i}$, angle of arrival (delay) $\tau_i = d_A \sin \alpha_i$, and phase noise contribution $\Delta\phi_{k,i}$. $x(t)$. The signal $x(t)$, of course, also experiences a different delay through the different paths, but this delay is insignificant relative to the bandwidth of $x(t)$. The non-linear model for multipath is rather complex, but if the linear model is valid, we get a simpler system model (again, after dc-removal)

$$\begin{aligned} r_k(t) &\approx x(t) \sum_{i=1}^{M} C_{k,i} s_k(\tau_i) j\frac{4\pi}{\lambda} + (\tilde{w})_k(t) \\ &= v_k x(t) + \tilde{w}_k(t), \end{aligned} \quad (7.6)$$

where v_k is a complex constant that summarizes all the factors affecting the signal at the nth antenna.

7.2.1 Physiological Signal Model

The signal displacement generated by the subject consists of respiration and heartbeat. The respiration is usually in the range 0.1–0.8 Hz and the heartbeat in the range 0.8–2 Hz. While the respiration is a stronger signal than the heartbeat, it is also more difficult to characterize and therefore to detect. In the current method, we therefore remove most of the respiration by high pass filtering. The heartbeat signal itself is a rather complicated signal. It is nearly periodic, but the period can vary from one beat to the next; this is called heart rate variability (HRV). HRV can be modeled as a random process [16] with strong periodicity. We consider the filtered received signal: bandpass filtered with a pass band of 0.8–2 Hz so that only the fundamental frequency of the heartbeat is received. The resulting signal is modeled as

$$s(t) = (A + \alpha(t)) \cos(\omega t + \theta(t) + \theta_0). \quad (7.7)$$

The amplitude variations $\alpha(t)$ and $\theta(t)$ are zero mean random processes modeling the HRV. Notice that this is the same kind of model used in [22] but that the frequency ω is unknown here. Dealing directly with the model (7.7) for signal processing is complicated, since it is a non-linear function of a random process $\theta(t)$ that is not completely characterized. We therefore must use some approximation for $\theta(t)$. As $\theta(t)$ is rapidly varying (the time between heartbeats can vary 10% from one heartbeat to the next), a linear approximation is not appropriate. Instead, we will use a piecewise constant approximation of $\varphi(t)$ (and $\alpha(t)$) so that

$$s(t) = A_i \cos(\omega t + \theta_i), \quad t \in [(i-1)T_0, iT_0], \tag{7.8}$$

where T_0 is some suitably chosen interval that can be optimized for performance, and A_i and φ_i are modeled as deterministic, unknown constants.

7.3 Single Person Signal Processing

7.3.1 Demodulation

Each antenna has two output signals, for a total of $2K$ signals. In a single person system, the first step is to combine these $2K$ signal into a single signal that best possible represents the heartbeat, a step we will call demodulation. The purpose is both graphical display of the signal (i.e., similar to the usual ECG signal), and to use in further signal processing. However, it should be noticed that the demodulated signal is not necessarily a sufficient statistic, so better performance can be obtained from using the $2K$ direct signals; this is in essence using the full MIMO advantage. However, using a single signal can considerably simplify signal processing.

There are two methods for demodulation. The first is *linear demodulation*. A linear demodulator is of the form

$$\hat{x}[n] = \sum_{k=1}^{K} a_k \Re(r_k[n]) + b_k \Im(r_k[n]) = \mathbf{d}^T \mathbf{r}[n] \tag{7.9}$$

where $r_k[n]$ is the sampled version of $r_k(t)$, $\mathbf{d} = [\, a_1 \; b_1 \; \ldots \; a_K \; b_K \,]^T$, and $\mathbf{r}[n] = [\, \Re(r_1[n]) \; \Im(r_1[n]) \; \ldots \; \Re(r_K[n]) \; \Im(r_K[n]) \,]^T$. As criterion for performance we will use mean square error (MSE)

$$\min_{\mathbf{d}} E\left[(\hat{x}[n] - x[n])^2\right]. \tag{7.10}$$

However, it can also be proven that the minimum MSE (MMSE) solution maximizes the signal-to-noise ration (SNR) at the output. In [23] we have proved that the optimum demodulator is projection onto the eigenvector corresponding to the maximum eigenvalue of the covariance matrix. This result is not surprising if the linear model (7.6) is assumed, but [23] shows it is also true for the non-linear model (without multipath).

The second method for demodulation is non-linear demodulation. This is mainly relevant for a single antenna, although non-linear modulation outputs from multiple antennas could be combined. So, assume a single antenna and let $r[n] = r_1[n]$ Clearly, from the model (7.5) the optimum non-linear demodulator is given by

$$\hat{x}[n] = \text{Arg}(r[n] - c)\frac{\lambda}{4\pi}. \tag{7.11}$$

Here, c is unknown dc offset of the signal. So, to implement this estimator c needs to be known, or estimated.

It is easy to see that, given c, the ML estimator of the remaining parameters is

$$\hat{x}(c)[n] = \text{Arg}(r[n] - c)\frac{\lambda}{4\pi}, \tag{7.12}$$

$$\hat{A}(c) = \frac{1}{N}\sum_{n=0}^{N-1}\Re\left\{(r[n] - c)\exp\left(-j\frac{4\pi}{\lambda}\hat{x}(c)[n]\right)\right\}. \tag{7.13}$$

The estimation problem for c can now be stated as:

$$\hat{c} = \underset{c \in \mathbb{R}}{\arg\min}\, d(c) \tag{7.14}$$

$$d(c) = \sum_{n=0}^{N-1}\left|r[n] - \hat{A}(c)\exp(j\hat{x}(c)[n]) - c\right|^2. \tag{7.15}$$

Unfortunately, a closed form expression for c does not exist, and the MLE is difficult to find numerically. In [17] we have developed a heuristic estimator for k, which is almost as good as the MLE for reasonable MSE.

The data in [17] shows that linear and non-linear demodulation is almost equivalent at low frequencies, such as the 2.4 GHz we use for experiments, but at higher frequencies, non-linear demodulation is better.

7.3.2 Detection of Heartbeat and Estimation of Heart Rate

In this section, we consider how to estimate the heart rate from the demodulated wireless signal and how to detect if a heartbeat is present. The signal after demodulation is still a very noisy signal. It does not have the well-defined peaks known from ECG signals, and methods from ECG signal processing therefore cannot be directly applied. We will first derive the maximum likelihood estimator (MLE) based on the model (7.8) for the heart rate, and corresponding to this the generalized likelihood ratio test (GLRT).

MLE Based on Demodulated Data

Consider estimation of average heart rate from the demodulated data. In that case the data is real, and we have a model

$$x[n] = s[n] + w[n], \tag{7.16}$$

where $w[n]$ is identically distributed zero mean real Gaussian noise with unknown variance σ^2. We assume the model (7.8), where $T_0 = NT$. We consider a *window* of data of length MN and divide this into M subwindows of length N. We can then write the received signal $s_m[n]$ during the mth subwindow as

$$s_m[n] = A_m \cos(\omega n + \theta_m) = a_m \cos(\omega n) + b_m \sin(\omega n), \qquad (7.17)$$

where the constant A_m includes all scaling due to body reflections, wireless propagation, demodulation, and heartbeat amplitude variations. For simplicity we use $\omega \in [0, \pi]$ for the discrete-time frequency. We can write the received signal during the mth subwindow as

$$x_m[n] = s_m[n] + w[n + mN]. \qquad (7.18)$$

The joint density function for the observation vector is, under the assumption of white Gaussian noise

$$f_\Phi(x) = \left(2\pi\sigma^2\right)^{-(MN/2)} \exp\left(-\frac{1}{2\sigma^2} \sum_{m=0}^{M-1} \sum_{n=0}^{N-1} (x_m[n] - s_m[n])^2\right), \qquad (7.19)$$

where $\Phi = [\omega, \alpha_0, \ldots, \alpha_{M-1}, \beta_0, \ldots, \beta_{M-1}, \sigma^2]$ denotes the vector of unknown parameters. Define the square error γ^2 and γ_m^2

$$\gamma_m^2 = \sum_{n=0}^{N-1} (x_m[n] - s_m[n])^2, \qquad (7.20)$$

$$\gamma^2 = \sum_{m=0}^{M-1} \gamma_m^2. \qquad (7.21)$$

To maximize the log-likelihood function is equivalent to minimize the square error γ^2, the summation of γ_m^2 over m. The maximization results in [23]

$$\hat{\omega} = \arg\max_\omega \frac{1}{M} \sum_{m=0}^{M-1} |X_m(\omega)|^2, \qquad (7.22)$$

$$\hat{a}_m = \frac{1}{N} \Re\{X_m(\hat{\omega})\}, \qquad (7.23)$$

$$\hat{b}_m = -\frac{1}{N} \Im\{X_m(\hat{\omega})\}, \qquad (7.24)$$

$$\hat{\sigma}^2 = \frac{1}{MN} \left(\sum_{m=0}^{M-1} \sum_{n=0}^{N-1} x_m^2[n] - \frac{N}{2} \sum_{m=0}^{M-1} |X_m(\hat{\omega})|^2\right). \qquad (7.25)$$

We notice that the ML estimator of the frequency is obtained from the combination of the DTFTs in each interval. Each of these can be calculated using FFTs, so the complexity of the algorithm is low.

For detection of heartbeat, we consider H_1 the hypothesis that a heartbeat is present, and H_0 that no heartbeat is present. Since heartbeat frequency and other parameter are unknown, this is a composite hypothesis test. A general method and in many cases optimum solution for this problem is the generalized likelihood ratio test (GLRT) detector [18]. The principle is to estimate the unknown parameters under each of the hypothesis using the MLE, and then calculating the resulting likelihood ratio. Above, we have estimated the parameters under the H_1 hypothesis. Under the H_0 hypothesis, we only need to estimate the noise variance:

$$\hat{\sigma}^2 = \frac{1}{MN} \sum_{m=0}^{M-1} \sum_{n=0}^{N-1} x_m^2[n], \quad (7.26)$$

$$f_{\mathbf{\Phi}}(x; H_0) = \left(2\pi\hat{\theta}_0^2\right)^{-(MN/2)} \exp\left\{-\frac{MN}{2}\right\}. \quad (7.27)$$

Now we can represent the generalized likelihood ratio for hypothesis H_1 and H_0 as

$$L_G(x) = \frac{f_{\mathbf{\Phi}}(x; H_1)}{f_{\mathbf{\Phi}}(x; H_0)} = \left(\frac{\hat{\sigma}_1^2}{\hat{\sigma}_0^2}\right)^{-(MN/2)}. \quad (7.28)$$

MLE Estimator Based on Direct Data

As mentioned, the demodulated data does not constitute a sufficient statistic. We therefore consider ML estimation directly from the received data. Thus, we utilize more directly the MIMO advantage. As for the demodulated case, we divide the data into M windows of length N samples. The received signal at the kth antenna during the mth window can then be written as

$$z_{k,m}[n] = s_{k,m}[n] + w_{k,m}[n] = x_{k,m}[n] + jy_{k,m}[n], \quad (7.29)$$

where $(x_{k,m}[n], y_{k,m}[n])$ is the received I and Q data, and $w_{k,m}[n]$ is a sequence of independent, identically distributed zero mean circular complex Gaussian noise with unknown variance σ^2. Let $A_{k,m}$ and $B_{k,m}$ denote the magnitudes for the I and Q channel data for the kth antenna. We can write these as:

$$A_{k,m} = C_{k,m} \cos(\psi_k), \quad (7.30)$$
$$B_{k,m} = C_{k,m} \cos(\psi_k). \quad (7.31)$$

In the multipath model, it is most reasonable to assume that the ψ_k is independent between antennas. We can now write the received signal as

$$x_{k,m}[n] = A_{k,m} \cos(\omega t_{m,n} + \theta_m) + \Re\{w_{k,m}[n]\} \quad (7.32)$$
$$= \cos(\psi_k)\left(C_{k,m} \cos(\theta_m) \cos(\omega t_{m,n})\right) \quad (7.33)$$
$$\quad - C_{k,m} \sin(\theta_m) \sin(\omega t_{m,n})) + \Re\{w_{k,m}[n]\},$$
$$y_{k,m}[n] = A_{k,m} \cos(\omega t_{m,n} + \theta_m) + \Im\{w_{k,m}[n]\} \quad (7.34)$$
$$= \sin(\psi_k)\left(C_{k,m} \cos(\theta_m) \cos(\omega t_{m,n})\right) \quad (7.35)$$
$$\quad - C_{k,m} \sin(\theta_m) \sin(\omega t_{m,n})) + \Im\{w_{k,m}[n]\}. \quad (7.36)$$

It can be noticed here that ψ_k depends only on k while θ_m depends only on m. This is the most accurate model, but it seems impossible to get explicit expressions for the MLE estimator. We therefore consider two cases. First, we assume that ψ_k also depends on m. It can then be proven [23] that we get the following explicit solution:

$$\hat{\omega} = \arg\max_{\omega} \sum_{m=1}^{M} \left(\frac{1}{N} \sum_{k=1}^{K} |X_{k,m}(\omega)|^2 + |Y_{k,m}(\omega)|^2 + \frac{1}{N}\sqrt{g_m^2 + 4f_m^2} \right), \tag{7.37}$$

$$\hat{\sigma}^2 = \sum_{m=1}^{M} \left(\sum_{k=1}^{K} \sum_{n=0}^{N-1} x_{k,m}^2[n] + \sum_{k=1}^{K} \sum_{n=0}^{N-1} y_{k,m}^2[n] \right.$$
$$\left. - \frac{1}{N} \sum_{k=1}^{K} |X_{k,m}(\hat{\omega})|^2 + |Y_{k,m}(\hat{\omega})|^2 - \frac{1}{N}\sqrt{g_m^2 + 4f_m^2} \right), \tag{7.38}$$

where

$$f_m = \sum_{k=1}^{K} \Re\{X_{k,m}(\omega)\} \Im\{X_{k,m}(\omega)\} + \Re\{Y_{k,m}(\omega)\} \Im\{Y_{k,m}(\omega)\}, \tag{7.39}$$

$$g_m = \sum_{k=1}^{K} \Re\{X_{k,m}(\omega)\}^2 + \Im\{X_{k,m}(\omega)\}^2 - \Re\{Y_{k,m}(\omega)\}^2 - \Im\{Y_{k,m}(\omega)\}^2. \tag{7.40}$$

The estimate of σ^2 can then be used in (7.28) to get the GLRT for this case.

Second, we assume that θ_m depends also on k. It can then be proved [23] that we get the following explicit solution:

$$\hat{\omega} = \arg\max_{\omega} \sum_{k=1}^{K} \left(\|X_k(\omega)\|^2 + \|Y_k(\omega)\|^2 \right.$$
$$\left. + \sqrt{\left(\|X_k(\omega)\|^2 - \|Y_k(\omega)\|^2 \right)^2 + 4\Re\{X_k(\omega)Y_k(\omega)^*\}^2} \right),$$

$$\sigma^2 = \sum_{k=1}^{K} \gamma_k,$$

$$\gamma_k^2 = \sum_{m=0}^{M-1} \sum_{n=0}^{N-1} x_{k,m}^2[n] + \sum_{m=0}^{M-1} \sum_{n=0}^{N-1} y_{k,m}^2[n],$$
$$- \frac{1}{N} \|X_k(\omega)\|^2 - \frac{1}{N} \|Y_k(\omega)\|^2,$$
$$- \frac{1}{N} \sqrt{\left(\|X_k(\omega)\|^2 - \|Y_k(\omega)\|^2 \right)^2 + 4\Re\{X_k(\omega)Y_k(\omega)^*\}^2}, \tag{7.41}$$

where

$$\|X_k(\omega)\|^2 = \sum_{m=0}^{M-1} |X_{k,m}(\omega)|^2,$$

$$\|Y_k(\omega)\|^2 = \sum_{m=0}^{M-1} |Y_{k,m}(\omega)|^2,$$

$$X_k(\omega)Y_k(\omega)^* = \sum_{m=0}^{M-1} X_{k,m}Y_{k,m}(\omega)^*. \quad (7.42)$$

Again, the estimate of σ^2 can then be used in (7.28) to get the GLRT for this case.

What should be noticed is that although the solutions look complex, they are expressed as simple combinations of discrete time Fourier transforms, and therefore they can be calculated extremely fast using FFTs.

Figures 7.2 and 7.3 show the detection performance of the detectors in the form of ROCs (receiver operating characteristic) based on measured data for a number of subjects. Figure 7.2 compares the detector based on demodulated data ((7.25) and (7.28)) with one based on direct data ((7.41) and (7.28)); the figure confirms that using direct data is more efficient. Figure 7.3 compares the performance of the direct data detector for various sizes of windows and subwindows.

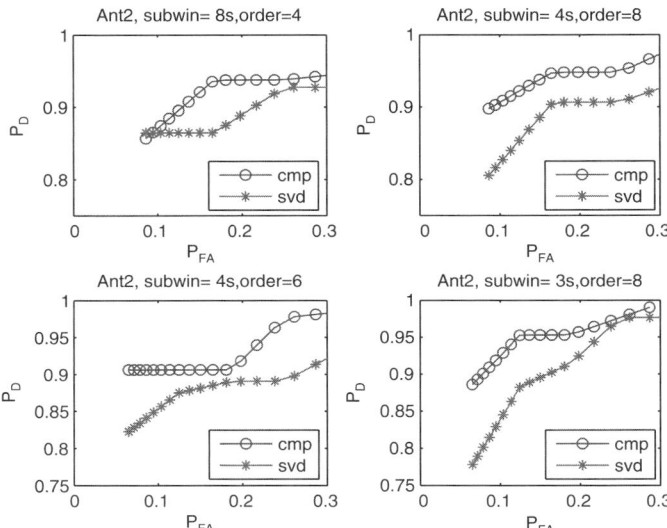

Fig. 7.2. ROC curves for the GLRT either based on demodulated data (SVD) or direct data (CMP)

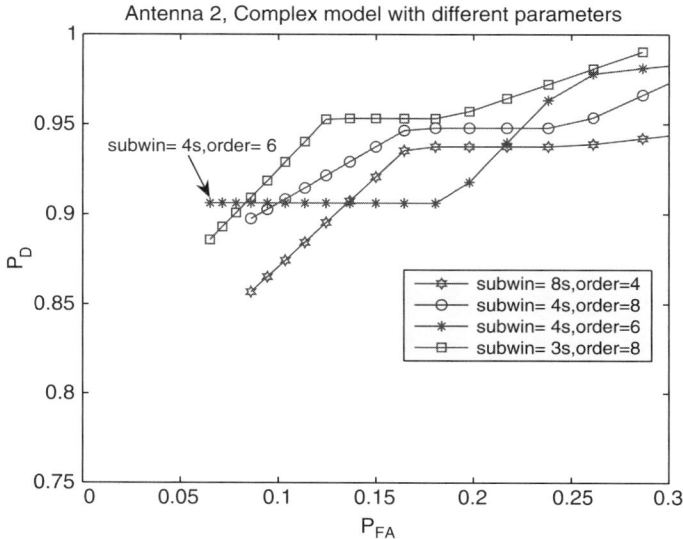

Fig. 7.3. Performance of the GLRT for direct data with different parameters

7.4 Multiple People Signal Processing

Consider the signal model (7.6). If there are d multiple sources, they add up linearly, resulting in

$$r_k(t) = \sum_{j=1}^{d} v_{k,j} x_j(t) + \tilde{w}_k(t) \tag{7.43}$$

In the sampled version, this can be written as the classical instantaneous linear model:

$$\mathbf{r}[n] = \mathbf{M}\,\mathbf{x}[n] + \mathbf{w}[n].$$

As the data is not known, nor the mixing matrix, blind source separation (BSS) methods are the natural choice. Note also that (1) the noise power is of the same magnitude or higher than the signal power, (2) the received IQ signal is complex, but the sources $s[n]$ are real, so we apply real BSS, and we work on:

$$\mathbf{y}[n] = \begin{bmatrix} \Re(\mathbf{r}[n]) \\ \Im(\mathbf{r}[n]) \end{bmatrix} = \begin{bmatrix} \Re(\mathbf{M}) \\ \Im(\mathbf{M}) \end{bmatrix} \mathbf{x}[n] + \begin{bmatrix} \Re(\mathbf{w}[n]) \\ \Im(\mathbf{w}[n]) \end{bmatrix},$$
$$\mathbf{y}[n] = \overline{\mathbf{M}}\mathbf{x}[n] + \overline{\mathbf{w}}[n], \tag{7.44}$$

where the size of the received signal is multiplied by two, thus allowing twice as many independent beamformers.

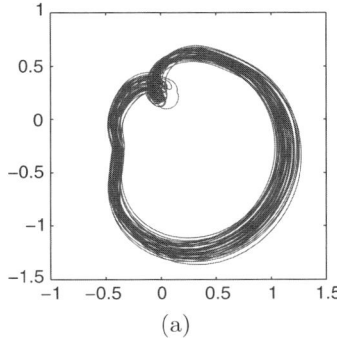 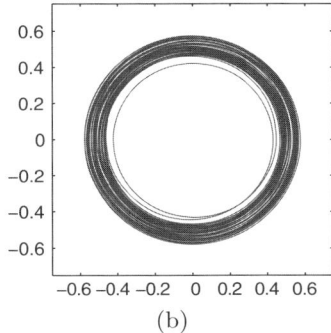

Fig. 7.4. Analytical representation of a finger pulse sensor signal after bandpass filtering in the range 0.03–30 Hz. (**a**) Full version, (**b**) Filtered version

The aim of multiple people signal processing is now to separate the multiple sources. Once they are separated, the single person signal processing from Sect. 7.3 can be utilized.

7.4.1 Heartbeat Signal

As a model for the heartbeat, we consider (7.7). Because of the phase variations $\theta(t)$ the signal is not strongly periodic. However, the amplitude $A+\alpha(t)$ is nearly constant, which means that $x(t)$ has an analytical representation which is almost a constant modulus signal. The analytical signal is obtained by adding the signal and the quadrature of its Hilbert transform [2].

Figure 7.4 shows the analytical signal of an ECG signal from, and its lowpass filtered version. The almost circular trajectory indicates that indeed the heartbeat signals are nearly constant modulus, and that after lowpass filtering, this property shows up even stronger.

In real-world applications, there might be many sources, but we are only interested in those that are heartbeat and/or respiration signals. We therefore need to use specific characteristics of these signals. One possibility is to use the signal that is quasi-periodic, as we used in the single person detector in Sect. 7.3. However, the signal is not very periodic, and we therefore believe that the constant modulus property is a stronger indicator. For this reason, we focus on BSS methods that use the constant modulus property.

7.4.2 Algorithm

Our algorithm is based on three steps:

(α) Use a band-pass filter over the range $[0.75;2]$ Hz.
(β) Generate the analytic version of our signal by taking it and then adding its Hilbert Transform, $\mathcal{H}\{r(t)\}$, in quadrature:

$$r_a(t) = r(t) + j\mathcal{H}\{r(t)\},$$

(γ) Use the ACMA algorithm on the data [21]. ACMA is a BSS algorithm that search for the beamformers $\mathbf{w}_i$, such that the estimated sources $\hat{x}_i = \mathbf{w}^H \mathbf{r}$ are constant-modulus: $|\hat{x}_i| \approx 1$.

Step α removes the heartbeat harmonics, the respiration and the low frequency interferences.

Doing these pre-processing steps insures that each source produces a *rank-one only matrix* and is constant-modulus, so that the assumptions needed by ACMA are respected.

The choice of the separation algorithm follows the next reasoning: RACMA [20] is then not applicable in our case, since at the end of step β, the sources are complex. ACMA [21] is the logical choice as it needs less samples than any HOS-based method, as ACI [4] or EFICA [9], is computationally faster, and our data is constant-modulus.

7.4.3 Results

As we do not have enough measurements with two subjects to draw statistically sound performance evaluations of our algorithm, we have to recourse to semi-synthesized and full simulations in this section.

We compare our method, PP-ACMA, to either RACMA, ICA, or EFICA. The later algorithms do not need the pre-processing to function, but we also evaluate EFICA with the pre-processed data in order to have a comparison that is fair to ACMA.

Simulations

To have better insight on the algorithm behavior, we conduct now the evaluation over fully simulated data, i.e., the heartbeat source signal are now generated by the method proposed in [16] with different average rate (so different subjects). Note that we can tune the signal noise ratio (SNR), the incoming average heart rate and time integration, and therefore fully explore the limits of the algorithms. Once the separation algorithm has delivered output signals, they are used to estimate the heart rate. The separation is declared a success if both measured hearbeats are within 0.08 Hz of their true values, and the output SINR is above 3 dB, otherwise it is a failure. We also compare to EFICA, which is either applied to the direct data or to the pre-processed data.

First, we consider two sources with an SNR of -5 dB, and an integration time of 6 s (so 120 samples). Figure 7.5 presents the success rate of the PP-ACMA, ICA, EFICA directly on data (denoted EFICA,R), and EFICA on the pre-processed data (denoted EFICA,C). We note that even two sources with the same average heart rate, due to the difference in HRV, the separation is still possible up to 50% of the cases with ACMA. In the other cases, ICA and EFICA (R) behaves also quite well, but not with a perfect 100% as the ACMA. Notice also that the pre-processed methods cannot work outside the frequency

7 Signal Processing Methods for Doppler Radar Heart Rate Monitoring 135

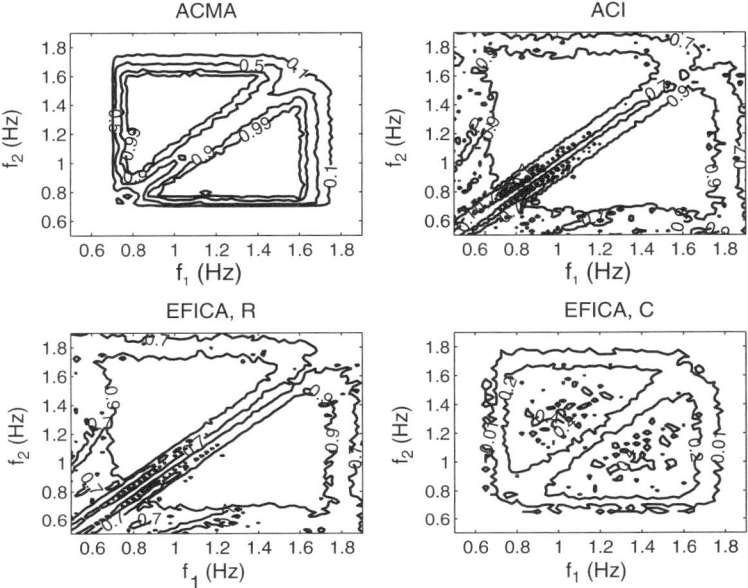

Fig. 7.5. Success rate of the algorithms as a function of average rate of the heartbeat

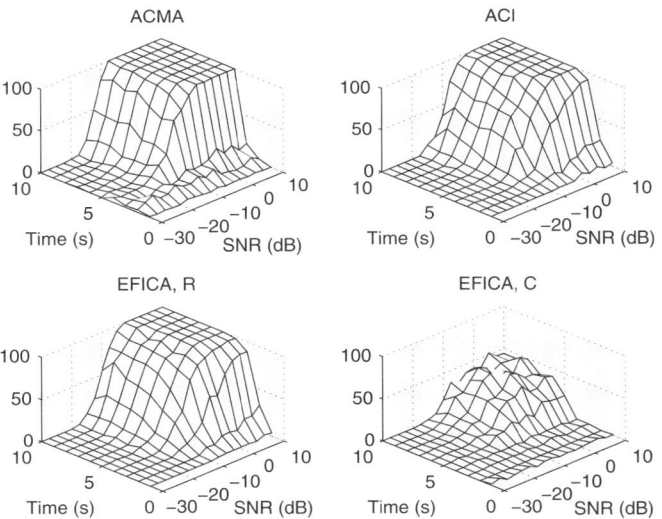

Fig. 7.6. Success rate of the algorithms as a function of SNR and the integration time

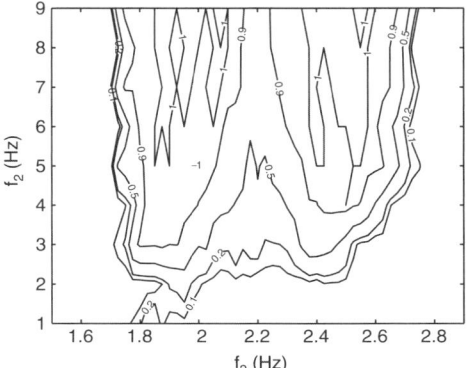

Fig. 7.7. Success rate of the PP-ACMA as a function of average rate of the second heartbeat and the integration time

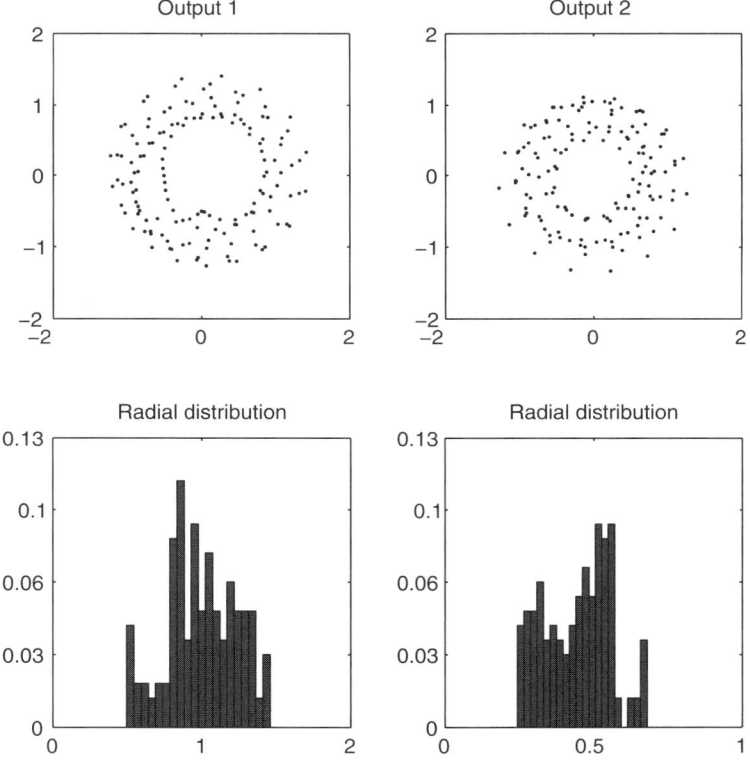

Fig. 7.8. Radial distribution of the two channels after separation with ACMA

7 Signal Processing Methods for Doppler Radar Heart Rate Monitoring

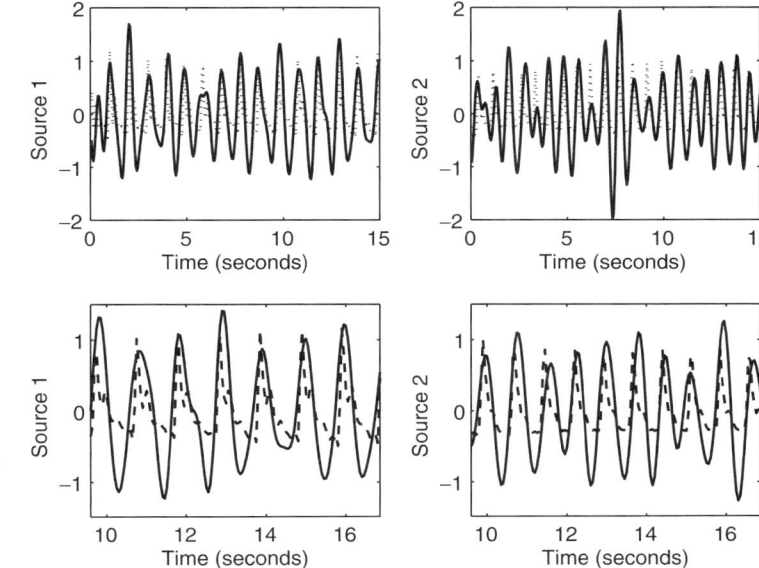

Fig. 7.9. The two separated sources in the time domain. The *solid curve* shows the reference measurements

range of the filters, but the frequency limits have been chosen according to reasonable heartbeat rate. Finally, we note that EFICA-C is not reliable in any case, this is due to the loss of diversity in the signal, which impairs the high-order statistics.

Next, we keep the frequencies fixed to 0.9 and 1.2 Hz, and we vary the SNR and the time of observation. We note in Fig. 7.6 that thanks to the pre-processing gain, acceptable success rate can be obtained till −15 dB SNR for PP-ACMA, and that 4 s is a minimum integration time (approximately four cycles). While ACI and EFICA-R needs 5 dB more, but 3 s is already enough for them. This counter-intuitive result can be explained by the fact that we do not strictly compare the algorithm over the same signals: indeed, for the PP-ACMA, the signal has been amputated from most of its harmonics and its spectral diversity, that ICA and EFICA-R are still using. Next to it, in a real scenario, where the pre-processing is mandatory, note that EFICA-C under-performs severely.

From now, we just concentrate on our scheme for our specific applications. We keep the first average frequency at 1.1 Hz, and we vary the other average frequency as well the time of integration. From Fig. 7.7, we confirm that given enough times of integration it is possible to separate two sources with equal average rate, as for instance for $T = 9$ s, the success rate is above 60%.

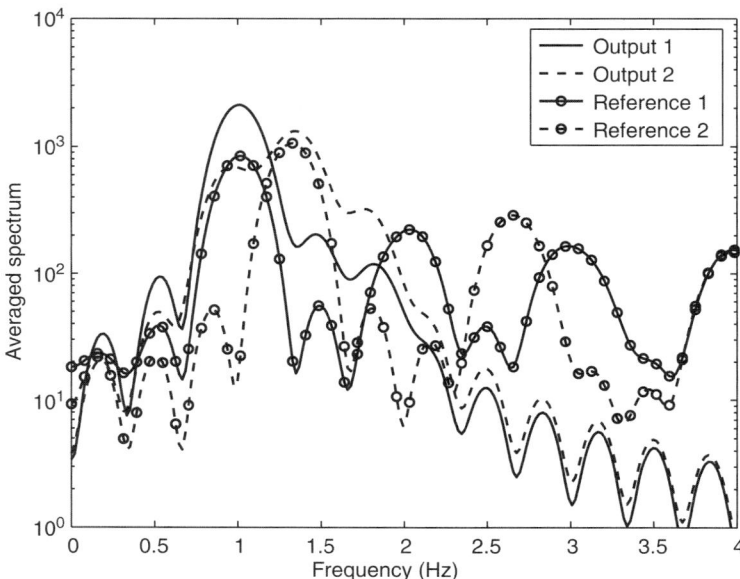

Fig. 7.10. The two separated sources in the frequency domain

Experimental Results

Finally, we will present some results from real measurements. The data is from a single experiment with two persons and two antennas. The ACMA algorithms was applied. Figure 7.8 shows the radial distribution after using ACMA. As can be seen, the CM property is well satisfied. Figures 7.9 and 7.10 show the two separated signals in the time domain, respectively, frequency domain. It can be seen that the two sources agree with the reference measurements, although there is some leakage from the stronger source to the weaker source.

7.5 Conclusion

In this chapter, we have outlined methods for signal processing of Doppler signals. The main problems we have focused on is detection of a signal and separation of multiple signals. We have outlined a number of methods for detection and separation and shown the efficacy on real signals. The key technology we considered was using multiple antenna transmitters and receivers. We first showed how this could be used for enhanced signal processing when a single person is present. We then showed how this could be used to monitor multiple persons.

Acknowledgement

This material is based upon work supported by the US National Science Foundation under Grant No. ECS0428975.

References

1. Boric-Lubecke, O., Ong, P.W., Lubecke, V.M.: 10 GHz Doppler radar sensing of respiration and heart movement. In: Proceedings of the IEEE 28th Annual Northeast Bioengineering Conference, pp. 55–56 (2002)
2. Bracewell, R.: The Fourier Transform and Its Applications, 3rd edn. McGraw-Hill, New York (1999)
3. Chen, K.M., Mirsa, D., Wang, H., Chuang, H.R., Postow, E.: An X-band microwave life detection system. IEEE Transactions on Biomedical Engineering **33**, 697–70 (1986)
4. Comon, P.: Independent component analysis, a new concept? Signal Processing, Special issue on Higher-Order Statistics **36**(3), 287–314 (1994)
5. Droitcour, A., Boric-Lubecke, O., Lubecke, V.M., Lin, J.: 0.25 μm CMOS and biCMOS single chip direct conversion Doppler radars for remote sensing of vital signs. In: IEEE ISSCC Digest of Technical Papers, pp. 348–349 (2002)
6. Droitcour, A.D., Boric-Lubecke, O., Lubecke, V.M., Lin, J., Kovacs, G.T.A.: Range correlation effect on ISM band I/Q CMOS radar for non-contact sensing of vital signs. In: IEEE MTT-S IMS2003 Digest, vol. 3, pp. 1945–1948 (2003)
7. Droitcour, A.D., Lubecke, V.M., Lin, J., Boric-Lubecke, O.: A microwave radio for Doppler radar sensing of vital signs. In: IEEE MTT-S IMS2001 Digest, vol. 1, pp. 175–178 (2001)
8. Hilton, M.F., Bates, R.A., Godfrey, K.R., et al.: Evaluation of frequency and time-frequency spectral analysis of heart rate variability as a diagnostic marker of the sleep apnea syndrome. Medical and Biological Engineering and Computing **37**(6), 760–769 (1999)
9. Koldovsk, Z., Tichavsk, P., Oja, E.: Dfficient variant of algorithm fastica for independent component analysis attaining the Cramer–Rao lower bound. IEEE Transactions on Neural Networks **17**(5), 1265–1277 (2006)
10. Lin, J.C.: Non-invasive microwave measurement of respiration. Proceedings of IEEE **63**, 1530 (1975)
11. Lin, J.C.: Microwave apexcardiography. IEEE Transactions MTT **27**, 618–620 (1979)
12. Lin, J.C.: Microwave sensing of physiological movement and volume change: A review. Bioelectromagnetics **13**, 557–565 (1992)
13. Lohman, B.B., Boric-Lubecke, O., Lubecke, V.M., Ong, P.W., Sondhi, M.M.: A digital signal processor for Doppler radar sensing of vital signs. In: 23rd Annual International Conference of the IEEE Engineering in Medicine and Biology Society (2001)
14. Lubecke, V., Boric-Lubecke, O., Awater, G., Ong, P.W., Gammel, P., Yan, R.H., Lin, J.C.: Remote sensing of vital signs with telecommunications signals. In: World Congress on Medical Physics and Biomedical Engineering (2000)

15. Lubecke, V., Boric-Lubecke, O., Beck, E.: A compact low-cost add-on module for Doppler radar sensing of vital signs using a wireless communications terminal. In: IEEE MTT-S International Microwave Symposium (2002)
16. McSharry, P.E., Clifford, G.D., Tarassenko, L., Smith, L.: A dynamical model for generating synthetic electrocardiogram signals. IEEE Transactions Biomedical Engineering **50**, 289–294 (2003)
17. Park, B.K., Vergara, A., Boric-Lubecke, O., Lubecke, V., Høst-Madsen, A.: Center tracking quadrature demodulation for a Doppler radar motion detector. IEEE Transactions on Microwave Theory and Techniques (2007). Submitted. Available at http://www.ee.hawaii.edu/~madsen/papers/index.html
18. Poor, H.V.: An Introduction to Signal Detection and Estimation. Springer, Berlin Heidelberg New York (1994)
19. Thakor, N.V., Zhu, Y.S.: Application of adaptive filtering to ECG analysis: Noise cancellation and arrhythmia detection. IEEE Transactions on Biomedical Enginering **38**, 785–794 (1991)
20. van der Veen, A.: Analytical method for blind binary signal separation. IEEE Transactions on Signal Processing **45**(4), 1078–1082 (1997)
21. van der Veen, A., Paulraj, A.: An analytical constant modulus algorithm. IEEE Transactions on Signal Processing **44**(5), 1136–1155 (1996)
22. Veeravalli, V., Poor, H.: Quadratic detection of signals with drifting phase. Journal of the Acoustic Society of America **89**, 811–819 (1991)
23. Zhou, Q., Petrochilos, N., Høst-Madsen, A., Boric-Lubecke, O., Lubecke, V.: Detection and monitoring of heartbeat using Doppler radar. IEEE Transactions on Signal Processing (2007). Submitted. Available at http://www.ee.hawaii.edu/~madsen/papers/index.html

8

Multimodal Fusion for Car Navigation Systems

Dragan Obradovic, Henning Lenz, Markus Schupfner, and Kai Heesche

The main tasks of car navigation systems are positioning, routing, and guidance. This chapter describes a novel, two-step approach to vehicle positioning founded on the appropriate combination of the in-car sensors, GPS signals, and a digital map. The first step is based on the application of a Kalman filter, which optimally updates the model of car movement based on the in-car odometer and gyroscope measurements, and the GPS signal. The second step further improves the position estimate by dynamically comparing the continuous vehicle trajectory obtained in the first step with the candidate trajectories on a digital map. This is in contrast with standard applications of the digital map where the current position estimate is simply projected on the digital map at every sampling instant. In addition to the positioning problem, this chapter addresses a fuzzy-logic based approach to guidance.

8.1 Introduction[1]

The need for accurate navigation devices is probably as old as man-made transportation systems. Nevertheless, the development of the current car navigation systems was enabled by the ongoing improvement of electronic devices as well as by the availability of new position information sources such as GPS and digital maps. In general, there are two concepts of navigation systems. The first type is given by centralized systems, where there is a continuous two-way communication with the vehicles requesting the navigation service. The information from the on-board vehicle sensors is transmitted to the navigation center, which estimates the car position and transmits the guidance commands back to the driver.

On the contrary, autonomous navigation systems process all the information on-board and calculate optimal route and the necessary guidance commands without participation of an external server. Due to the lower costs,

---

[1] This chapter is based on [9], with the permission from IEEE.

the autonomous navigation systems have become the standard in the car industry.

The car navigation systems have to perform three distinctive tasks: positioning, routing, and guidance. Positioning relies on the available sensory and the digital map information. The car build-in sensors used for this purpose are the odometer and the gyroscope. The odometer provides information about the traveled distance while the gyroscope calculates the heading (orientation) change with a given time interval. Both of these sensors are, due to the financial reasons, of limited accuracy and subject to drift. Additional information sources used for car positioning are the GPS signals and the digital map of the roads.

The global positioning system (GPS) is a system of at least 24 satellites with synchronized atomic clocks, which continuously transmit the time and the needed satellite identification information. The GPS receiver that is available within the vehicle detects the number of satellites whose signals are strong enough to be processed. The receiver delivers the vehicle position and the information about its velocity as two separate outputs. Until May 2000, the GPS accuracy was intentionally deteriorated to make the civilian use of the system less accurate than its military applications. This property is called selective availability and, as stated above, is currently turned off. Nevertheless, the decision to reactivate is subject to a yearly review by the US government. In addition, the coverage (the number of visible satellites) of the GPS system is not evenly distributed, and in some areas (urban, mountainous regions, etc.) the number of visible satellites is further reduced.

Digital maps contain information about the road network including the road properties (highway, one-way street, etc.). The road representation within the digital map is piecewise linear. Nowadays the digital maps used in standard navigation systems provide additional information about the type and location of different services of interest to the car passengers such as hotels, shopping mall locations, etc.

The information from the on-board sensors, the GPS and the digital map has to be combined appropriately to maximize the accuracy of the estimated vehicle position and heading. This is a standard sensor fusion problem where different sensors are of different accuracy and whose properties are also changing over time. Typically, the sensor fusion is performed stepwise. In the first step, the odometer and gyroscope information is combined within the process called dead reckoning. The result of the dead reckoning process is the estimated vehicle trajectory. In a further step, the dead reckoning position is projected on the digital map. This step is called map matching. If the dead reckoning position is in-between several roads, several projections will be made. Each projection is then seen as a possible alternative whose likelihood is estimated over time based on the GPS information and the trajectory time evolution. If the discrepancy between the matched position and the GPS signal is too high, the so-called "GPS reset" is performed, i.e., the position is assigned to the current GPS estimate.

Not only that the sensor fusion algorithms have to deliver an accurate position estimate, but they also have to accomplish this task under very stringent computational and memory constraints. Hence, the final solution is always a tradeoff between the optimality of the mathematical approach and the implementation limitations.

Routing is the calculation of the route between locations A and B that is optimal with respect to the selected criterion. Once the optimal route and the current position are known, the guidance algorithm issues recommendations to the driver so that the vehicle remains on the selected route.

8.2 Kalman Filter-Based Sensor Fusion for Dead Reckoning Improvement

If the initial vehicle location is known, the odometer and gyroscope measurements can be used to reconstruct the traveled trajectory. The process of incremental integration of the vehicle trajectory relative to a known location is called dead reckoning [13] and is illustrated in Fig. 8.1 [9]. The dead reckoning equations are as follows:

$$
\begin{aligned}
x_i &= x_0 + \sum_{j=1}^{i} l_j \cos(\varphi_j), \\
y_i &= y_0 + \sum_{j=1}^{i} l_j \sin(\varphi_j), \\
\varphi_i &= \varphi_0 + \sum_{j=1}^{i} \Delta\varphi_j.
\end{aligned}
\tag{8.1}
$$

As it can be seen, dead reckoning is the simplest way of determining the vehicle position. Unfortunately, the accuracy of this method is very low since the incremental information is noisy and the error accumulates over the time. The standard way of dealing with the inaccuracy of the dead reckoning is to project the so determined vehicle position on the roads depicted in the digital map and use GPS to evaluate the certainty of different projections. The approach described in this chapter is different. The novelty of the herein presented approach is that the dead reckoning trajectory is corrected by the

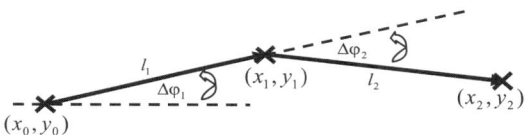

Fig. 8.1. Dead reckoning as vector addition

GPS measurement every time when the "quality" of the GPS reception is acceptable. Hence, the GPS information is combined with the odometer and gyroscope measurements before the digital map is used. This sensor fusion is achieved by a suitable Kalman filter implementation.

There are several possible ways in setting up the sensor fusion as a Kalman filter problem [1, 5, 6, 10, 12, 13]. The "complete" Kalman filter representation would use the dead reckoning equations as state equations and the GPS as measurements. In this case the states would be the actual positions. Although this is a common sense approach, it has the following unfavorable characteristics:

The state equations are non-linear. The non-linearity stems from the presence of the sine and cosine function as well as from the transformation needed to adjust the plane assumption of the dead reckoning to the curvature of the Earth.

The additive Gaussian noises in the two state equations are not uncorrelated. The actual independent measurements are the incremental traveled length and the orientation. Since both longitude and latitude positions are functions of the latter two variables, they are correlated.

Consequently, the decision was made to use the "decentralized" Kalman filter [13] implementation, where the odometer and gyroscope data will be enhanced with the available GPS measurements. A decentralized Kalman filter, differently from the standard Kalman filter implementation, is applied only to the part of the dynamics of the positioning system. The complete system dynamics in this case would have had the actual position coordinates of the vehicle as its states. In the current implementation, the Kalman filter is applied only to the angle and traveled distance measurements, which are after the estimation used to compute the vehicle position. The odometer and gyroscope drift is modeled with extra parameters P^{gyro} and P^{odo}. In the state-space representation of the Kalman filter (8.2) these two parameters appear as two auxiliary states driven by additive noise. Due to the fact that the gyroscope drift is more severe than the odometer drift, and due to the computation and memory limitation of the system, the following Kalman filter is implemented:

$$\begin{cases} \varphi_{i+1} = \varphi_i + (\Delta\varphi_i)_{\text{gyro}} + P_i^{\text{gyro}} + \vartheta_i^1 \\ P_{i+1}^{\text{gyro}} = P_i^{\text{gyro}} + \vartheta_i^2 \\ P_{i+1}^{\text{odo}} = P_i^{\text{odo}} + \vartheta_i^3 \end{cases} \text{state equations,}$$

$$\begin{cases} y_i^{(1)} = \varphi_i + \eta_i^{(1)} \\ y_i^{(2)} = (l_i)_{\text{odo}} + P_i^{\text{odo}} + \eta_i^{(2)} \end{cases} \text{measurement equations.}$$

(8.2)

The heading and traveled distance measurements $y^{(1)}$ and $y^{(2)}$ are obtained from the received GPS information. The GPS receiver delivers independently the position estimate as well as the velocity and the heading. The Kalman filter presented herein relies only on the GPS velocity and heading information, while the GPS position estimate is not explicitly used. The limitation of the usage of the GPS information as a "teacher" comes from the fact that the quality of the velocity and heading (as well as the position) estimate vary based on the number of visible satellites and their positions. Consequently, a novel proprietary algorithm of Siemens AG [11] is applied that changes the covariance matrices of the additive measurement noise $\eta^{(1)}$ and $\eta^{(2)}$ based on the GPS signal quality. The worse the GPS reception, the more noisy the measurements and, as a result, the smaller the correction of the odometer and gyroscope information. In the extreme case when the GPS information is completely unreliable, no correction of the gyroscope and odometer information is performed. Due to the changes of the GPS quality, the Kalman filter is implemented with a suitable forgetting factor which forces the statistics to weight the past information less and less over time [6].

The resulting corrected heading and traveled distance estimates are further used to determine the vehicle position according to the following formula:

$$\begin{pmatrix} \text{Long}_{k+1} \\ \text{Lat}_{k+1} \end{pmatrix} = \begin{pmatrix} \text{Long}_k \\ \text{Lat}_k \end{pmatrix} + \psi \left\{ \frac{\sin(\Delta\varphi_k/2)(1+P_k^{\text{odo}})(l_k)_{\text{odo}}}{\Delta\varphi_k/2} \begin{pmatrix} \cos(\varphi_k+\Delta\varphi_k/2) \\ \sin(\varphi_k+\Delta\varphi_k/2) \end{pmatrix} \right\}, \tag{8.3}$$

where Ψ stands for the transformation from the planar to the WGS84 coordinate system [9], which takes into account the geometry of the Earth surface. The argument of the function is the dead reckoning position with the Kalman filter updated gyroscope and odometer information. The dead reckoning formula are slightly changed from (8.1) to better accommodate driving on a circle [8]. According to the fact that dead reckoning is a discrete time process, the orientation at the sampling instants and the orientation between two samples differ: The orientation of the traveling car in between two sample points is given by the mean of the two orientations at the two sample instants. Therefore, the averaged orientation has to be used for dead reckoning. For practical reasons, this averaged orientation can be determined by the orientation of a first sample instant plus half of the orientation change within the following time step. The differences between (8.1) and (8.3) become clearly visible, when a car is driving in a circle as it happens in parking garages and as is illustrated in Fig. 8.2 [9].

In addition to using an appropriate heading during the dead reckoning, the traveled length has also to be adjusted to avoid inaccuracies. The car physically drives on the circle, such that the length of a circle section is measured by the odometer. On the other hand, the discrete dead reckoning uses straight lines, which create a shortcut between two sample instants. Accordingly, the length measured by the odometer has to be corrected by the term $\sin(\Delta\varphi/2)/(\Delta\varphi/2)$ as applied in (8.3) where $\lim_{\Delta\varphi\to 0}[\sin(\Delta\varphi/2)/(\Delta\varphi/2)] = 1$.

A typical dead reckoning improvement after the Kalman filter implementation is depicted in Fig. 8.3 [9]. The dashed line shows the dead reckoning path without the Kalman filter while the dotted line shows the Kalman filter

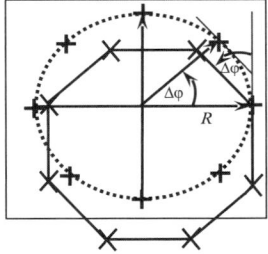

Fig. 8.2. The *plus* signs represent true positions on a *circle*. *Crosses* represent the dead reckoning positions based on (8.1). Using the dead reckoning positions based on (8.3), the true and dead reckoning positions coincide

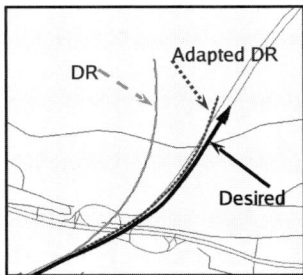

Fig. 8.3. Dead reckoning paths without (*dashed*, light grey) and with (*dotted*, dark grey) Kalman filter implementation. The black *solid line* depicts the true trajectory

produced dead reckoning. The true driven path is depicted by a solid line. As seen, the Kalman filter corrected dead reckoning improves significantly the accuracy of the original dead reckoning since the GPS information is used.

The computational burden of the Kalman filter implementation in the navigation system was acceptable. The total number of floating-point operations per time step was 188.

Once the Kalman filter adapted dead reckoning is available, it should be further combined with the digital map information. The process of comparing the calculated vehicle trajectory against known roads is called map matching [4, 13]. Section 8.3 describes a novel pattern matching based method for map matching.

8.3 Map Matching Improvement by Pattern Recognition

The Kalman filter adaptation of the odometer and gyroscope information not only provides their optimal estimates but also the standard deviation of their errors. This error information is propagated in the positioning equations (8.1) leading to the error distribution in the determined position, i.e., the region where the vehicle true position might be. This position error distribution is then used in the standard navigation systems for determining the candidate positions on the road map. If there are more than one road intersecting the region with possible vehicle positions obtained from the Kalman filter updated dead reckoning, several candidate on-road positions will be determined. In the standard positioning approach each candidate position is then separately evaluated based on different criteria such as the heading, connectivity with respect to the candidates in the previous time step, closeness to the dead reckoning position estimate, etc. The candidate position with the highest likelihood is adopted as the true position at the current time instant. The problem with this approach is that the evaluation of the candidates is based only on the instantaneous information that is prone to error.

The novelty of the herein presented approach to map matching is that the historical information over several time integration steps in dead reckoning is compared in a suitable way to determine the optimal map matching position. The underlying idea is to extract features from both dead reckoning trajectory and digital map that could be compared. The problem is that the resolutions of the two representations are different. The road representation in the digital map is piecewise linear while the dead reckoning trajectory is smooth due to the small time step in odometer and gyroscope time sampling. If the goal is to represent each of these trajectories as a sequence of curve and straight-line patterns, a method is needed to initially extract these features and to compare them in reliable manner.

8.3.1 Generation of Feature Vectors by State Machines

State machines are used herein to process the available information to determine the apparent features "straight" and "turn." The features themselves are derived by summation of values over a defined interval. The feature "straight" corresponds to the distance between the estimated vertices of two successive turns. The feature "turn" gives the change in heading in a curve.

During the development of the rules for state transitions, it has been found that intermediate states have to be defined in addition to the original states "straight" and "turn" (see Figs. 8.4 and 8.5 [9]). The intermediate states take care of the inaccuracies in the sensor and map data and, therefore, they enhance the reliability of the pattern matching process.

There are two critical aspects during the development of state machines, the selection of states and the definition of the rules for the state transitions. The rules are typically defined as a comparison of a variable of interest with the appropriate threshold (see Figs. 8.4 and 8.5 [9]). Therefore, the thresholds have a major impact on the instance when a transition takes place and they simultaneously define if a feature is considered as significant or not. The evaluation criteria for feature generation applied in the state machines are: the current heading change, the accumulated heading change since start of

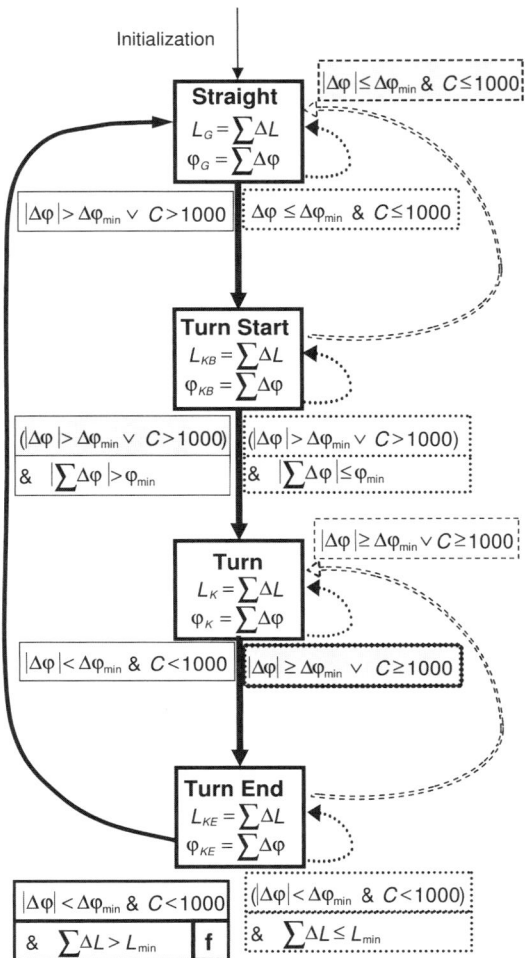

Fig. 8.4. State machine for sensor data, where $\Delta\varphi_{\min} = 2°$, $\varphi_{\min} = 30°$, and $L_{\min} = 10\,\text{m}$. Here, $C = 1{,}000$ means $C = 1{,}000°/1\,\text{km} = 1\,\text{deg}\,\text{m}^{-1}$

the turn, the curvature (inverse radius of the turn),[2] and the distance traveled while the state machines remain in the current state.

Since both the types of paths (sensor and map) are available with a different resolution, separate state machines are chosen for each of them. One of the main differences between the analysis of the sensor and map data is that the sensor data are continuous (smooth) while the digital map data are piecewise linear. Hence, two different state machines were derived, one for each path type.

---

[2] Instead of the radius, the curvature is considered to avoid singular behavior for straight roads.

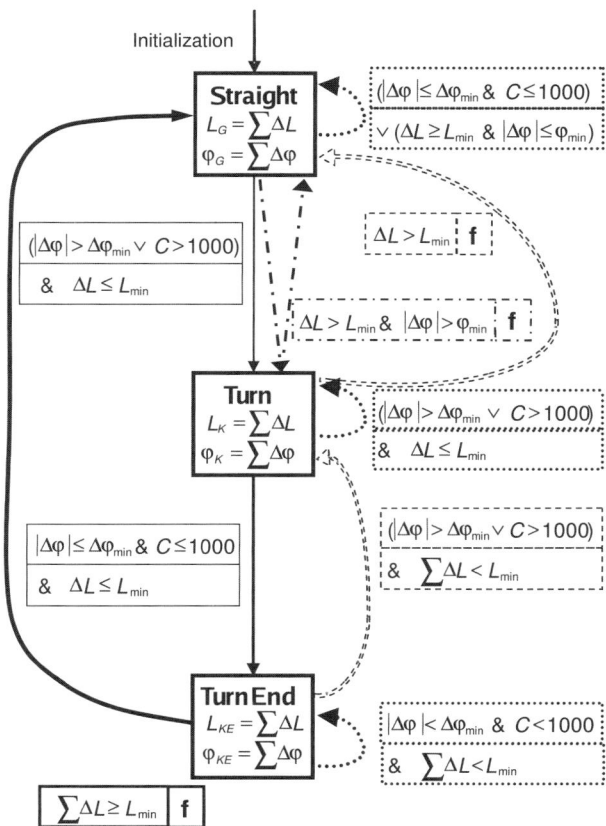

Fig. 8.5. State machine for sensor data, where $\Delta\varphi_{\min} = 4°$, $\varphi_{\min} = 30°$, and $L_{\min} = 35\,\text{m}$. Here, $C = 1{,}000$ means $C = 1{,}000°/1\,\text{km} = 1°/\text{m}$

Figure 8.4 depicts the state machine for sensor data with four possible states and the rules for the transitions. $\Delta\varphi$ is the heading change in the current time step, $\sum \Delta\varphi$ is the accumulated heading change up to the current time step, $C = \Delta\varphi/\Delta L$ is the curvature where $\sum \Delta L$ describes distance traveled while the state machine remains in the current state. A curvature of 1,000 means a heading change of $1{,}000\,\text{deg}\,\text{km}^{-1} = 1\,\text{deg}\,\text{m}^{-1}$, where the corresponding turn has a radius of about 60 m.

The features 'straight' with the length L and 'turn' with the heading change φ are simultaneously provided pair wise with the transition from 'turn end' to 'straight'. This transition is indicated by 'f' in Figs. 8.4 and 8.5. With the indices TS for turn start, T for turn, TE for turn end, and S for straight, the following relations hold:

$$L = L_{\text{TE}} + L_{\text{S}} + (L_{\text{TS}} + L_{\text{T}})/2, \tag{8.4}$$

$$\varphi = \varphi_{\text{TS}} + \varphi_{\text{T}}. \tag{8.5}$$

The given thresholds have been defined based on the analysis of test drives. For the adjustment of the thresholds the following aspects should be considered:

- $\Delta\varphi$ or C too large: slowly driven turns are not recognized.
- $\Delta\varphi$ or C too small: the state 'turn' is never left again.
- $\varphi_{\min}$ determines the heading change for a turn considered as significant, i.e., the heading change which is considered as relevant feature.
- $L_{\min}$ determines the minimum distance between two turns to be considered as separate turns and not just a single joined turn.

In Fig. 8.5 the state machine for map data is depicted – with the three possible states and the rules for the transitions. The state machine, which analyzes the candidate trajectories on the digital map, has one state less (turn start) than the state machine for sensor data. The curve beginning is not needed due to the piecewise linear road representation. The curve end state is still needed since a curve can be represented by a sequence of incremental angle changes, which should be ultimately summed up together.

In general, the state machine for map data uses different rules and different thresholds compared with the rules and thresholds of the state machine for sensor data (e.g. 4° heading change compared to 2°). This difference is caused by the difference in resolution of sensor and map data. Similar to sensor data, the tuning of thresholds for the map data has been achieved based on test data – and this has been performed such that sequences of features of an equal length for sensor and map data are achieved.

The features of sensor data are always produced as pair. The transitions where a feature of map data is produced are indicated by 'f' in Figs. 8.4 and 8.5. Features are provided only, if the accumulated heading change is larger than $\varphi_{\min}$ (e.g. 30°).

8.3.2 Evaluation of Certainties of Road Alternatives Based on Feature Vector Comparison

The features determined by state machines are stored in two-dimensional vectors. For the sensor signals there is one vector, for the map there are as many vectors as there are possible map alternatives. In the following, the term path refers to the fact that feature vectors represent a course, i.e., the pattern.

The likelihood (certainty) of each possible map trajectory is calculated by comparing its feature vector with the features extracted from the dead reckoning trajectory. The comparison is based on the suitable norm of the distance between feature vectors with different weighting on the differences between the straight-line lengths and the curve angles. The error is additive, i.e., each new pair (curve, straight-line) is compared and the corresponding error is added to the already existing evaluation. The comparison starts only when four features have been already identified.

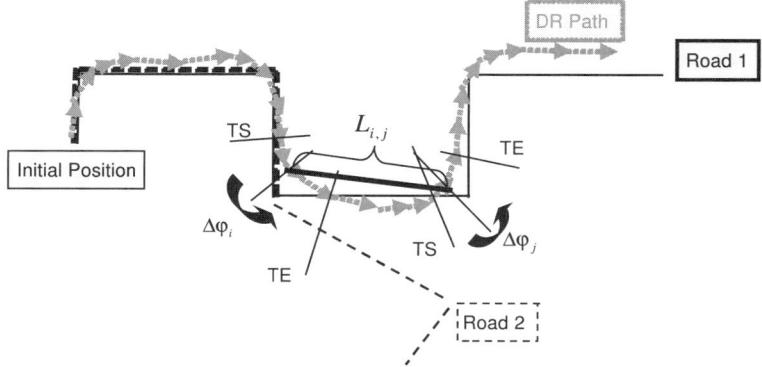

Fig. 8.6. Example of the pattern recognition for evaluation of two possible road trajectories (*solid* and *dashed lines*) based on the digital map. The (adapted, cf. above) dead reckoning path is given by a *dotted line* of *arrows*. States and features corresponding to the dead reckoning path are indicated. In this example, map alternative 1 is obviously true

Figure 8.6 [9] illustrates a comparison between a dead reckoning (DR) trajectory and the two possible trajectories on the digital map. The extracted curves with angle changes and the corresponding straight-line length between them are enough to determine that the road 1 is a much more likely candidate than the road 2 candidate. The same figure illustrates also the dynamic nature of the pattern recognition approach. Between the start position and the curve φ_i, there was only one candidate. At that point, the digital map shows an intersection with two possible successive roads. Both of these roads became candidates with the same initial evaluation inherited from the original "parent" candidate. Both of these candidates are further evaluated based on their own features.

A special care had to be taken to guarantee robustness of the pattern generation and their comparison. Every time when there is a feature in one domain and not in the other, e.g., in the dead reckoning path but not on the candidate road, the comparison is delayed for some time (or traveled distance) to see if the corresponding feature can be detected. If, in spite of everything, the corresponding feature cannot be detected, this is interpreted as a mismatch between the dead reckoning and the candidate trajectory. As a consequence, the certainty of this candidate trajectory is decreased (i.e., its cumulative error is increased).

Figure 8.7 [9] shows a typical situation with and without the pattern matching algorithm (test drive A). As visible in the left half of Fig. 8.7, positioning without pattern recognition is not stable. It jumps from one road alternative to another and even shows an "off-road" candidate as the most probable one. The correct position is identified at the end of the shown trajectory. On the other hand, the positioning system with the pattern matching (right half of Fig. 8.7) follows smoothly the exact trajectory in a stable fashion.

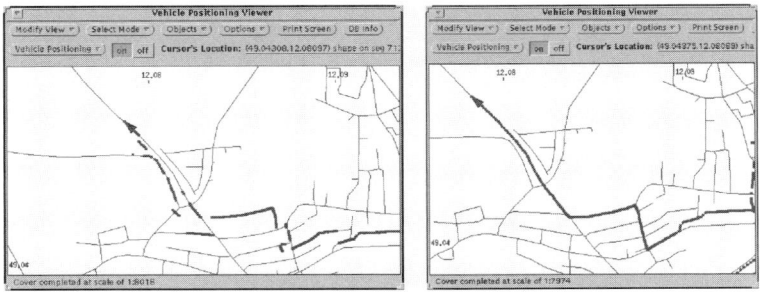

Fig. 8.7. Test drive A without (*left*) and with (*right*) pattern recognition

Fig. 8.8. Test drive B without (*left*) and with (*right*) pattern recognition

Another typical situation is depicted by the test drive B (see Fig. 8.8 [9]). The driving direction is from top to bottom. Again the positioning without pattern recognition is not stable – it jumps from one road alternative to another and it takes the complete test drive B to identify the correct position. At the start of the shown drive, the pattern matching approach enables the navigation system to keep the correct map alternative and to identify this map alternative as the true position despite of an initializing error. Accordingly, the precise position becomes clear after a few meters (see top of right half of Fig. 8.8), when the pattern matching is applied. During test drive B the positioning system with the pattern matching follows smoothly the exact trajectory in a stable fashion.

A third situation is given by test drive C (see Fig. 8.9 [9]). The driving direction is from above right to left bottom. Without pattern matching the correct map alternative has been deleted and there are positioning problems when the highway is entered, at the highway intersection and behind it. Using the pattern matching, the correct map alternative is kept as second alternative. When the highway is entered, the system recognizes that the second alternative should be the first and changes the position correspondingly. While doing this a small length error appears which causes small positioning problems at the highway intersection. In total, the accuracy of the positioning improves significantly when the pattern matching is applied.

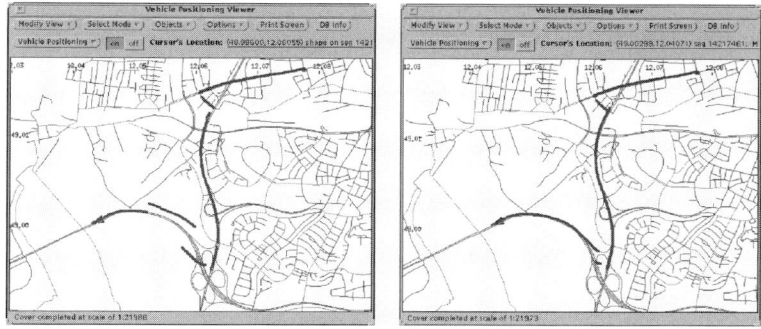

Fig. 8.9. Test drive C without (*left*) and with (*right*) pattern recognition

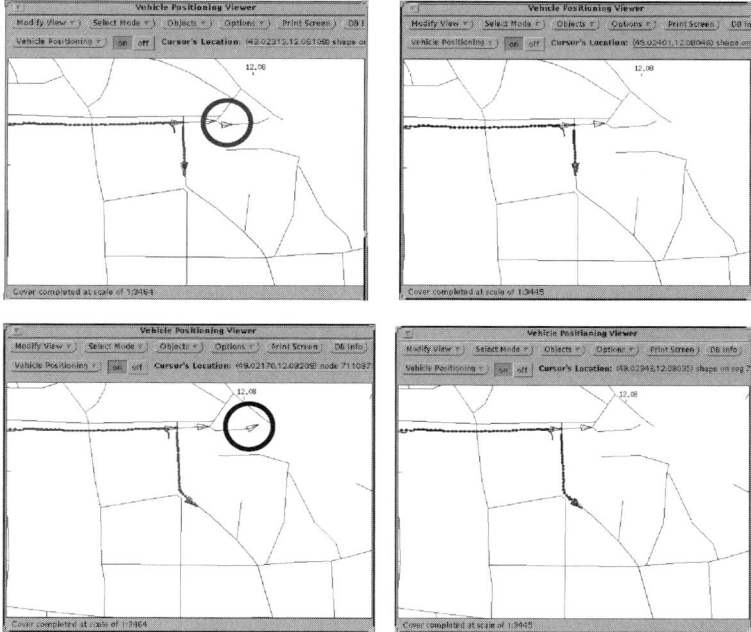

Fig. 8.10. Test drive D without (*left, top* and *bottom*) and with (*right, top* and *bottom*) pattern recognition

Figures 8.10 and 8.11 [9] illustrates typical situations, where the pattern matching helps the navigation system to delete unproductive map alternatives faster: without pattern matching the map alternative, which is highlighted by the circle and which does not contain the feature "right turn", is kept over the whole test drive D (left half of Fig. 8.10, top and bottom). With the pattern matching, this map alternative is not followed (right half of Fig. 8.10, top and bottom). In test drive E (see Fig. 8.11) there are ten map alternatives kept

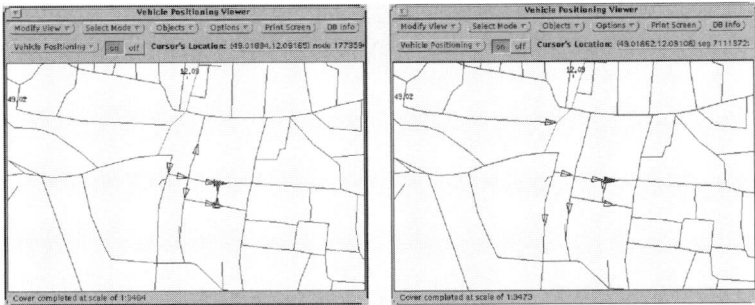

Fig. 8.11. Test drive E without (*left*) and with (*right*) pattern recognition

without pattern matching, while there are only eight map alternatives kept with pattern matching.

The overall advantages of the pattern matching approach are as follows:

- Usage of the historical information in the form of a pattern sequence (curve, straight line) contrary to the standard use of instantaneous information (current dead reckoning position, projections on the digital map, etc.). The historical information is herein combined with the classical projection algorithms and it can override the projection based solution.
- Robustness: the feature extraction is performed by integration of the distance and heading change, which leads to noise averaging. In addition, the feature vector matching is performed iteratively leading to the gradual change of the certainty (mismatch).
- Feature vectors generation is independent from the projection on the digital map performed by the standard positioning algorithm.

Other approaches to utilizing digital map data in the car positioning process can be found in [2, 3, 7, 14].

8.4 Fuzzy Guidance

As well as determining the position and planning the route a navigation system has to guide along the route. The instructions or maneuvers are presented to the driver visually as arrows and bar charts or by means of a voice output. Here, eight different maneuvers are used to illustrate the different approaches to maneuver generation: straight (S), left (L), right (R), slight left (SL), slight right (SR), hard left (HL), hard right (HR) and u-turn (U). In simple maneuver generation approaches, each of these commands is assigned to a given angle slot as it is depicted in Fig. 8.12. The actual generation of clear maneuvers out of these basic commands is a challenge particularly at complex route intersections with multiple possible alternative routes, where each route exhibits similar changes of angle. Figure 8.13 gives an example, where these fixed

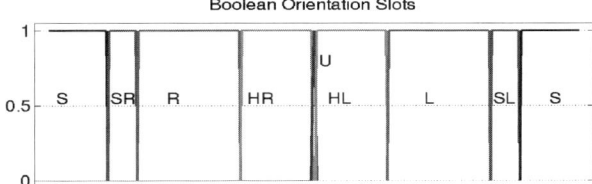

Fig. 8.12. Boolean orientation slots

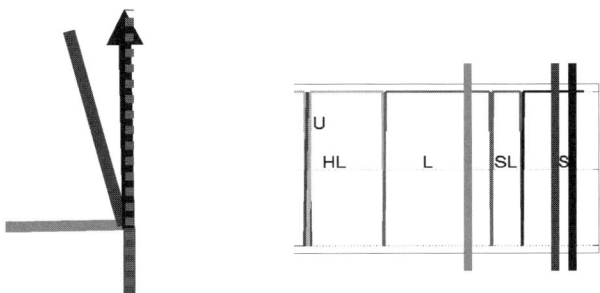

Fig. 8.13. Example for critical maneuver with fixed angle slots

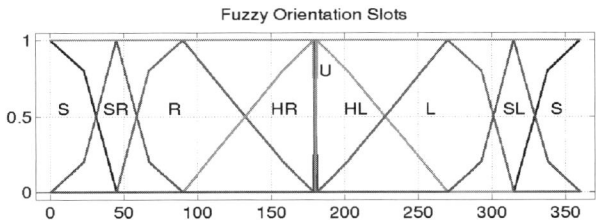

Fig. 8.14. Fuzzy maneuver membership functions

angle slots are used to generate a maneuver for the given routes. The simple approach with fixed angle slots fails since two routes occupy the same slot S. To resolve this problem a more complex logic was originally applied to select the appropriate command for this rather simple example.

The approach proposed here overcomes the restrictions of the fixed angle slots by applying fuzzy membership functions for each command. Figure 8.14 shows an example for the distribution of the maneuver membership function over angle. In contrast to the fixed angle slot approach the fuzzy membership functions performs no hard transitions between maneuvers but it performs a smooth transition with overlap between the different maneuvers. This overlap automatically solves the critical route example from Fig. 8.13. The second route is automatically remapped from the S to the SL maneuver, as it is shown

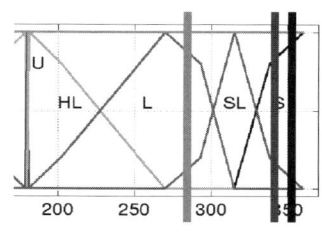

Fig. 8.15. Example for critical maneuver with fuzzy angle slots (based on fuzzy membership functions)

in Fig. 8.15. This remapping takes place because the maneuver generation in the first step calculates the degree of membership $\mu_{m,i}$ for each maneuver m and route i and assigns each maneuver to the route with the highest degree of membership. In addition to this fuzzy angle classification a further evaluation variable is introduced for selecting the appropriate maneuver. This further variable describes a significance or certainty $S_{m,i}$ of a maneuver and is produced from the quotient of the match of the maneuver under consideration $\mu_{m,i}$ for a street i and the sum of the membership values with regard to maneuver m:

$$S_{m,i} = \frac{\mu_{m,i}}{\sum_{j=1}^{N} \mu_{m,j}} \tag{8.6}$$

with $m \in \{\text{S}, \text{SR}, \text{R}, \text{HR}, \text{U}, \text{HL}, \text{L}, \text{SL}\}$

This value for certainty simplifies the selection of an understandable maneuver. A threshold >50% has to be chosen to guarantee the certainty of a maneuver for a specific route. In the example in Fig. 8.15, the maneuver S would be applied for a different route than maneuver SL, i.e., each possible maneuver gets a clear meaning.

If, for a given route, a maneuver with a certainty above the threshold cannot be found, then this means the number of different maneuvers (S, SL, SR, etc.) is too small to consider the intersection at hand. An approach counting the number of streets and producing maneuvers like 'take the second right street' might be considered instead. Also, a rule-based approach, which evaluates the neighbors of each route to select the best maneuver, could be applied. However, in the practical examples considered such an extension was not required. For any route at least one maneuver with a certainty above the selected threshold has been found. If there was more than one certain maneuver the one with the maximum degree of fuzzy membership is selected producing well understandable maneuvers for the driver.

8.5 Conclusions

This chapter has presented two novel sensor and information source fusion methods implemented in the Siemens car navigation systems, as well as a fuzzy-logic based guidance methodology. The first sensor fusion is implemented via Kalman filter, which updates the odometer and gyroscope signals needed for dead reckoning by using the appropriate GPS measurements. The so calculated dead reckoning trajectory was extensively tested in the whole set of test drives where it was shown that the resulting improvement in accuracy exceeds 20% in average.

The second novelty presented in this chapter is the evaluation of different position candidates on the road map based on the historical information. The historical information is presented in the form of feature vectors consisting of the curve and straight-line elements. Appropriately designed state machines perform the feature extraction from both the dead reckoning and digital map. The certainty (mismatch) of the different position alternatives is iteratively updated every time a new feature is generated.

A further contribution of this chapter is the presentation of a fuzzy-logic based approach to guidance. Using this approach the maneuver generation process produces more reliable, i.e., better understandable instructions.

The navigation system including the Kalman filter improved dead reckoning, the pattern recognition algorithm and the fuzzy-logic based guidance is a standard commercial product of Siemens AG and since 2001 has been implemented in Opel, Porsche, Alfa Romeo, and Lancia models. In January 2002, the German car magazine Auto Bild has evaluated the performance of different navigation systems implemented in ten different cars. An important focus was the evaluation of the positioning accuracy. The Siemens navigation system implemented in Opel cars and in Alfa Romeo cars was graded with 8 out of 10 possible points. The same evaluation was given for the navigation systems in these cars: Jaguar S-Type, Mercedes ML 55 AMG, Toyota Yaris. However, the other five systems got grades below such that the average evaluation of the ten systems on the positioning accuracy was merely 5.7 points. Considering other aspects like time of route calculation and routing accuracy, the Siemens navigation system implemented in Opel cars was awarded the first place (test winner).

References

1. Barshan, B., Durrant-Whyte, H.: Inertial navigation systems for mobile robots. IEEE Transactions on Robotics and Automation **11**(3), 328–342 (1995)
2. Blewitt, G., Taylor, G.: Mapping dilution of precision (MDOP) and map-matched GPS. International Journal on Geographical Information Science **16**(1) (2002)

3. Edelkamp, S., Schroedl, S.: Route planning and map inference with global positioning trees. Computer Science in Perspective Springer, Berlin Heidelberg New York (2003)
4. French, R.: Map matching origins, approaches and applications. In: 2nd International Symposium on Land Vehicle Navigation, pp. 93–116 (1989)
5. Grewal, M., Hendersom, V., Miyasko, R.: Application of Kalman Filtering to the calibration and alignment of inertial navigation systems. IEEE Transactions on Automatic Control **36**(1), 4–13 (1991)
6. Jazwinski, A.: Stochastic Processes and Filtering Theory. Academic, New York (1970)
7. Kim, W., et al.: Efficient use of digital road map in various positioning for ITS. In: IEEE Positioning and Navigation Symposium. San Diego, USA (2000)
8. Lenz, H., et al.: Determination of sensor path by dead reckoning and using the time discretization. Technical Report, Siemens Internal Report (2000)
9. Obradovic, D., Lenz, H., Schupfner, M.: Fusion of sensor data in Siemens car navigation systems. IEEE Transactions on Vehicular Technology **56**(1) (2007)
10. Scott, C.: Improved positioning of motor vehicles through secondary information sources. Ph.D. thesis, University of Technology, Sydney (1996)
11. Siemens AG proprietary algorithm. Siemens VDO internal report (1999)
12. Sukkarieh, S., Nebot, E., Durrant-Whyte, H.: A high integrity IMU/GPS navigation loop for autonomous land vehicle applications. IEEE Transactions on Robotics and Automation **15**(3) (1999)
13. Zao, Y.: Vehicle Location and Navigation Systems. Artech House, USA (1997)
14. Zewang, C., et al.: Development of an algorithm for car navigation system based on dempster-shafer evidence reasoning. In: IEEE 5th Conference on Intelligent Transportation Systems. Singapore (2002)

Part III

Information Fusion in Imaging

9

Cue and Sensor Fusion for Independent Moving Objects Detection and Description in Driving Scenes

Nikolay Chumerin and Marc M. Van Hulle

In this study we present an approach to detecting, describing and tracking independently moving objects (IMOs) in stereo video sequences acquired by on-board cameras on a moving vehicle. In the proposed model only three sensors are used: stereovision, speedometer and light detection and ranging (LIDAR). The IMOs detected by vision are matched with obstacles provided by LIDAR. In the case of a successful matching, the descriptions of the IMOs (distance, relative speed and acceleration) are provided by adaptive cruise control (ACC) LIDAR sensor, or otherwise these descriptions are estimated based on vision. Absolute speed of the IMO is evaluated using its relative velocity and ego-speed provided by the speedometer. Preliminary results indicate the generalization ability of the proposed system.

9.1 Introduction

The detection of the *independently moving objects* (IMOs) can be considered as an exponent of the obstacle detection problem, which plays a crucial role in traffic-related computer vision. Vision alone is able to provide robust and reliable information for autonomous driving or guidance systems in real-time but not for the full spectrum of real-world scenarios. The problem is complicated by ego-motion, camera vibrations, imperfect calibrations, complex outdoor environments, insufficient camera resolutions and other limitations. The fusion of information obtained from multiple sensors can dramatically improve the detection performance [2–5, 8, 9, 11, 12, 15–18, 27–30].

In Table 9.1, we present a chronological list of studies which are related to sensor fusion in traffic applications and which are relevant to the considered topic. Various sensors can be used for traffic applications: video (color or grayscale) cameras in different setups (monocular, binocular or trinocular), IR (infrared) cameras, light detection and ranging (LIDAR), radio detection and ranging (radar), global positioning system/differential GPS (GPS/DGPS) as

Table 9.1. Sensor fusion for traffic applications papers short overview

Study	Sensors	Cues/features	Techniques used
Handmann et al. [11]	Monocular color vision, radar	Color, edges, texture (local image entropy), (up to 3) obstacle positions	MLP
Stiller et al. [29]	Stereo vision, radar, LIDARs, DGPS/INS	Horizontal edges, stereo disparity, optical flow, 2D range profile, global ego-position and ego-orientation	Kalman filter
Becker and Simon [2]	Stereo vision, DGPS, vehicle sensors (IMUs), LIDARs, radar	Local ego-position and ego-orientation (w.r.t. lane), global ego-position and ego-orientation, ego-speed, ego-acceleration, steering angle, 2D range profile	Kalman filter
Kato et al. [16]	Monocular vision, radar	Kanade–Lucas–Tomasi feature points, range data	Frame-to-frame feature points coupling based on range data
Fang et al. [8]	Stereo vision, radar	Edges, stereo disparity, depth ranges	Depth-based target edges selection and contour discrimination
Steux et al. [28]	Monocular color vision, radar	Shadow position, rear lights position, symmetry, color, 2D range profile	Belief network
Hofmann et al. [12]	Monocular color vision, monocular grayscale vision, radar, ACC-radar	Lane position and width, relative ego-position and ego-orientation (w.r.t. road), radar-based obstacles	extended Kalman filter
Laneurit et al. [18]	Vision, GPS, odometer, wheel angle sensor, LIDAR	Relative ego-position and ego-orientation (w.r.t. road), global ego-position and ego-orientation, steering angle, path length, LIDAR-based obstacle profile	Kalman filter
Sergi [26]	Vision, LIDAR, DGPS	Video stream, global ego-position and ego-orientation, LIDAR-based obstacle profile	Kalman filter
Sole et al. [27]	Monocular vision, radar	Horizontal and vertical edges, 'pole like' structures, radar target,	Matching
Blanc et al. [5]	IR camera, radar, LIDAR	IR images, range profile	Kalman filter, matching
Labayrade et al. [17]	Stereo vision, LIDAR	Stereo disparity, "v-disparity", lighting conditions, road geometry, obstacle positions	Matching, Kalman filter, belief theory based association -
Thrun et al. [31]	Monocular color vision, GPS, LIDARs, radars, accelerometers, gyroscopes	Color images, global ego-position and ego-orientation, ego-speed, short-range profile (LIDARs), long-range obstacles (radars)	Unscented Kalman Filter

well as data from vehicle inertial measurement unit (IMU) sensors: accelerometer, speedometer, odometer and angular rate sensors (gyroscopes). There are a number of approaches to fusion characterization [7, 10, 24, 33] but, most frequently, fusion is characterized by the abstraction level:

1. Low (signal) level fusion combines raw data provided directly from sensors, without any preprocessing or transformation.
2. Intermediate (feature) level fusion aggregates features (e.g. edges, corners, texture) extracted from raw data before aggregation.
3. High (decision) level fusion aligns decisions proposed by different sources.

Depending on the application, several different techniques are used for fusion. Matching of the targets detected by different sensors is often used for obstacle detection. Extensions of the Kalman filter (KF) [14] (e.g., extended Kalman filter (EKF) and unscented Kalman filter (UKF) [13]) are mostly involved in estimation and tracking of obstacle parameters, as well as in ego-position and ego-motion estimation.

The flow diagram of the proposed model is shown in Fig. 9.1. To detect independent motion, we propose to extract visual cues and subsequently fuse them using appropriately trained multi-layer perceptron (MLP). Object recognition is used as cooperative stream which helps to delineate and classify IMOs. If detected by vision IMO appears within sweep of the ACC LIDAR, we use distance, speed and acceleration of the IMO provided by ACC system. Otherwise these descriptions are estimated based on vision.

To validate the model we have used the data obtained in the frameworks of the DRIVSCO and ECOVISION European Projects. In recording sessions a modified Volkswagen Passat B5 was used as a test car. It was equipped by Hella KGaA Hueck and Co.

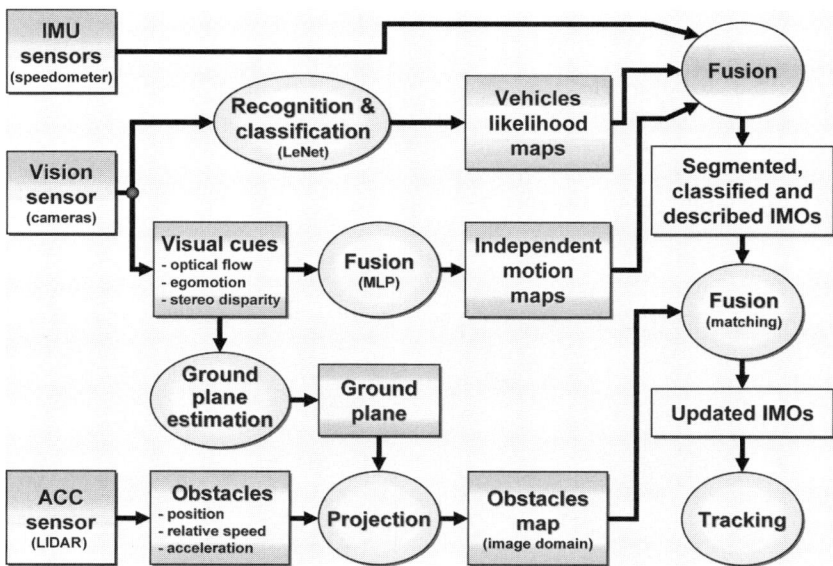

Fig. 9.1. Flow-diagram of the proposed model

9.2 Vision Sensor Data Processing

For vision-based IMO detection, we used an approach proposed by Chumerin and Van Hulle [6]. This method is based on the processing and subsequent fusing of two cooperative streams: the *independent motion detection stream* and the *object recognition stream*. The recognition stream deals with static images (i.e., does not use temporal information) and therefore cannot distinguish between independently moving and static (i.e., with respect to the environment) objects, but which can be detected by the independent motion stream.

9.2.1 Vision Sensor Setup

In the recording sessions, we used a setup with two high resolution progressive scan color CCD cameras (see Table 9.2). The camera rig was mounted inside the cabin of the test car (see Fig. 9.2) at 1.24 m height above the ground, with 1.83 m from the front end and 17 cm displacement from the middle of the test car towards the driver's side. Both cameras were oriented parallel to each other and to the longitudinal axis of the car and look straight ahead into

Table 9.2. Video sensor specifications

Sensor parameter	Value
Manufacturer	JAI PULNiX Inc.
Model	TMC-1402Cl
Field of View	$53° \times 42.4°$ (horizontal × vertical)
Used resolution	$1{,}280 \times 1{,}024$
Used frequency	25 fps
Color	RGB Bayer pattern
Interocular distance	330 mm
Focal length	12.5 mm
Optics	Pentax TV lenses

Fig. 9.2. Setup of the cameras in the car

the street. Before each recording session, the cameras were calibrated. Raw color (Bayer pattern) images and some crucial ACC parameters were stored via CAN-bus for further off-line processing. In the model, we used rectified grayscale images downscaled to a 320 × 256 pixels resolution.

9.2.2 Independent Motion Stream

The problem of *independent motion* detection can be defined as the problem of locating objects that move independently from the observer in his field of view. In our case, we build so-called *Motion modelling independent motion map* where each pixel encodes the likelihood of belonging to an IMO. For each frame we build an independent motion map in two steps: visual cues extraction and classification.

As visual cues we consider: *stereo disparity* (three components – for current, previous and next frame), *optical flow* (two components) and *normalized coordinates*[1] (two components). The optical flow and stereo disparity are computed using multiscale phase-based optical flow and stereo disparity algorithms [23, 25]. Unfortunately, there are no possibilities to estimate reliably all these cues for every pixel in the entire frame. This means that the motion stream contains incomplete information, but this gap will be bridged after fusion with the recognition stream.

We consider each pixel as a multidimensional vector with visual cues as components. We classify all the pixels (which have every component properly defined) in two classes: IMO or background. We have tried a number of setups for classification, but the optimal performance was obtained with a multi-layer perceptron (MLP) with three layers: a linear (4–8 neurons), a nonlinear layer (8–16 neurons), and one linear neuron as output (see Fig. 9.3).

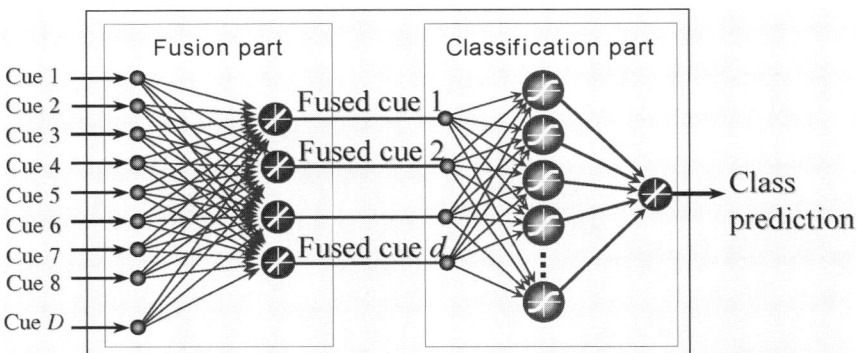

Fig. 9.3. MLP used as classifier in independent motion stream

---

[1] By a normalized coordinate system on a frame we mean the rectangular coordinate system with origin in the center of the frame, where the upper-left corner is (−1, −1) and the lower-right corner is (1, 1).

For training purposes, we labeled the pixels in every frame of a number of movies into background and different IMOs, using a propriety computer-assisted labeling tool (see Fig. 9.4).

After training, the MLP can be used for building an IMO likelihood map I for the entire frame:

$$I(x,y) = p\left(\text{IMO}|(x,y)\right), \tag{9.1}$$

where x, y are pixel coordinates. Figure 9.5 shows an example of an IMO likelihood map obtained using the proposed approach.

Fig. 9.4. myLabel – a tool for labeling video sequences. The labeling is similar to what is done in graphical editors like Photoshop or Gimp. Each label mask is represented by a separate layer with its own color (on the figure instead of colors we have used contours). The user can easily edit label masks in a pixel-wise manner as well as change their colors, transparencies and visibilities. myLabel allows semi-automatic labeling by interpolating the labels between two labeled frames

Fig. 9.5. (*Left*) Frame number 342 of motorway3 sequence. (*Right*) Matrix I, output of the motion stream for the same frame. Value $I(x,y)$ is defined as probability of pixel (x,y) being part of an IMO

9.2.3 Recognition Stream

For the recognition of vehicles and other potentially dangerous objects (such as bicycles and motorcycles, but also pedestrians), we have used a state-of-the-art recognition paradigm – the convolutional network LeNet, proposed by LeCun and colleagues [19]. We have used the CSCSCF configuration of LeNet (see Fig. 9.6) comprising six layers: three convolutional layers (C0, C1, C2), two subsampling layers (S0, S1) and one fully connected layer (F). As an input, LeNet receives a 64×64 grayscale image. Layer C0 convolves the input with ten 5×5 kernels, adds (ten) corresponding biases, and passes the result to a squashing function[2] to obtain ten 60×60 feature maps.

In layer S0, each 60×60 map is subsampled to a 30×30 map, in such a way that each element of S0 is obtained from a 2×2 region of C1 by summing these four elements, by multiplying with a coefficient, adding a bias, and by squashing the end-result. For different S0 elements, the corresponding C1's 2×2 regions do not overlap. The S0 layer has ten coefficient-bias couples (one couple for each feature map). Computations in C1 are the same as in C0 with the only difference in the connectivity: each C1 feature map is not obtained by a single convolution, but as a sum of convolutions with a set of previous (S0) maps (see Table 9.3). Layer S1 subsamples the feature maps of C1 in the same manner as S0 subsamples the feature maps of C0. The final convolutional layer C2 has kernels sized 13×13 and 180 feature maps which are fully connected to all 16 S1 feature maps. It means that the number of C2 kernels is $16 \times 180 = 2{,}880$, and the corresponding connectivity matrix should have all cells shaded. The output layer consists of seven neurons, which are fully connected to C2's outputs. It means that each neuron in F (corresponding to a particular class *background, cars, motorbikes, trucks, buses, bicycles* and *pedestrians*) just squashes the biased weighted sum of all C2's outputs.

LeNet scans (see Fig. 9.7) the input image with a sliding window and builds seven matrices, $R_0, \ldots, R_6$, which are regarded as likelihood maps for

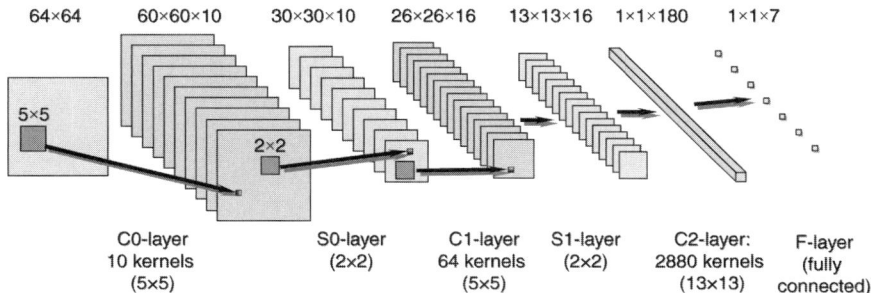

Fig. 9.6. LeNet – a feed-forward convolutional neural network, used in the recognition stream

---
[2] $f(x) = A \tanh(Sx)$, $A = 1.7159$ and $S = 2/3$ according to [19].

Table 9.3. S0-C1 connectivity matrix. A shaded cell which belongs to the ith column and jth row indicates that the jth feature map of S0 participates in the computation of the ith feature map of C1. For example, to compute the fourth feature map of C1, one has to find a sum of convolutions of S0 feature maps 0, 8 and 9 with corresponding kernels. The number of kernels in C1 (the number of shaded cells in the table) is 64

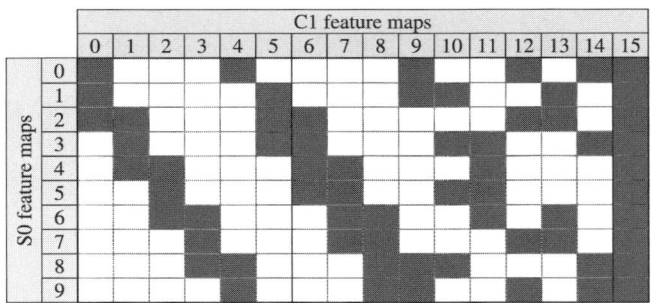

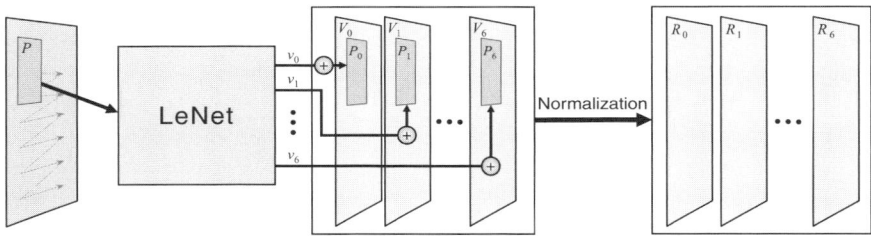

Fig. 9.7. For each position of the 64×64 sliding window, we feed the corresponding image patch P to LeNet to obtain seven values $v_0, \ldots, v_6$ which are related to the class likelihoods of P. Then we update the regions P_i associated with P in intermediate matrices V_i by adding v_i ($i = 0, \ldots, 6$). After scanning the entire image, we normalize V_i to obtain matrices R_i. Normalization here means that we divide each element of V_i by its number of updates (different elements are updated different number of times) and then linearly remap all obtained values into the interval $[0, 1]$

the considered classes (see Fig. 9.8). To make the recognition more scale-invariant, we process the input image in two scales 320×256 and 640×512, but the resulting (mixed) matrices R_i are sized 320×256. Note that, for further processing, the most important map is R_0, which corresponds to the background class and the so-called *non-background* map is obtained as $(1 - R_0)$. The rest of the maps $(R_1, \ldots, R_6)$ are used only in IMO classification.

9.2.4 Training

For training both vision streams, we used two rectified stereo video sequences, each consisting of 450 frames. We have labeled IMOs in all left frames of the sequences. These labels were used for training the motion stream classifier.

Fig. 9.8. Result of recognition for frame number 342 of motorway3 sequence. (**a**) Input grayscale image. (**b**) Input image overlaid with the output of the recognition stream (all the classes except background), showing, from left to right: motorcycle, two cars and truck. (**c**)–(**f**) Likelihood maps for the classes *background*, *cars*, *trucks* and *motorbikes*, respectively. Pixel intensities correspond to the likelihood values

We have used small batches with the increasing size version of the BFGS Quasi-Newton algorithm for the independent motion classifier training. Samples for each batch were randomly taken from all the frames of all the scenes. Training was stopped after reaching 0.04 (MSE on training set) performance.

To train LeNet, we have prepared a data set of 64 × 64 grayscale images (approximately 67,500 backgrounds, 24,500 cars, 2,500 motorbikes, 6,200 trucks, 1,900 bicycles, 78 buses, and 3,500 pedestrians). Although this may seem an unbalanced data set, it should reflect the proportion of different object classes present in real driving scenes, especially in rural ones. Furthermore, we have doubled the data set by including horizontally flipped versions of all the samples. Images were taken mainly from publicly available object recognition databases (LabelMe[3], VOC[4]). We have randomly permuted samples in the entire data set and then split the latter into training (90%) and testing (10%) sets. A stochastic version of the Levenberg–Marquardt algorithm with diagonal approximation of the Hessian [19] was used for LeNet training. Training was stopped after reaching a misclassification rate on the testing set of less than 1.5%. To increase the robustness of the classification, we have run the training procedure several times, every time by adding a small (2%) amount of uniform noise and by randomly changing the intensity (97–103%) of each training sample.

---

[3] http://labelme.csail.mit.edu/
[4] http://www.pascal-network.org/challenges/VOC/

9.2.5 Visual Streams Fusion

Fusion of the visual streams for a particular frame is achieved in three steps.

1. Masking (elementwise multiplication) of the independent motion map I by the mask M of the most probable locations of the IMOs in the frame (see Fig. 9.9):
$$F_1(x,y) = I(x,y)M(x,y). \tag{9.2}$$

2. Masking of the previous result F_1 by the non-background map $(1 - R_0)$:
$$F_2(x,y) = F_1(x,y)(1 - R_0(x,y)). \tag{9.3}$$

3. Masking of the previous result F_2 by the likelihood maps $R_1, \ldots, R_6$ of each class, which results in six maps $L_1, \ldots, L_6$ (one for each class, except the background):
$$L_k(x,y) = F_2(x,y)R_k(x,y), \quad k = 1, \ldots, 6. \tag{9.4}$$

The first step is necessary for rejecting regions of the frame where the appearance of the IMOs is implausible. After the second step we obtain crucial information about regions which have been labeled as non-backgrounds (vehicles or pedestrians) and which, at the same time, contain independently moving objects. This information is represented as the saliency map F_2, which we will further use for IMO detection/description and in the tracking procedure. The third step provides us the information needed in the classification stage.

Fig. 9.9. Matrix M, masking regions of possible IMO appearance in a frame. This matrix has been chosen heuristically based on the logical assumptions that in normal circumstances IMOs are not supposed to appear "in the sky" or beneath the ego-vehicle

9.3 IMO Detection and Tracking

For detecting an IMO, we have used a simple technique based on the detection of the local maximas in the maps defined in (9.3). We have performed a spatio-temporal filtering (i.e. for ith frame we apply smoothing of a three-dimensional array – a concatenation of the $(i-2)$th, $(i-1)$th, ith, $(i+1)$th and $(i+2)$th two-dimensional maps along the third time-dimension). Then we search for local maximas in the entire (ith) filtered frame and consider them as the IMO centers $\mathbf{x}_k$ for this frame.

For tracking IMOs, we have introduced a parameter called *tracking score*. For a particular IMO, we increase this parameter when, in the next frame, only in a small neighborhood of the IMO center there is a good candidate for the considered IMO in the next frame, namely the IMO with the same class label, and approximately with the same properties (size, distance and relative speed in depth). Otherwise, the tracking score is decreased. An IMO survives while the tracking score is above a fixed threshold. The tracking score works as a momentum and allows the system to keep tracking an IMO even when there are no sufficient data in the next few frames.

9.4 Classification and Description of the IMOs

As soon as we are able to detect IMOs, it becomes possible to classify them and to retrieve their properties (size, absolute speed in depth, relative speed in depth, time to contact, absolute acceleration, etc).

We define the *class* c_k of the kth IMO as:

$$c_k = \arg\max_{1 \leq c \leq 6} \{L_c(\mathbf{x}_c)\}, \tag{9.5}$$

where $\mathbf{x}_k = (i_k, j_k)$ is the center of the kth IMO (in image domain D) and L_c are the maps, defined in (9.4).

Let σ_k be the *size* of the kth IMO. For σ_k estimation, we search for the spread of the appropriately scaled circular Gaussian blob, locally best fitting L_{c_k} in $\mathbf{x}_k$. Analytically this can be expressed as a search for the argument of the first minimum of the function:

$$\Delta_k(\sigma) = \int_{D_k} \left| L_{c_k}(\mathbf{x}_k) e^{-||\mathbf{x}_k - \mathbf{x}||^2/\sigma^2} - L_{c_k}(\mathbf{x}) \right| d\mathbf{x}, \tag{9.6}$$

where D_k is a neighborhood of $\mathbf{x}_k$ in image D. In our simulations we have used $D_k = D$, but the choice of D_k could easily be optimized.

The IMO's *distance* estimation is a crucial point in the retrieval process. Using an averaged (in a small neighborhood of the IMO's center) disparity and known calibration parameters of the two cameras, we have computed the distance to the IMO. To compensate for instabilities in the distance estimations, we have used a robust linear regression based on the previous five estimates.

The present-day motor vehicles are being equipped with an increasing number of electronic devices, including control units, sensors, actuators, etc. All these devices communicate with each other over a data bus. During recording sessions, we have stored the ego-speed provided by test car's speedometer.

The *relative speed in depth*, we estimated as the derivative (with respect to time) of the distance using robust linear regression based on the last five estimations of the distance. To estimate the *time to contact*, we have divided the averaged distance by the averaged relative speed in depth. Using the precise value of the ego-motion speed from the CAN-bus data, and simply by adding it to the relative speed in depth we have also obtained the *absolute speed in depth* of the considered IMO.

The derivative of the absolute speed in depth can be considered as an estimation of the *acceleration* (it is true only in the case when the ego-heading is collinear to the heading of the IMO). An example of IMO tracking and the retrieved properties is shown in Fig. 9.13.

9.5 LIDAR Sensor Data Processing

The ACC system of the used test car was able to detect and track up to ten obstacles, when in the range of the LIDAR sensor. In addition to position, the ACC can also provide information about relative lateral extent and speed of the tracked obstacle.

9.5.1 LIDAR Sensor Setup

We used data recorded by the test car equipped with the LIDAR sensor manufactured by Hella KGaA Hueck and Co (see Table 9.4 for specifications). The sensor was mounted in the test car at 30 cm height above ground, with 18 cm from the front-end and 50 cm from the middle of the car to the driver's side (see Fig. 9.10). The ACC system analyzes raw LIDAR data and tracks up to ten targets within a distance of up to 150 m. The tracking data are updated and available for recording via the CAN-bus (Flex-ray) every 60 ms.

Table 9.4. LIDAR sensor specifications

Sensor parameter	Value
Manufacturer	Hella KGaA Hueck and Co
Model	IDIS 1.0
Field of view	$12° \times 4°$ (horizontal × vertical)
Range	up to 200 m
Description	12 fixed horizontally distributed beams, each beam observes a $1° \times 4°$ angular cell

9 Cue and Sensor Fusion for IMOs Detection and Description   173

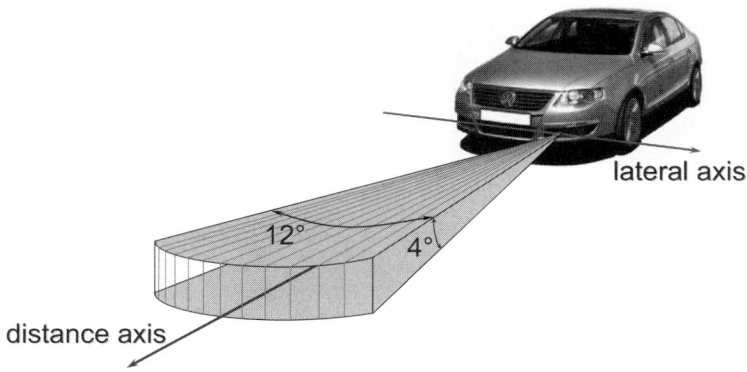

Fig. 9.10. ACC LIDAR configuration

Each tracked target is described by its distance, lateral position (left and right edges), relative velocity and acceleration.

9.5.2 Ground Plane Estimation

The LIDAR provides the depth and lateral position of the detected obstacles. This information is not sufficient for the correct projection of the obstacles onto the video frame. To estimate the missing vertical components (in the frame domain) of the IMOs, we assume that all IMOs are located near the dominant ground plane. Here, we use a strong assumption of road planarity, which is not met in all driving scenarios and could introduce bias. However, in our model, the positions of the LIDAR-based obstacles are used only to verify (confirm) vision-based obstacles, so that the bias caused by the non-planarity of the road is to a large extend unimportant.

To estimate the ground plane, we estimate the *disparity plane*, then map the set of points from the disparity domain into a 3D world domain, and finally fit a plane through the projected set.

Before the disparity plane estimation, we intersect the disparity map with the predefined road mask (see Fig. 9.11, left panel). By this step, we filter out the majority of pixels which do not belong to the ground plane and are outliers in the disparity plane linear model:

$$\Delta : D = \alpha x + \beta y + \gamma, \qquad (9.7)$$

where (x, y) are pixel coordinates and D is disparity.

The disparity plane parameters α, β and γ are estimated using iteratively reweighted least-squares (IRLS) with weight function proposed by Beaton and Tukey [1] and tuning parameter $c = 4.6851$. 7 iterations.

For the ground plane parameters estimation, we choose a set of nine points (3×3 lattice) in the lower half of the frame (see Fig. 9.11, right panel). Disparities for these points are determined using the estimated disparity plane (9.7).

Fig. 9.11. (*Left*) Predefined road mask. (*Right*) Example of the ground plane estimation. *Red points* represent points used for ground plane estimation (see text)

Given the disparities and camera calibration data, we project the selected points into a 3D world coordinate system. In addition, we add two so-called *stabilization points* which correspond to the points where the front wheels of the test car are supposed to touch the road surface. For the inverse projection of the stabilization points, we use parameters of the *canonical disparity plane*: it is a disparity plane which corresponds to the horizontal ground plane observed by cameras in a quiescent state. The parameters of the canonical disparity plane and positions of the stabilization points were obtained based on the test car geometry and camera setup position and orientation in the test car. The full set of 11 points is then used for IRLS fitting of the ground plane in a world coordinate system:

$$\pi : aX + bY + cZ + d = 0, \tag{9.8}$$

where (X, Y, Z) are pixel coordinates in the 3D world coordinate system connected to the left camera. Here were assume that $a^2 + b^2 + c^2 = 1$ (otherwise one can divide all coefficients by $\sqrt{a^2 + b^2 + c^2}$) and $b > 0$. In this case, vector $\mathbf{n} = (a, b, c)^{\mathrm{T}}$ represents the normal unity vector of the ground plane and coefficient d represents the distance from the camera to the ground plane. During the disparity plane estimation, we use the estimation from the previous frame for weight initialization in IRLS; for the first frame, for the same purpose, we use the parameters of the canonical disparity plane. We assume that the ground plane is estimated correctly if the following conditions are met:

$$\|\mathbf{n}_t - \mathbf{n}_0\| < \theta_0 \text{ and } \|\mathbf{n}_t - \mathbf{n}_{t-1}\| < \theta_1, \tag{9.9}$$

where $\mathbf{n}_k$ is normal vector for kth frame, and $\mathbf{n}_0$ is canonical normal vector. Thresholds $\theta_0 = 0.075$ and $\theta_1 = 0.015$ were chosen empirically. If the estimated ground plane does not satisfy (9.9), the previous estimation is used.

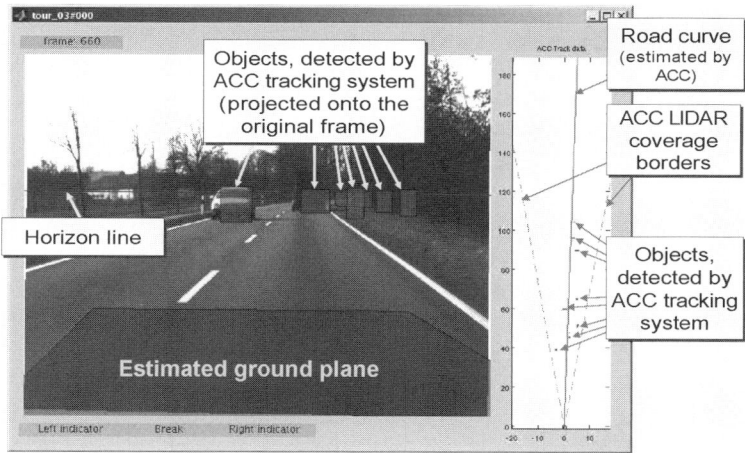

Fig. 9.12. ACC obstacles projection. Left part contains the grayscale version of current frame, overlaid by the horizon line, the ground plane segment and projected ACC (LIDAR) obstacles. Right part represents obstacles 2D range profile, provided by ACC system

9.5.3 LIDAR Obstacles Projection

Projection of the LIDAR-based obstacles into the (left) frame is based on the ground plane position, the obstacle positions, the camera projective matrix (from calibration data) and the position and orientation of the LIDAR sensor with respect to the camera. Only the height of the obstacles is not available. We have set the height of all the obstacles to a fixed value of 1.5 m. The result of the LIDAR obstacles projection is shown in Fig. 9.12.

9.6 Vision and LIDAR Fusion

The fusion of the vision-based IMOs with LIDAR-based obstacles is based on a simple matching process.

1. For the current IMO I_k, we look for candidates from the LIDAR obstacles O_l by means of the high intersection ratio:

$$r_{kl} = \#(I_k \cap O_l)/\#(I_k), \qquad (9.10)$$

where $\#(\cdot)$ is number of pixels of the set in the brackets. If ratio $r_{kl} > 0.5$, then obstacle O_l is an IMO I_k candidate and considered for further verification. If all obstacles were rejected, IMO I_k is not updated and process continues from step 4.

2. All the obstacles O_{k_m} with distances d_{k_m} satisfying the following condition:

$$\frac{|d_{k_m} - d_k^*|}{d_k^*} > 0.15, \tag{9.11}$$

where d_k^* denotes the distance of the IMO I_k, are rejected. Like in the previous step, if all obstacles were rejected, IMO I_k is not updated and the process continues from step 4.
3. Among the remaining obstacles, we choose the best matching candidate O_{k_i} for the IMO I_k with minimal depth deviation $|d_{k_i} - d_k^*|$. Distance, relative velocity and acceleration of IMO I_k are updated using corresponding values of the obstacle O_{k_i}. The absolute velocity of the IMO I_k is re-estimated in accordance with the new value of the relative speed. The obstacle O_{k_i} is eliminated from the search process. If all the obstacles were rejected, IMO I_k is not updated.
4. The process finishes if all IMOs are checked, otherwise the next IMO is selected for matching and the process continues from step 1.

9.7 Results

Due to the lack of ground truth benchmarks, systems similar to the proposed one are tested mainly in a qualitative sense. For the evaluation of the presented system, we have used two complex real-world video sequences (different from the training sequences). ACC LIDAR tracking data has been provided only for one of them. Nevertheless, even without this important information the system has shown the ability to detect, track and describe IMOs (see Fig. 9.13)

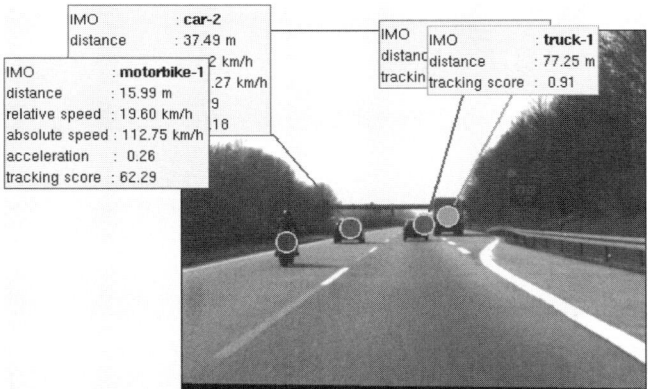

Fig. 9.13. Vision-based IMO detection, classification, description and tracking result

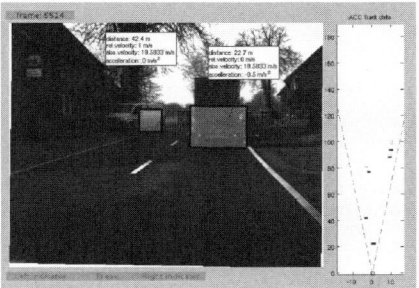 

Fig. 9.14. Results of the final fusion. Light (*pink*) bars with thick edges represent the detected IMOs, whereas the LIDAR obstacles rejected by the fusion procedure are shown as dark (*green*) bars

relying only on the data from the vision sensor and the IMU (speedometer). This testing scenario is important because during rapid turns or when the IMOs are out of reach of the LIDAR sensor, the system has to rely on the vision sensor only. In this case, the quality of the properties estimation suffers from the noise presented in the visual cues.

The experiments with the ACC LIDAR tracking data (see Fig. 9.14) have shown a significant improvement of the accuracy of the IMOs properties extraction especially in low curvature driving scenarios (when the ACC latencies have a negligible influence on the LIDAR-based obstacle localization).

One of the most important results is the fact that both networks have shown to be acceptable of generating qualitative results even on completely new testing sequences, which attests to the generalization ability of the used networks.

9.8 Conclusions and Future Steps

A high level sensor fusion model for IMO detection, classification and tracking has been proposed. The model incorporates three independent sensors: vision, LIDAR and speedometer. Vision plays the most important role in the model, whereas LIDAR data are used for confirming the IMO detection and for updating the IMO properties. The speedometer is used only for the IMOs absolute speed in depth estimation.

The existing model is still not a real-time system, but we see a number of ways to increase its speed. Both visual streams of the model have feed-forward architectures, which can be easily implemented in hardware such as Field-Programmable Gate Arrays (FPGAs). Moreover, as far as the streams are independent, they can be implemented as separate FPGAs, working in parallel. To speed up the entire model, we propose to switch from LeNet-based object recognition to faster and more task-specific recognition paradigm (e.g. [32] or [20]). Another way to increase the speed of the model could be the transition from an MLP-based fusion of the visual cues to a hard-coded fusion of the visual cues with ego-motion (e.g. [22]). As another future step of the model development, we envisage the incorporation of KF-based approaches [13, 21] for IMO tracking.

Finally, as more abstract descriptions of the IMOs are generated, this is expected to facilitate the development of models of driving behavior, in response to IMOs, which is one of the goals of the DRIVSCO project[5]. Indeed, *stopping, following, lane changing*, and so on, are important aspects of driving behavior, and which are key elements of driver-assistance systems that support collision free driving and increased traffic safety in general. These are heavily researched topics in academia, research institutes, and the automotive industry.

Acknowledgments

The first author is supported by the European Commission (NEST-2003-012963). The second author is supported by the Excellence Financing program (EF 2005) and the CREA Financing program (CREA/07/027) of the K.U. Leuven, the Belgian Fund for Scientific Research – Flanders (G.0248.03, G.0234.04), the Flemish Regional Ministry of Education (Belgium) (GOA 2000/11), the Belgian Science Policy (IUAP P5/04), and the European Commission (NEST-2003-012963, STREP-2002-016276, and IST-2004-027017).

References

1. Beaton, A., Tukey, J.: The fitting of power series, meaning polynomials, illustrated on band-spectroscopic data. Technometrics **16**(2), 147–185 (1974)
2. Becker, J., Simon, A.: Sensor and navigation data fusion for an autonomous vehicle. Intelligent Vehicles Symposium, 2000. IV 2000. Proceedings of the IEEE pp. 156–161 (2000)
3. Bertozzi, M., Broggi, A., Cellario, M., Fascioli, A., Lombardi, P., Porta, M.: Artificial vision in road vehicles. Proceedings of the IEEE **90**(7), 1258–1271 (2002)
4. Bertozzi, M., Broggi, A., Fascioli, A.: Vision-based intelligent vehicles: State of the art and perspectives. Robotics and Autonomous Systems **32**(1), 1–16 (2000)
5. Blanc, C., Trassoudaine, L., Le Guilloux, Y., Moreira, R.: Track to track fusion method applied to road obstacle detection. Procedings of the Seventh International Conference on Information Fusion (2004)
6. Chumerin, N., Van Hulle, M.: An approach to on-road vehicle detection, description and tracking. Proceedings of the 2007 17th IEEE Signal Processing Society Workshop on Machine Learning for Signal Processing (2007). (in press)

---

[5] http://www.pspc.dibe.unige.it/drivsco/

7. Dasarathy, B.: Sensor fusion potential exploitation-innovative architectures and illustrative applications. Proceedings of the IEEE **85**(1), 24–38 (1997)
8. Fang, Y., Masaki, I., Horn, B.: Depth-based target segmentation for intelligent vehicles: fusion of radar and binocular stereo. Intelligent Transportation Systems, IEEE Transactions on **3**(3), 196–202 (2002)
9. Gandhi, T., Trivedi, M.: Vehicle surround capture: Survey of techniques and a novel omni video based approach for dynamic panoramic surround maps. Intelligent Transportation Systems, IEEE Transactions on **7**(3), 293–308 (2006)
10. Hall, D., Llinas, J.: An introduction to multisensor data fusion. Proceedings of the IEEE **85**(1), 6–23 (1997)
11. Handmann, U., Lorenz, G., Schnitger, T., Seelen, W.: Fusion of different sensors and algorithms for segmentation. IV'98, IEEE International Conference on Intelligent Vehicles 1998 pp. 499–504 (1998)
12. Hofmann, U., Rieder, A., Dickmanns, E.: Radar and vision data fusion for hybrid adaptive cruise control on highways. Machine Vision and Applications **14**(1), 42–49 (2003)
13. Julier, S., Uhlmann, J.: A new extension of the kalman filter to nonlinear systems. International Symposium Aerospace/Defense Sensing, Simulation and Controls **3** (1997)
14. Kalman, R.: A new approach to linear filtering and prediction problems. Journal of Basic Engineering **82**(1), 35–45 (1960)
15. Kastrinaki, V., Zervakis, M., Kalaitzakis, K.: A survey of video processing techniques for traffic applications. Image and Vision Computing **21**(4), 359–381 (2003)
16. Kato, T., Ninomiya, Y., Masaki, I.: An obstacle detection method by fusion of radar and motion stereo. Intelligent Transportation Systems, IEEE Transactions on **3**(3), 182–188 (2002)
17. Labayrade, R., Royere, C., Gruyer, D., Aubert, D.: Cooperative fusion for multi-obstacles detection with use of stereovision and laser scanner. Autonomous Robots **19**(2), 117–140 (2005)
18. Laneurit, J., Blanc, C., Chapuis, R., Trassoudaine, L.: Multisensorial data fusion for global vehicle and obstacles absolute positioning. Intelligent Vehicles Symposium, 2003. Proceedings. IEEE pp. 138–143 (2003)
19. LeCun, Y., Bottou, L., Bengio, Y., Haffner, P.: Gradient-based learning applied to document recognition. pp. 2278–2324 (1998)
20. Leibe, B., Schiele, B.: Scale invariant object categorization using a scale-adaptive mean-shift search. DAGM'04 pp. 145–153 (2004)
21. van der Merwe, R.: Sigma-point kalman filters for probabilistic inference in dynamic state-space models. Ph.D. thesis, University of Stellenbosch (2004)
22. Pauwels, K., Van Hulle, M.: Segmenting independently moving objects from egomotion flow fields. Isle of Skye, Scotland (2004)
23. Pauwels, K., Van Hulle, M.: Optic flow from unstable sequences containing unconstrained scenes through local velocity constancy maximization. pp. 397–406. Edinburgh (2006)
24. Pohl, C.: Review article multisensor image fusion in remote sensing: concepts, methods and applications. International Journal of Remote Sensing **19**(5), 823–854 (1998)
25. Sabatini, S., Gastaldi, G., Solari, F., Diaz, J., Ros, E., Pauwels, K., Van Hulle, M., Pugeault, N., Krueger, N.: Compact and accurate early vision processing in the harmonic space. Barcelona (2007)

26. Sergi, M.: Bus rapid transit technologies: A virtual mirror for eliminating vehicle blind zones. University of Minnesota ITS Institute Final Report (2003)
27. Sole, A., Mano, O., Stein, G., Kumon, H., Tamatsu, Y., Shashua, A.: Solid or not solid: vision for radar target validation. Intelligent Vehicles Symposium, 2004 IEEE pp. 819–824 (2004)
28. Steux, B., Laurgeau, C., Salesse, L., Wautier, D.: Fade: a vehicle detection and tracking system featuring monocular color vision and radar data fusion. Intelligent Vehicle Symposium, 2002. IEEE **2** (2002)
29. Stiller, C., Hipp, J., Rossig, C., Ewald, A.: Multisensor obstacle detection and tracking. Image and Vision Computing **18**(5), 389–396 (2000)
30. Sun, Z., Bebis, G., Miller, R.: On-road vehicle detection: a review. Pattern Analysis and Machine Intelligence, IEEE Transactions on **28**(5), 694–711 (2006)
31. Thrun, S., Montemerlo, M., Dahlkamp, H., Stavens, D., Aron, A., Diebel, J., Fong, P., Gale, J., Halpenny, M., Hoffmann, G., et al.: Stanley: The robot that won the darpa grand challenge. Journal of Field Robotics **23**(9), 661–692 (2006)
32. Viola, P., Jones, M.: Rapid object detection using a boosted cascade of simple features. Proceedings of CVPR **1**, 511–518 (2001)
33. Wald, L.: Some terms of reference in data fusion. IEEE Transactions on Geoscience and Remote Sensing **37**(3) (1999)

10

Distributed Vision Networks for Human Pose Analysis

Hamid Aghajan, Chen Wu, and Richard Kleihorst

Multi-camera networks offer potentials for a variety of novel human-centric applications through provisioning of rich visual information. Local processing of acquired video at the source camera facilitates operation of scalable vision networks by avoiding transfer of raw images. Additional motivation for distributed processing stems from an effort to preserve privacy of the network users while offering services in applications such as assisted living. Yet another benefit of processing the images at the source is the flexibility it offers on the type of features and the level of data exchange between the cameras in a collaborative processing framework. In such a framework data fusion can occur across the three dimensions of 3D space (multiple views), time, and feature levels.

In this chapter collaborative processing and data fusion mechanisms are examined in the context of a human pose estimation framework. For efficient collaboration between the cameras under a low-bandwidth communication constraint, only concise descriptions of extracted features instead of raw images are communicated. A 3D human body model is employed as the convergence point of the spatiotemporal and feature fusion. The model also serves as a bridge between the vision network and the high-level reasoning module, which can extract gestures and interpret them against the user's context and behavior models to arrive at system-level decisions. The human body model also allows the cameras to interact with one another on the initialization of feature extraction parameters, or to evaluate the relative value of their derived features.

10.1 Introduction

In a camera network, access to multiple sources of visual data allows for more comprehensive interpretations of events and activities. Vision-based sensing fits well within the notion of pervasive sensing and computing environments enabling novel interactive user-based applications that do not require users to wear sensors. Having access to interpretations of posture and gesture elements obtained from visual data over time enables higher-level reasoning modules to deduce the user's actions, context, and behavior models, and decide upon suitable actions or responses to the situation.

The increasing interest in understanding human behaviors and events in a camera context has heightened the need for gesture analysis of image sequences. Gesture recognition problems have been extensively studied in human computer interfaces (HCI), where often a set of pre-defined gestures are used for delivering instructions to machines [7, 11]. However, "passive gestures" predominate in behavior descriptions in many applications. Some traditional application examples include surveillance and security applications, while novel classes of applications arise in emergency detection in elderly care and assisted living [1, 26], video conferencing [16, 17], creating human models for gaming and virtual environments [9], and biomechanics applications analyzing human movements [15].

A variety of methods have focused on 3D model reconstruction from a *single* camera [4, 21]. They have demonstrated great capability to track human gestures in many situation. However, there lies the intrinsic inability of monocular cameras to fully explain 3D objects. Ambiguities result from either perspective views of the cameras or self-occlusion of the human body [25]. In multiple camera scenarios, ambiguity can be reduced by finding the best match for all views or reconstructing the 3D visual hull [3, 14, 23].

In this chapter we introduce a model-based data fusion framework for human posture analysis which employs multiple sources of vision-based information obtained from the camera network. The fusion framework spans the three dimensions of space (different camera views), time (each camera collecting data over time), and feature levels (selecting and fusing different feature subsets). Furthermore, the chapter outlines potentials for interaction between the distributed vision network and the high-level reasoning system.

The structure of the vision-based processing operation has been designed in such a way that the lower-level functions as well as other in-node processing operations will utilize feedback from higher levels of processing. While feedback mechanisms have been studied in active vision areas, our approach aims to incorporate interactions between the vision and the high-level reasoning or artificial intelligence (AI) modules as the source of active vision feedback. To facilitate such interactions, we introduce a human model as the convergence point and a bridge for the two sides, enabling both to incorporate the results of their deductions into a single merging entity. For the vision network, the human model acts as the embodiment of the fused visual data contributed by the multiple cameras over observation periods. For the AI-based functions, the human model acts as the carrier of all the sensed data from which gesture interpretations can be deduced over time through rule-based methods or mapping to training data sets of interesting gestures. Figure 10.1 illustrates this concept in a concise way.

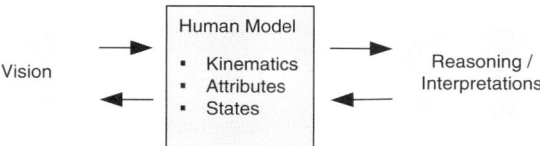

Fig. 10.1. The human model bridges the vision module and the reasoning module, as the interactive embodiment

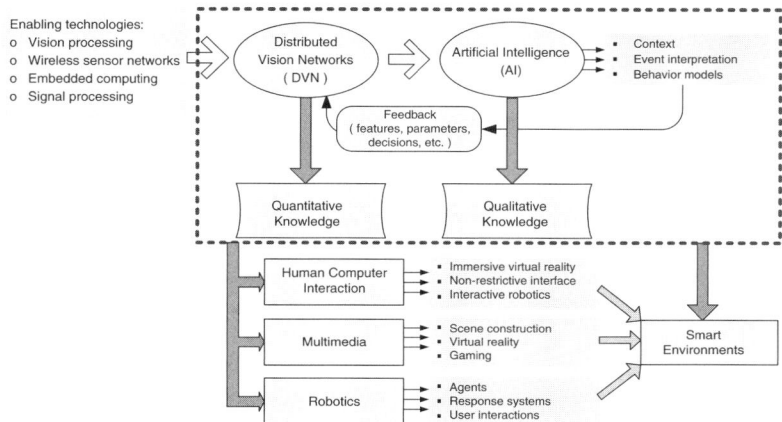

Fig. 10.2. The relationship between vision networks and high-level AI reasoning, and a variety of novel applications enabled by both

10.2 A Unifying Framework

Our notion of the role a vision network can play in enabling novel intelligent applications derives from the potential interactions between the various disciplines outlined in Fig. 10.2. The vision network offers access to quantitative knowledge about the events of interest such as the location and other attributes of a human subject. Such quantitative knowledge can either complement or provide specific qualitative distinctions for AI-based functions. On the other hand, we may not need to extract all the detailed quantitative knowledge available in visual data since often a coarse qualitative representation may be sufficient in addressing the application [2]. In turn, qualitative representations can offer clues to the features of interest to be derived from the visual data allowing the vision network to adjust its processing operation according to the interpretation state. Hence, the interaction between the vision processing module and the reasoning module can in principle enable both sides to function more effectively. For example, in a human gesture analysis application, the observed elements of gesture extracted by the vision module can assist the AI-based reasoning module in its interpretative tasks, while the

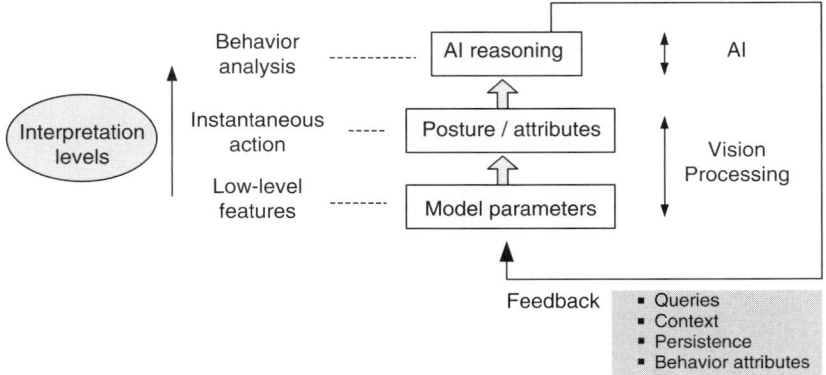

Fig. 10.3. Interpretation focuses on different levels from vision to behavior reasoning

deductions made by the high-level reasoning system can provide feedback to the vision system from the available context or behavior model knowledge.

Figure 10.3 shows the relationship between the low-level vision processing, which occurs in the camera nodes, the instantaneous state resulting from camera collaboration in the visual domain, and the high-level behavior interpretation which is performed in the AI module. The feedback elements provided by the AI module help the vision processing system to direct its processing effort toward handling the more interesting features and attributes.

10.3 Smart Camera Networks

The concept of feedback flow from higher-level processing units to the lower-level modules also applies when considering the vision network itself. Within each camera, temporal accumulation of features over a period of time can, for example, enable the camera to examine the persistence of those features, or to avoid re-initialization of local parameters. In the network of cameras, spatial fusion of data in any of the forms of merged estimates or a collective decision, or in our model-based approach in the form of updates from body part tracking, can provide feedback information to each camera. The feedback can, for example, be in the form of indicating the features of interest that need to be tracked by the camera, or as initialization parameters for the local segmentation functions. Figure 10.4 illustrates the different feedback paths within the vision processing unit.

Our work aims for intelligent and efficient vision-based interpretations in a camera network. One underlying constraint of the network is the rather low bandwidth. Therefore, for efficient collaboration between cameras, we expect concise descriptions instead of raw image data as outputs from local processing in a single camera. This process inevitably removes certain details in images

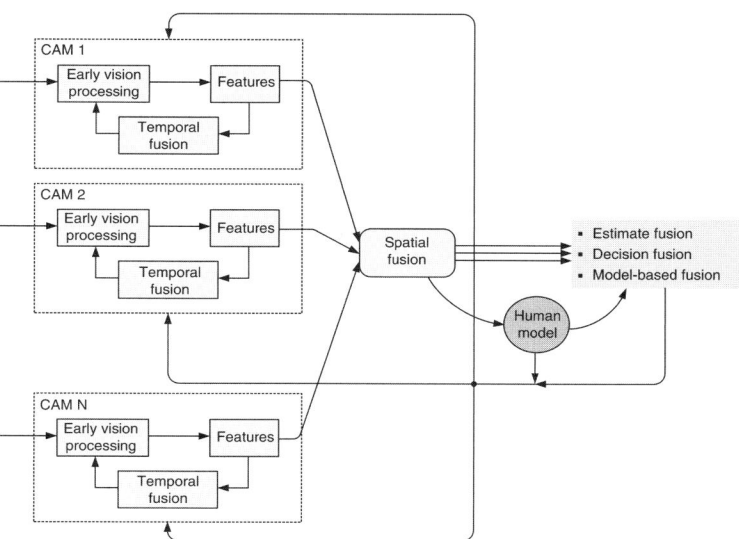

Fig. 10.4. Different feedback paths within distributed vision processing units

of a single camera, which requires the camera to have some "intelligence" on its observations (smart camera), i.e., some knowledge of the subject. This derives one of the motivations for opportunistic data fusion between cameras, which compensates for partial observations in individual cameras. So the output from opportunistic data fusion (a model of the subject) is fed to local processing. On the other hand, outputs of local processing in single cameras enable opportunistic data fusion by contributing local descriptions from multiple views. It is the interactive loop that brings in the potential for achieving both efficient and adequate vision-based analysis in the camera network.

10.4 Opportunistic Fusion Mechanisms

The opportunistic fusion framework for gesture analysis is conceptually illustrated in Fig. 10.5. At each observation time the cameras derive relevant features from their own observations, which are exchanged at a possibly lower rate with other cameras. This results in spatial fusion of the features towards describing the body model. As time progresses, the model is modified based on the new observations, while at each new observation instance, the most recent composition of the model is input to both the spatial fusion module as well as the in-node feature extraction module. The spatial fusion module employs the knowledge about the model to initialize any search operation for tracking the movement of the body parts. The in-node feature extraction module uses the information accumulated in the model to initialize segmentation functions.

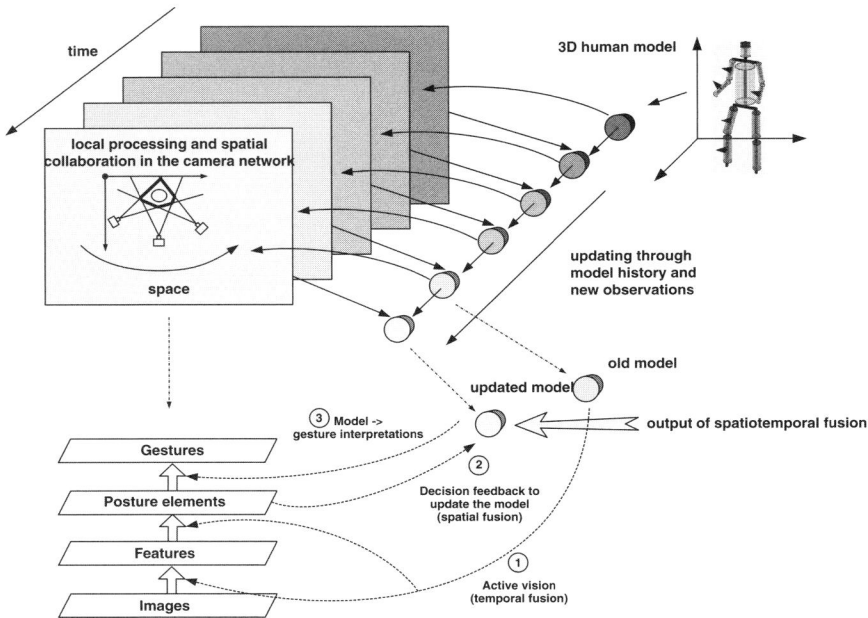

Fig. 10.5. Spatiotemporal fusion for human gesture analysis

The spatial fusion operation can be composed of layers of processing from low-level vision to extract features, to intermediate levels in which partially processed data is exchanged between the cameras, and to higher-level vision processing functions that are employed to make deductions about the observations. The lower part of Fig. 10.5, which is detailed in Fig. 10.6 illustrates this layered operation. As an example of how functions in different layers are defined, in the bottom layer of a gesture analysis operation, image features are extracted from local processing. No explicit collaboration between cameras is done in this stage since communication is not expected until images/videos are reduced to short descriptions. Distinct features (e.g., colors) specific for the subject are registered in the current model M_0 and are used for analysis. This allows for targeting relevant features in the early vision operations (arrow ① in Fig. 10.5). After local processing, data is shared between cameras to derive a new estimate of the model. Decisions from spatial fusion of cameras are used to update the model (arrow ② in Fig. 10.5). Therefore, for every update of the model, the method combines spatial (multiple views), temporal (the previous model) and feature levels (choice of image features in local processing from new observations and subject-specific attributes in the model). Finally, the new model is used for high-level gesture deductions in the certain application scenario (arrow ③ in Fig. 10.5).

An implementation for the 3D human body posture estimation is illustrated in Fig. 10.7. Local processing in single cameras includes segmentation

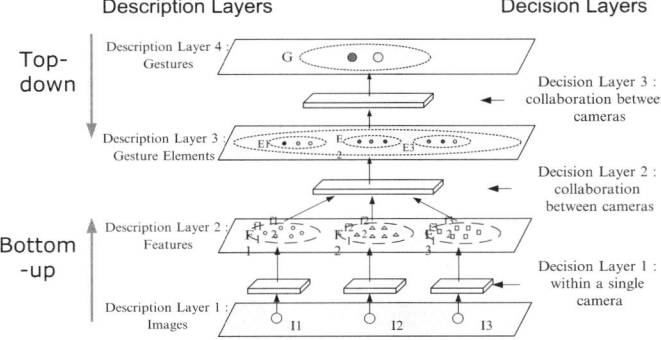

Fig. 10.6. The layered and collaborative architecture of the gesture analysis system. I_i stands for images taken by camera i; F_i is the feature set for I_i; E_i is the posture element set in camera i; and G is the set of possible gestures

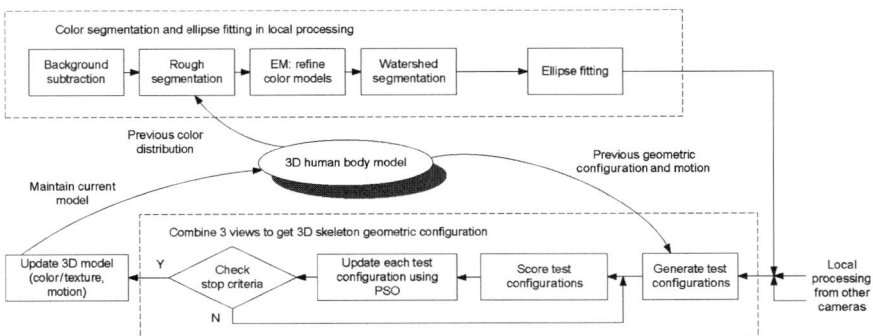

Fig. 10.7. Algorithm flowchart for 3D human skeleton model reconstruction

and ellipse fitting for a concise parameterization of segments. For spatial collaboration, ellipses from all cameras are merged to find the geometric configuration of the 3D skeleton model. These operations are described in more detail in Sects. 10.7 and 10.8.

10.5 Human Posture Estimation

All posture estimation approaches are essentially based on matching between the body model and image observations. The approaches can be generally categorized in two types, *top-down* and *bottom-up*. The common goal is to reconstruct a 3D model from 2D images.

In the top-down approach, a 3D model configuration is first assumed, then a similarity evaluation function is calculated between projections of the 3D

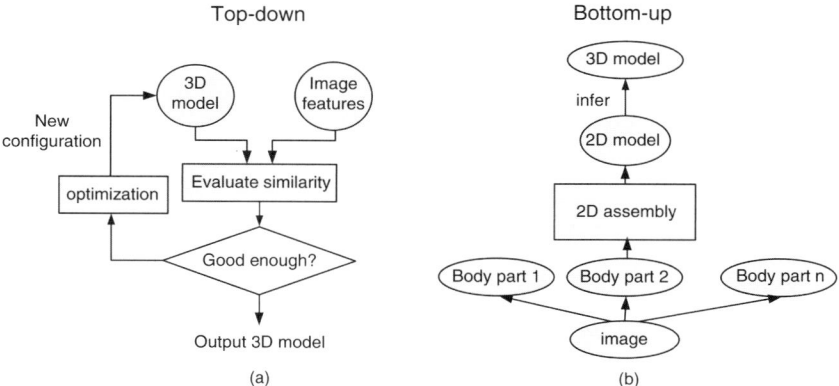

Fig. 10.8. Top-down and bottom-up approaches for human body model reconstruction

model and the images (see Fig. 10.8a). In [8] the model is projected into four views and the edge contours are compared with image edge maps. The search space is decomposed into body parts and a non-greedy algorithm is applied to look for the best match. In [10] the 3D model is evolved through particle swarm optimization (PSO) while the evaluation function compares silhouettes from the image and silhouettes from model projections. Another group of work which uses shape context or template matching also belongs to the top-down approach. In [5] templates representing underlying 3D postures are used to detect humans in the image. The templates are generated through motion captured data.

In the bottom-up approach, instead of the assumption-validation scheme, different features are first sought in the images, then they assemble into the 3D model (see Fig. 10.8b). One category of methods first reconstructs the visual hull from a set of calibrated cameras. Mikic et al. [14] start from the silhouettes to reconstruct the voxel data. At the beginning the model is acquired based on template fitting and growing to the voxel data, then a tracker is applied for motion tracking. Similarly, in [18] 3D stereo data are taken as features for model fitting. A different type of work looks for different body parts and then assembles them to recover the posture [6, 24]. Sigal et al. in [24] detect limb segments candidates in the image, on which a 2D loose-limbed body model is searched. The 3D body model is estimated based on a Mixture of Experts model to represent the conditionals of the 3D body state given a 2D body state.

Both the top-down and bottom-up approaches have their advantages and weaknesses. In the top-down approach, knowing the correspondence between projections and the 3D model, one can find the optimal 3D configuration as long as matching scores are available. Top-down approaches also naturally handle occlusions since that is taken care of by the projection process. Association between projected contours or silhouettes and body parts is known as well. The complexity of the top-down approach lies in the searching technique. A number of methods such as partitioned sampling, annealing, particle swarm filtering, etc., have been adopted. Due to the high dimensionality involved in the human model, usually a hierarchical search is implemented based on a subset of body parts. In addition, each of the 3D configurations needs to be projected to each image plane, resulting in much computational load. An alternative would be to save these projections as templates, which may require big capacity to store the templates and a quick access to retrieve them.

Switching between the image and the 3D model is much less of a demand in the bottom-up approach. However, after the detection of body part candidates it will be a complicated task to assemble them into a certain 3D configuration within the model constraints. Another issue is the difficulty in detection of occlusions without an assumption of the 3D configuration.

10.6 The 3D Human Body Model

Fitting human models to images or videos has been an interesting topic for which a variety of methods have been developed. Usually assuming a dynamic model (such as walking) [4, 22] will greatly help to predict and validate the posture estimates. But tracking can easily fail in case of sudden motion or other movements that differ much from the dynamic model. Therefore, we always need to be aware of the balance between the limited dynamics and the capability to discover more diversified postures. For multi-view scenarios, a 3D model can be reconstructed by combining observations from different views [3, 13]. Most methods start from silhouettes in different cameras and find the points occupied by the subject. Then they fit a 3D model with principle body parts in the 3D space [14]. This approach is relatively "clean" since the only image components it is based on are the silhouettes. But at the same time the 3D voxel reconstruction is sensitive to the quality of the silhouettes and accuracy of camera calibrations. It is not difficult to find situations where background subtraction for silhouettes suffers for quality or is almost impossible (clustered, complex background, and when the subject is wearing clothes with similar colors to the background).

Another aspect of the human model fitting problem is the choice of image features. All human model fitting methods are based on some image features as targets to fit the model. Most of them are based on generic features such as silhouettes or edges [13, 20]. Some use skin color but are prone to failure since lighting often has a big influence on color and skin color varies from person to person.

In our work, we aim to incorporate appearance attributes adaptively learned by the network for initialization of segmentation. Another emphasis of our work is that images from a single camera are first reduced to short descriptions, which are exchanged in the network to reconstruct a 3D human model. Hence, concise descriptions are the expected outputs from the in-node processing operation.

The 3D model maps to the Posture Elements layer in the layered architecture for gesture analysis (Fig. 10.6) we proposed in [29]. However, here it not only assumes spatial collaboration between cameras, but also connects decisions from previous observations with current observations.

As such, the 3D human body model embodies up-to-date information from both current and historical observations of all cameras in a concise way (Fig. 10.5). It has the following components: (1) Geometric configuration: body part lengths, angles. (2) Color or texture of body parts. (3) Motion of body parts. All the three components are updated from the three dimensions of space, time, and feature levels. The 3D human model takes up two roles. One is as an intermediate step for high-level application-pertinent gesture interpretation , and the other is to create a feedback path from spatiotemporal and feature fusion operations to the low-level vision processing in each camera. It is true that for a number of gesture applications a human body model may not be needed to interpret the gesture. There is existing work for hand gesture recognition [28] where only part of the body is analyzed. Some gestures can also be detected through spatiotemporal motion patterns of some body parts [12, 19]. However, as the set of gestures to differentiate in an application expands, it becomes increasingly difficult to devise methods for gesture recognition based on only a few cues. A 3D human body model provides a unified interface for a variety of gesture interpretations. On the other hand, instead of being a passive output to represent decisions from spatiotemporal and feature fusion, the 3D model implicitly enables more interactions between the three collaboration dimensions by being actively involved in vision analysis. For example, although predefined appearance attributes are generally not reliable, adaptively learned appearance attributes can be used to identify the person or body parts. Those attributes are usually more distinguishable than generic features such as edges. An example of the communication model between five cameras to reconstruct the person's model is shown in Fig. 10.9. The circled numbers represent the sequence of events.

10.7 In-Node Feature Extraction

The goal of local processing in a single camera is to reduce raw images/videos to simple descriptions so that they can be efficiently transmitted between cameras. The output of the algorithm will be ellipses fitted to the segments representing body parts, and the mean color of the segments. As shown in the upper part of Fig. 10.7, local processing includes image segmentation for the subject and ellipse fitting to the extracted segments.

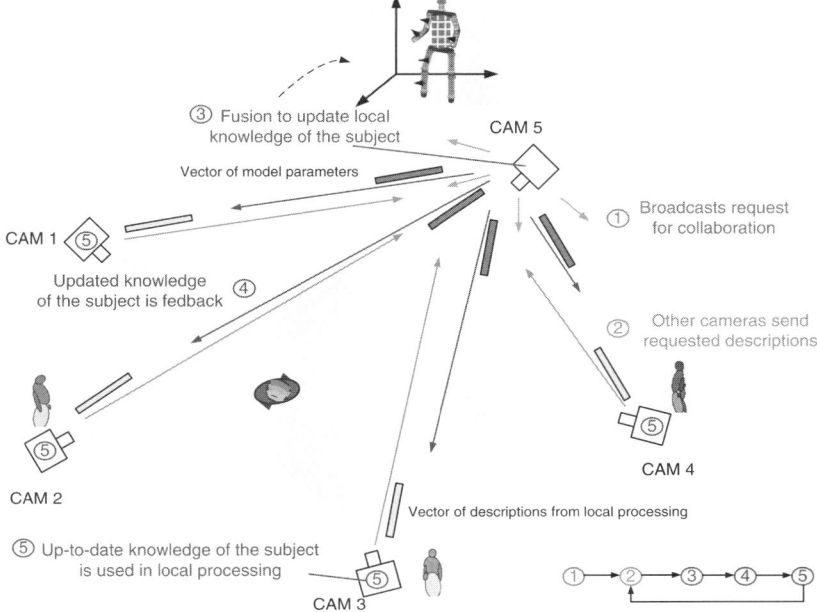

Fig. 10.9. Communication for collaboration in the camera network

We assume the subject is characterized by a distinct color distribution. A preliminary estimate of the foreground area is obtained through background subtraction. Pixels with high or low illumination are also removed since for those pixels chrominance may not be reliable. Then an improved segmentation for the foreground is achieved either based on K-means on the chrominance of the foreground pixels or color distributions from the current model. The color distribution maintained by the model may not be accurate for all cameras due to illumination variations. Also the subject's appearance may change due to movement or lighting conditions. Therefore, the color distribution of the model is only used for a rough segmentation in initialization of the segmentation scheme. Then an Expectation Maximization (EM) algorithm is used to refine the color distribution for the current image. The initial estimated color distribution plays an important role because it can prevent EM from being trapped in local minima.

Suppose the color distribution is a mixture of N Gaussian modes, with parameters $\Theta = \{\theta_1, \theta_2, \ldots, \theta_N\}$, where $\theta_l = \{\mu_l, \Sigma_l\}$ are the means and covariance matrices of the modes. Mixing weights of the different modes are $A = \{\alpha_1, \alpha_2, \ldots, \alpha_N\}$. The EM algorithm aims to find the probability of each pixel x_i belonging to a certain mode θ_l: $\Pr(y_i = l | x_i)$.

The basic EM algorithm takes each pixel independently, without considering the fact that pixels belonging to the same mode are usually spatially close to each other. In [27] perceptually organized EM (POEM) is introduced. In POEM, influence of neighbors is incorporated by a weighting measure

$$w(x_i, x_j) = e^{-\frac{\|x_i - x_j\|}{\sigma_1^2} - \frac{\|s(x_i) - s(x_j)\|}{\sigma_2^2}} \tag{10.1}$$

where $s(x_i)$ is the spatial coordinate of x_i. Then "votes" for x_i from the neighborhood are given by

$$V_l(x_i) = \sum_{x_j} \alpha_l(x_j) w(x_i, x_j), \text{ where } \alpha_l(x_j) = Pr(y_j = l | x_j). \tag{10.2}$$

Then modifications are made to EM steps. In the E step, $\alpha_l^{(k)}$ is changed to $\alpha_l^{(k)}(x_i)$, which means that for every pixel x_i, mixing weights for different modes are different. This is partially due to the influence of neighbors. In the M step, mixing weights are updated by

$$\alpha_l^{(k)}(x_i) = \frac{e^{\eta V_l^{(x_i)}}}{\sum_{k=1}^{N} e^{\eta V_k^{(x_i)}}} \tag{10.3}$$

η controls the "softness" of neighbors' votes. If η is as small as 0, then mixing weights are always uniform. If η approaches infinity, the mixing weight for the mode with the largest vote will be 1.

After refinement of the color distribution with POEM, we set pixels with high probability (e.g., bigger than 99.9%) that belong to a certain mode as markers for that mode. Then a watershed segmentation algorithm is implemented to assign labels for undecided pixels. Finally for every segment an ellipse is fitted to it to obtain a concise parameterization for the segment. Figure 10.10 includes examples of the segmentation and ellipse fitting results for a few frames.

10.8 Collaborative Posture Estimation

Human posture estimation is essentially treated as an optimization problem in which we aim to minimize the distance between the posture and the sets of ellipses from the multiple cameras. There can be several different ways to find the 3D skeleton model based on observations from multi-view images. One method is to directly solve for the unknown parameters through geometric calculations. In this method one needs to first establish correspondences between points/segments in different cameras, which is itself a hard problem. Common observations for points are rare for human problems, and body parts may take on very different appearances from different views. Therefore, it is difficult to resolve ambiguity in the 3D space based on 2D observations. A second method would be to treat this as an optimization problem, in which we find optimal θ_i's and ϕ_i's to minimize an objective function (e.g., the

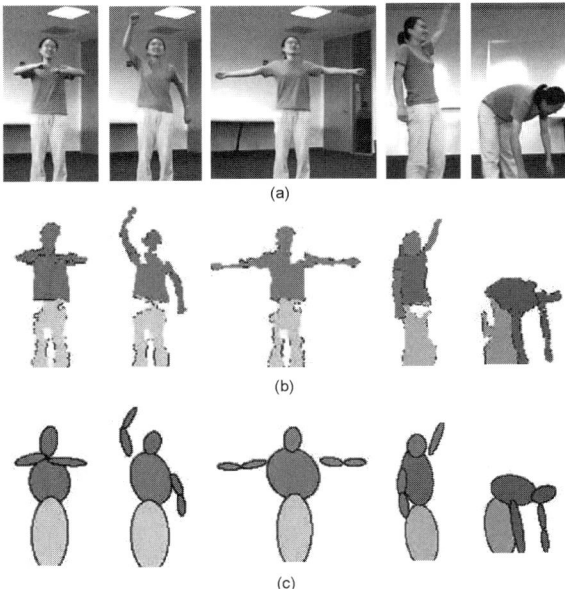

Fig. 10.10. Examples for in-node segmentation results. (**a**) Original frames. (**b**) Segmentation results. (**c**) Ellipse fitting results

difference between projections due to a certain 3D model and the actual segments) based on properties of the objective function. However, if the problem is highly nonlinear or non-convex, it may be difficult or time consuming to solve. Therefore, searching strategies which do not explicitly depend on the objective function formulation are desired.

Motivated by [10, 18], particle swarm optimization (PSO) is used for our optimization problem. The lower part of Fig. 10.7 shows the estimation process. Ellipses from local processing of single cameras are merged together to reconstruct the skeleton (Fig. 10.11). Here, we consider a simplified problem in which only arms change in position while other body parts are kept in the default location. Elevation angles (θ_i) and azimuth angles (ϕ_i) of the left/right, upper/lower parts of the arms are specified as the parameters (Fig. 10.11b). The assumption is that projection matrices from 3D skeleton to 2D image planes are known. This can be achieved either from locations of cameras and the subject, or it can be calculated from some known projective correspondences between the 3D subject and points in the images, without knowing the exact locations of the cameras or the subject.

PSO is suitable for posture estimation as an evolutionary optimization mechanism. It starts from a group of initial particles. During the evolution the particles are directed to the optimal position while keeping some randomness to explore the search space. Suppose there are N particles (test configurations)

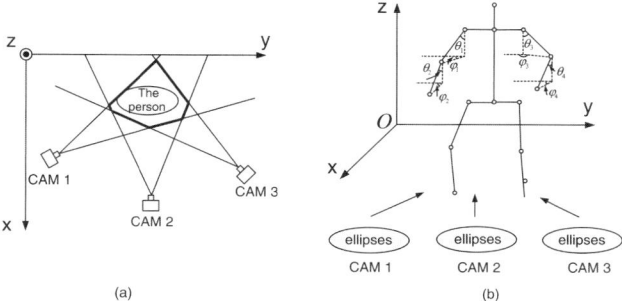

Fig. 10.11. 3D skeleton model fitting. (a) Top view of the experiment setting. (b The 3D skeleton reconstructed from ellipses from multiple cameras

x_i, each being a vector of θ_i's and ϕ_i's. The velocity of x_i is denoted by v_i. Assume the best position of x_i up to now is $\hat{x}_i$, and the global best position of all x_i's up to now is g. The objective function is $f(\cdot)$ for which we wish to find the optimal position x to minimize $f(x)$. The PSO algorithm is as follows:

1. Initialize x_i and v_i. The value of v_i is usually set to 0, and $\hat{x}_i = x_i$. Evaluate $f(x_i)$ and set $g = \arg\min f(x_i)$.
2. While the stop criterion is not satisfied, do for every x_i:
   - $v_i \leftarrow \omega v_i + c_1 r_1 (\hat{x}_i - x_i) + c_2 r_2 (g - x_i)$
   - $x_i \leftarrow x_i + v_i$
   - If $f(x_i) < f(\hat{x}_i)$, $\hat{x}_i = x_i$; If $f(x_i) < f(g)$, $g = x_i$
   - *The stop criterion:* After updating all N x_i's once, the increase in $f(g)$ falls below a threshold, then the algorithm exits.

Here ω is the "inertia" coefficient, while c_1 and c_2 are the "social" coefficients. r_1 and r_2 are random vectors with each element uniformly distributed on [0,1]. Choice of ω, c_1, and c_2 controls the convergence process of the evolution. If ω is big, the particles have more inertia and tend to keep their own directions to explore the search space. This allows for more chances of finding the "true" global optimal if the group of particles is currently around a local optimal. While if c_1 and c_2 are big, the particles are more "social" with the other particles and go quickly to the best positions known by the group. In our experiment, $N = 16$, $\omega = 0.3$, and $c_1 = c_2 = 1$.

Similar to other search techniques, PSO will be likely to converge to a local optimum without carefully choosing the initial particles. In the experiment, we assume that the 3D skeleton will not go through a big change in a time interval. Therefore, at time t_1 the search space formed by the particles is centered around the optimal solution of the geometric configuration at time t_0. That is, time consistency in postures is used to initialize particles for searching. Some examples showing images from three views and the posture estimates are shown in Fig. 10.12.

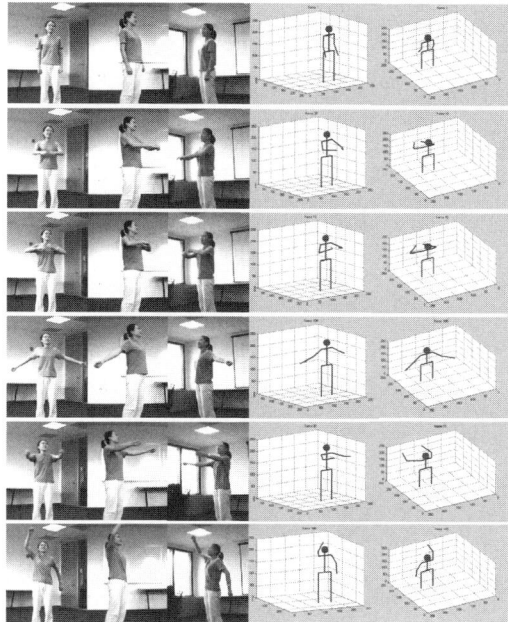

Fig. 10.12. Experiment results for 3D skeleton reconstruction. Original images from three camera views and the skeletons

10.9 Towards Behavior Interpretation

An appropriate classification is essential towards a better understanding of the variety of passive gestures. Therefore, we propose a categorization of the gestures as follows:

- Static gestures, such as standing, sitting, lying
- Dynamic gestures, such as waving arms, jumping
- Interactions with other people, such as chatting
- Interactions with the environment, such as dropping or picking up objects

Figure 10.6 illustrated the layered processing architecture defining collaboration stages between the cameras and the levels of vision-based processing from early vision towards discovery of the gesture elements.

To illustrate the process of achieving high-level reasoning using the collaborative vision-based architecture, we consider an application in assisted living in which the posture of the user (which could be an elderly or a patient) is monitored during daily activities for detection of abnormal positions such as lying down on the ground. Each of the cameras in the network employs local vision processing on its acquired frames to extract the silhouette of the person. A second level of processing employs temporal smoothing combined with shape fitting to the silhouette and estimates the orientation and the aspect

196   H. Aghajan et al.

ratio of the fitted (e.g. elliptical) shape. The network's objective at this stage is to decide on one of the branches in the top level of a tree structure (see Fig. 10.13) between the possible posture values of *vertical, horizontal,* or *undetermined*. To this end, each camera uses the orientation angle and the aspect ratio of the fitted ellipse to produce an *alert level*, which ranges from −1 (for safe) to 1 (for danger). Combining the angle and the aspect ratio is based on the assumption that nearly vertical or nearly horizontal ellipses with aspect ratios away from 1 provide a better basis for choosing one of the *vertical* and

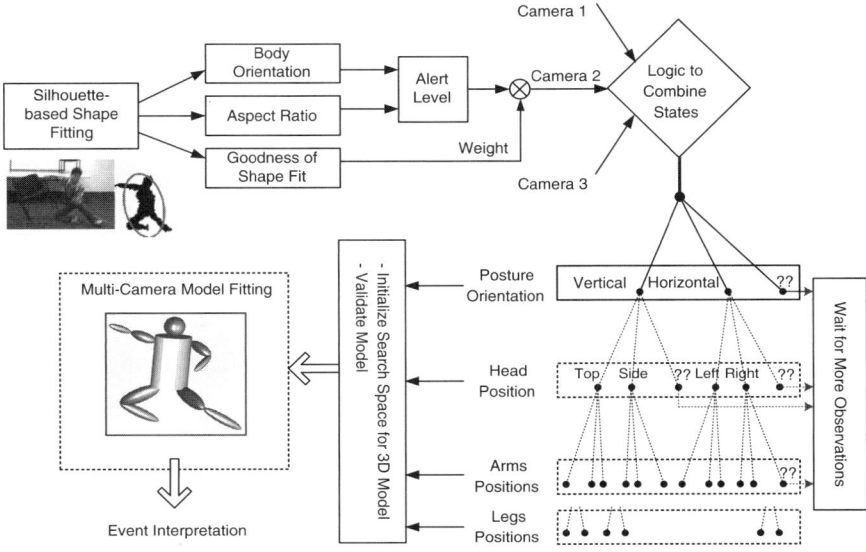

Fig. 10.13. A tree-based reasoning technique for fall detection. Qualitative descriptions can trace down the branches for specific event detection. The specific deductions can also be fed back to initialize posture reconstruction

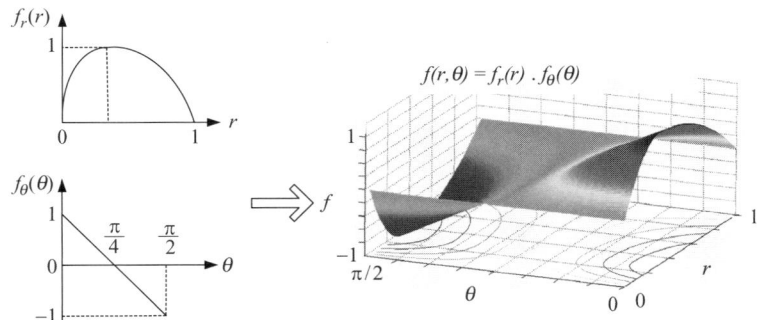

Fig. 10.14. The alert level functions based on the aspect ratio and the orientation angle of fitted ellipses

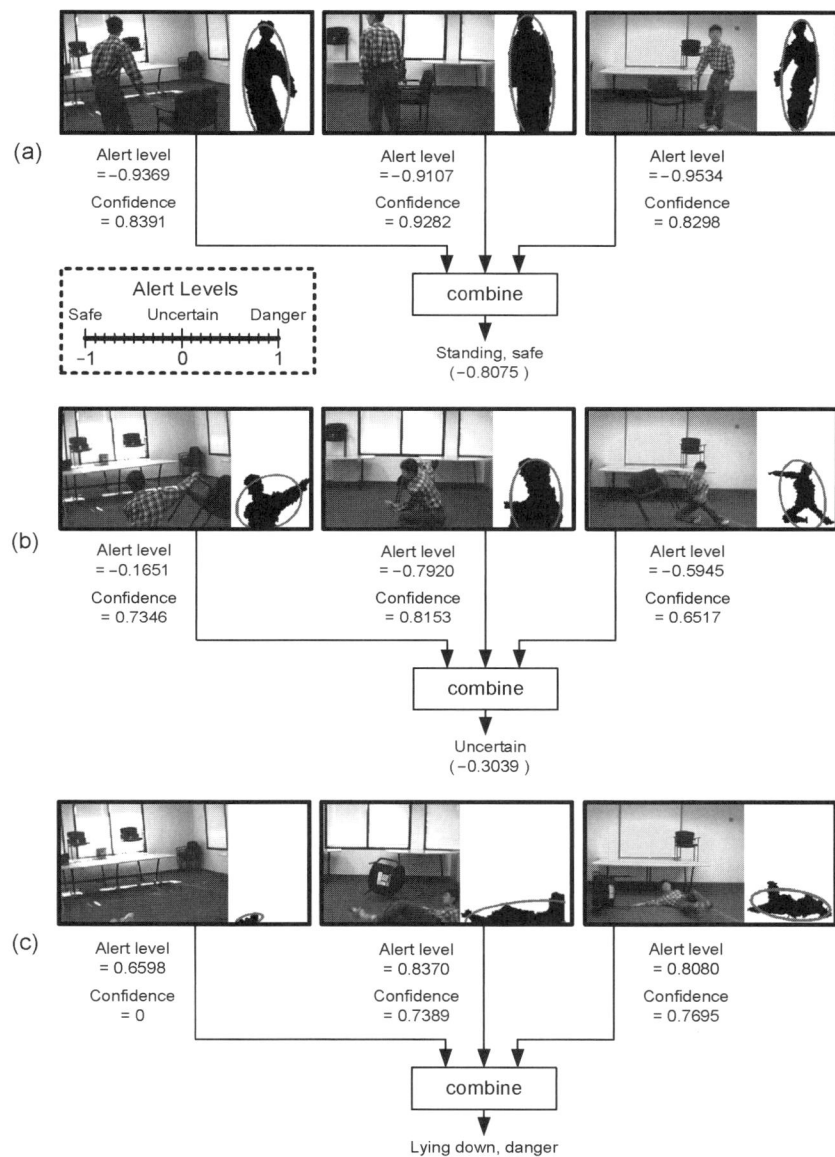

Fig. 10.15. Three sets of examples from three cameras of different views for fall detection. (**a**) Standing; (**b**) falling; (**c**) lying on the ground. Alert levels and their confidence levels are shown. After combining observations from the three cameras a final score is given indicating whether the person is standing (safe) or lying (danger)

horizontal branches in the decision tree than when the aspect ratio is close to one or when the ellipse has for example, a 45-degree orientation.

Figure 10.14 illustrates an example of the alert level function combining the orientation and aspect ratio attributes in each camera. The camera broadcasts the value of this function for the collaborative decision making process. Along with the alert level, the camera also produces a figure of merit value for the shape fitted to the human silhouette. The figure of merit is used as a weighting parameter when the alert level values declared by the cameras are combined.

Figure 10.15 presents cases in which the user is walking, falling, and lying down. The posture detection outcome is superimposed on the silhouette of the person for each camera. The resulting alert levels and their respective weights are shared by the cameras, from which the overall alert level shown in the figure is obtained.

The analysis presented here is an example of instantaneous interpretation of visual observations based on comparison with generic reference models. The aim of our continued work is to introduce two factors in the analysis to render the decision making process specific to the user. One is to replace the generic models with those obtained from long-term observations of the user, and the other is to connect the set of instantaneous observations into a gestural sequence which can then be compared with the database of gestures that have importance to be detected for the user. Such a database can be composed from long-term observations and trained decisions.

10.10 Conclusions

A framework of data fusion in distributed vision networks is proposed. Motivated by the concept of opportunistic use of available information across the different processing and interpretation levels, the proposed framework has been designed to incorporate interactions between the vision module and the high-level reasoning module. Such interactions allow the quantitative knowledge from the vision network to provide specific qualitative distinctions for AI-based problems, and in turn, allows the qualitative representations to offer clues to direct the vision network to adjust its processing operation according to the interpretation state. Two vision-based fusion algorithms were presented, one based on reconstructing the full-parameterized human model and the other based on a sequence of direct deductions about the posture elements in a fall detection application.

The current work includes incorporation of body part motion into the full-parameterized human body model allowing the model to carry the gesture elements in interactions between the vision network and the high-level reasoning module. Other extensions of interest include creating a link from the human model to the reduced qualitative description set for a specific application, and utilizing deductions made by the AI system as a basis for active vision in multi-camera settings.

References

1. Aghajan, H., Augusto, J., Wu, C., McCullagh, P., Walkden, J.: Distributed vision-based accident management for assisted living. In: ICOST 2007. Nara, Japan
2. Gottfried, B., Guesgen, H.W., Hübner, S.: Designing Smart Homes, chap. Spatiotemporal Reasoning for Smart Homes, pp. 16–34. Springer, Berlin Heidelberg New York (2006)
3. Cheung, K.M., Baker, S., Kanade, T.: Shape-from-silhouette across time: Part ii: Applications to human modeling and markerless motion tracking. International Journal of Computer Vision **63**(3), 225–245 (2005)
4. Deutscher, J., Blake, A., Reid, I.: Articulated body motion capture by annealed particle filtering. International Journal of Computer Vision **II**, 126–133 (2000)
5. Dimitrijevic, M., Lepetit, V., Fua, P.: Human body pose detection using bayesian spatio-temporal templates. Computer Vision and Image Understanding **104**(2), 127–139 (2006). DOI http://dx.doi.org/10.1016/j.cviu.2006.07.007
6. Felzenszwalb, P.F., Huttenlocher, D.P.: Efficient matching of pictorial structures. In: CVPR (2000)
7. Ye, G. Corso, J.J., Hager, G.D.: Real-Time Vision for Human-Computer Interaction, chap. 7: Visual Modeling of Dynamic Gestures Using 3D Appearance and Motion Features, pp. 103–120. Springer, Berlin Heidelberg New York (2005). URL gyeHCI2005.pdf
8. Gavrila, D.M., Davis, L.S.: 3-D model-based tracking of humans in action: A multi-view approach. In: CVPR (1996)
9. Hilton, A., Beresford, D., Gentils, T., Smith, R., Sun, W., Illingworth, J.: Whole-body modelling of people from multi-view images to populate virtual worlds. Visual Computer International Journal of Computer Graphics **16**(7), 411–436 (2000)
10. Ivecovic, S., Trucco, E.: Human body pose estimation with pso. In: IEEE Congress on Evolutionary Computation, pp. 1256–1263 (2006)
11. Kwolek, B.: Visual system for tracking and interpreting selected human actions. In: WSCG (2003)
12. Liu, Y., Collins, R., Tsin, Y.: Gait sequence analysis using frieze patterns. In: Proceedings of the 7th European Conference on Computer Vision (ECCV'02) (2002)
13. Ménier, C., Boyer, E., Raffin, B.: 3d skeleton-based body pose recovery. In: Proceedings of the 3rd International Symposium on 3D Data Processing, Visualization and Transmission, Chapel Hill, USA (2006). URL http://perception.inrialpes.fr/Publications/2006/MBR06
14. Mikic, I., Trivedi, M., Hunter, E., Cosman, P.: Human body model acquisition and tracking using voxel data. International Journal of Computer Vision **53**(3), 199–223 (2003). DOI http://dx.doi.org/10.1023/A:1023012723347
15. Muendermann, L., Corazza, S., Andriacchi, T.: The evolution of methods for the capture of human movement leading to markerless motion capture for biomechanical applications. Journal of NeuroEngineering and Rehabilitation **3**(1) (2006)
16. Patil, R., Rybski, P.E., Kanade, T., Veloso, M.M.: People detection and tracking in high resolution panoramic video mosaic. In: Proceedings of IEEE/RSJ International Conference on Intelligent Robots and Systems (IROS), vol. 1, pp. 1323–1328 (2004)

17. Robertson, C., Trucco, E.: Human body posture via hierarchical evolutionary optimization. In: BMVC'06, p. III:999 (2006)
18. Robertson, C., Trucco, E.: Human body posture via hierarchical evolutionary optimization. In: BMVC'06, p. III:999 (2006)
19. Rui, Y., Anandan, P.: Segmenting visual actions based on spatio-temporal motion patterns. pp. I:111–118 (2000)
20. Sidenbladh, H., Black, M.: Learning the statistics of people in images and video **54**(1–3), 183–209 (2003)
21. Sidenbladh, H., Black, M.J., Fleet, D.J.: Stochastic tracking of 3d human figures using 2d image motion. In: ECCV '00: Proceedings of the 6th European Conference on Computer Vision-Part II, pp. 702–718. Springer, Berlin, Heidelberg, New York (2000)
22. Sidenbladh, H., Black, M.J., Sigal, L.: Implicit probabilistic models of human motion for synthesis and tracking. In: ECCV '02: Proceedings of the 7th European Conference on Computer Vision-Part I, pp. 784–800. Springer, London, UK (2002)
23. Sigal, L., Bhatia, S., Roth, S., Black, M.J., Isard, M.: Tracking loose-limbed people. In: CVPR (2004)
24. Sigal, L., Black, M.J.: Predicting 3d people from 2d pictures. In: IV Conference on Articulated Motion and Deformable Objects (2006)
25. Sminchisescu, C., Triggs, B.: Kinematic jump processes for monocular 3d human tracking. In: CVPR (2003)
26. Tabar, A.M., Keshavarz, A., Aghajan, H.: Smart home care network using sensor fusion and distributed vision-based reasoning. In: ACM Multimedia Workshop on VSSN (2006)
27. Weiss, Y., Adelson, E.: Perceptually organized em: A framework for motion segmentaiton that combines information about form and motion. Technical Report 315, M.I.T Media Lab (1995). URL citeseer.ist.psu.edu/article/weiss95perceptually.html
28. Wilson, A.D., Bobick, A.F.: Parametric hidden markov models for gesture recognition. IEEE Transactions on Pattern Analysis and Machine Intelligence **21**(9), 884–900 (1999). URL citeseer.ist.psu.edu/wilson99parametric.html
29. Wu, C., Aghajan, H.: Layered and collaborative gesture analysis in multi-camera networks. In: ICASSP (2007)

11

Skin Color Separation and Synthesis for E-Cosmetics

Norimichi Tsumura, Nobutoshi Ojima, Toshiya Nakaguchi, and Yoichi Miyake

E-cosmetic function for digital images is introduced based on physics and physiologically based image processing. A practical skin color and texture analysis/synthesis technique is developed for this E-cosmetic function. Shading on the face is removed by a simple color vector analysis in the optical density domain as an inverse lighting technique. The image without shading is analyzed by extracting hemoglobin and melanin components using independent component analysis. Based on the image processing for the extracted components, we synthesized the various appearances of facial images changed due to these extracted components.

11.1 Introduction

The appearance of human skin is an essential attribute in various imaging applications such as image communication, cosmetic recommendations [16], medical diagnosis, and so on. Skin appearance is mainly caused by the color and texture of the skin, and people are very sensitive to any change of the skin appearance. Reproduced skin appearance depends on imaging device, illuminants and environments. With recent progress of color management technology, imaging devices and the color of an illuminant can be calibrated by device profiles [10, 11]. However, high-fidelity reproduction is not always effective in the practical imaging systems used for facial imaging; therefore, it is required that additional functions for color and texture reproduction [7]. These functions could be labeled "E-cosmetic," "digital cosmetic," and also "E-make [14]".

In Europe and America, it seems that the application of lipstick and makeup with some color is the general preference. On the other hand, in Asia, it is preferred that the application of a cosmetic for skin care, such as whitening essence, and the natural change of skin color. In this chapter, the "E-cosmetic"

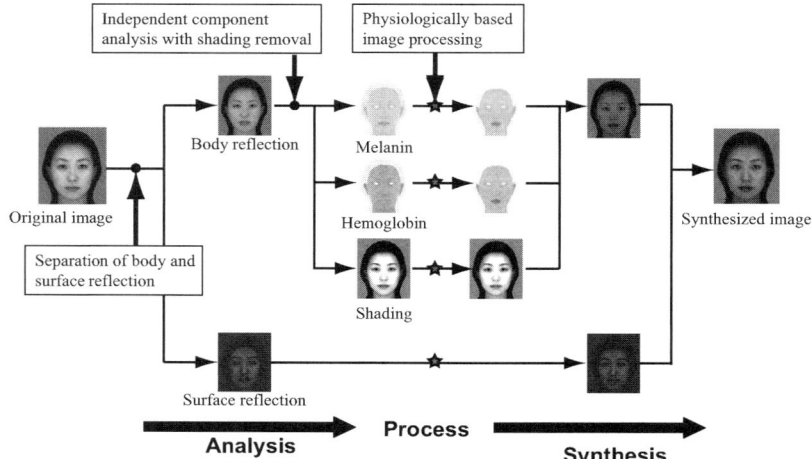

Fig. 11.1. Flow of the image based skin color and texture analysis/synthesis

function focuses on Asian people. The inverse lighting technique [1, 12, 13, 17] is the key technology for E-cosmetic, since it is necessary to obtain a unique reflectance of the skin by discounting illuminants for E-cosmetic.

We introduce a useful tool for skin appearance control for digital images, build on physics and physiologically based image processing [21–23]. We develop a practical skin color and texture analysis/synthesis technique for this function. Figure 11.1 shows the schematic flow of our technique. The surface reflection and body reflection are separated by polarization filters in front of the camera and light source by using the algorithm proposed by Ojima et al. [15]. The component maps of hemoglobin, melanin and shading information are extracted from a single skin color image for body reflection taken by digital camera. Based on the analysis of 123 skin textures in our database, we can control the texture of melanin continuously and physiologically.

In Sect. 11.2, the basic idea is introduced in applying the independent component analysis to extract the hemoglobin and melanin components from a skin color image with homogeneous shading. In Sect. 11.3, the practical analyzing model for skin images are introduced for inhomogeneous shading that has occurred in capturing the whole face, and the validation of the analysis is confirmed in Sect. 11.4. In Sect. 11.5, the pyramid-based texture analysis/synthesis technique was used for the spatial processing of texture. In Sect. 11.6, we extended the method for melanin texture analysis/synthesis to control the texture continuously and physiologically based on the analysis of 123 skin textures in our database. Finally, this capture is concluded in Sect. 11.7.

11.2 Image-Based Skin Color Analysis and Synthesis

The basic idea is introduced based on our previous paper [21] in applying the independent component analysis to extract the hemoglobin and melanin components from skin color images with homogeneous shading.

Let $s_1(x,y)$ and $s_2(x,y)$ denote the quantities of the two chromophores: hemoglobin and melanin on the image coordinates (x,y) in the digital color image. The unit vectors for the pure color of two chromophores are denoted by $\mathbf{a}_1$ and $\mathbf{a}_2$, respectively. There are two assumptions: (1) $\mathbf{a}_1$ and $\mathbf{a}_2$ are different from each other. (2) The compound color vector $\mathbf{v}(x,y)$ on the image coordinates (x,y) can be calculated by the linear combination of the pure color vectors with the quantities $s_1(x,y)$ and $s_2(x,y)$ as

$$\mathbf{v}(x,y) = s_1(x,y)\mathbf{a}_1 + s_2(x,y)\mathbf{a}_2. \qquad (11.1)$$

The elements in the compound color vector indicate the pixel values of the corresponding channels. Here only two color channels are considered for the simplicity of the explanation. Let us denote $\mathbf{A} = [\mathbf{a}_1, \mathbf{a}_2]^t$ as the constant 2×2 mixing matrix, whose column vectors are pure color vectors, $\mathbf{s}(x,y) = [s_1(x,y), s_2(x,y)]$ the quantity vector on the image coordinates (x,y), and $[\cdot]^t$ represents transposition. Equation (11.1) can be rewritten in vector and matrix formulation as follows:

$$\mathbf{v}(x,y) = \mathbf{A}\mathbf{s}(x,y). \qquad (11.2)$$

In independent component analysis, it is also assumed that the values $s_1(x,y)$ and $s_2(x,y)$ of the quantity vector are mutually independent for the image coordinates (x,y). Figure 11.2a illustrates the mixing process, and Fig. 11.2b is an example of the probability density distribution (PDF) of $s_1(x,y)$ and $s_2(x,y)$, they are mutually independent. Figure 11.2c shows the probability density distribution of $c_1(x,y)$ and $c_2(x,y)$, which are elements of the compound color vector $\mathbf{v}(x,y)$. Note that the observed color signals $c_1(x,y)$ and $c_2(x,y)$, are not mutually independent.

By independent component analysis, the relative quantity and pure color vectors of each chromophore, which will be changed depending on each skin optical property, can be extracted from the compound color vectors without a priori information about the quantity and color vector. Let us define the following equation by using the separation matrix $\mathbf{H}$ and the separated vector $\mathbf{e}(x,y)$, as shown in Fig. 11.2d

$$\mathbf{e}(x,y) = \mathbf{H}\mathbf{v}(x,y), \qquad (11.3)$$

where $\mathbf{H} = [\mathbf{h}_1, \mathbf{h}_2]$ is the separation matrix, and $\mathbf{e}(x,y) = [e_1(x,y), e_2(x,y)]^t$ is the extracted signal vector. By finding out the appropriate separation matrix $\mathbf{H}$, we can extract the mutually independent signals $e_1(x,y)$ and $e_2(x,y)$ from the compound color vectors in the image. There are many methods for finding the separation matrix $\mathbf{H}$ (e.g., [8]). In this chapter, we employ

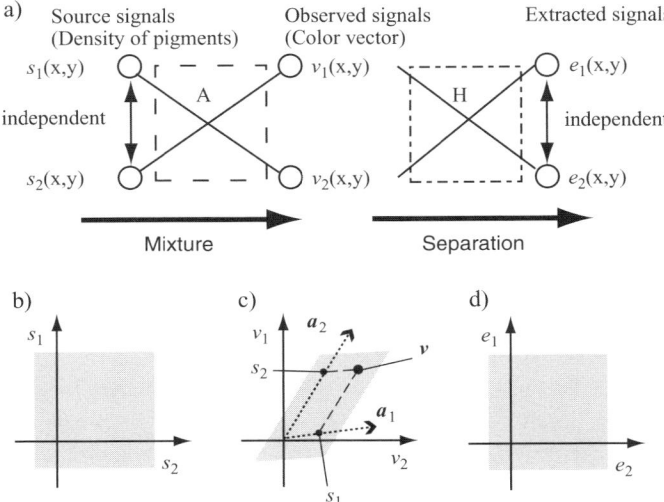

Fig. 11.2. Mixture and separation of independent signals

an optimization techniques based on the fixed-point method [9] to find the separation matrix **H**.

The extracted independent vector $\mathbf{e}(x, y)$ is given by

$$\mathbf{e}(x, y) = \mathbf{R\Lambda s}(x, y), \qquad (11.4)$$

where **R** is the permutation matrix which may substitute the elements of the vectors for each other, and $\mathbf{\Lambda}$ is the diagonal matrix that relates the absolute quantities to relative quantities. Substituting (11.2) and (11.3) into (11.4) yields

$$\mathbf{HAs}(x, y) = \mathbf{R\Lambda s}(x, y). \qquad (11.5)$$

If we take (11.5) for the arbitrary quantity vector, the matrix **HA** would be equivalent to the matrix **RΛ**, and the mixing matrix **A** is obtained by using the inverse matrix of **H** as follows:

$$\mathbf{A} = \mathbf{H}^{-1}\mathbf{R\Lambda}. \qquad (11.6)$$

Note that we obtained only the relative quantities and directions of the compound color vectors. However, it is enough since absolute values are not required in our application of color image separation and synthesis.

If the number of chromophores is larger than the number of channels, it is difficult to extract the independent components by relying on the reduction of signals. By contrast, if the number of chromophores is smaller than the number of channels, we can make the number of channels equal to the number of chromophores by using principal component analysis. This technique is also used in our analysis. White balance calibration of a digital camera is not necessary because a priori information is not used in independent component analysis. The gamma characteristic of a digital camera will also be automatically approximated as linear in the restricted variation of skin color. Therefore, this technique is very effective for images taken by commercially available digital cameras.

11.3 Shading Removal by Color Vector Space Analysis: Simple Inverse Lighting Technique

The directional light causes shading on the face and the previous technique results in a wrong estimate for the density of chromophore. Therefore, this technique is not enough to our practical application. This section describes a simple inverse lighting technique for shading removal by color vector space analysis [23]. First we model the imaging process, and then based on this model, the shading removal technique is described.

11.3.1 Imaging Model

Figure 11.3 illustrates the schematic model of the imaging process with a two-layered skin model. Note that we find melanin and hemoglobin predominantly in the epidermal and dermal layers, respectively. Part of the incident light is reflected on the surface as a Fresnel reflection, and other parts penetrate into the epidermis layer and dermis layers and then diffusely reflected from the surface. Assuming that the modified Lambert–Beer law [6] is applicable in the skin layer for incident light, we can write the body reflection as (11.7). In near-infrared range, the modified Lambert–Beer law is applicable by using the mean path length of photons in the medium as the depth of the medium (in the conventional Lambert–Beer law). The modified Lambert-Beer law is assumed in (11.7):

$$L(x,y,\lambda) = ce^{-\rho_m(x,y)\sigma_m(\lambda)l_m(\lambda) - \rho_h(x,y)\sigma_h(\lambda)l_h(\lambda)} E(x,y,\lambda), \qquad (11.7)$$

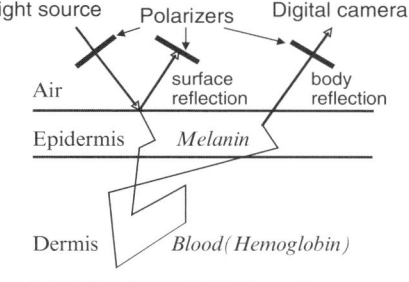

Fig. 11.3. Schematic model of the imaging process

where λ is the wavelength, $E(x, y, \lambda)$ and $L(x, y, \lambda)$ are the spectral irradiance from light source and spectral radiance to imaging devices on the surface position (x, y), respectively, and $\rho_m(x, y)$, $\rho_h(x, y)$, $\sigma_m(\lambda)$, $\sigma_h(\lambda)$ are the chromophore densities and spectral cross-sections of melanin and hemoglobin, respectively, and c is a constant value determined from the gain of the camera. For the modified Lambert–Beer law [6], $l_m(\lambda)$ and $l_h(\lambda)$ are the mean path lengths of photons in the dermis and epidermis layers, respectively.

As is mentioned before, the surface reflection and body reflection are separated by polarization filters in front of the camera and light source by using the algorithm proposed by Ojima et al. [15]. The surface reflection is added to the synthesized images to obtain the final image. If surface reflection is modified before adding, not only facial skin color but also appearance of the face (such as glossiness) would be changed. Any other separation techniques [18–20] for surface and body reflection can be used for this purpose. Therefore, the sensor response v_i ($i = R, G, B$) from the digital camera for diffuse component can be obtained as follows:

$$v_i(x, y) = \int L(x, y, \lambda) s_i(\lambda) d\lambda$$
$$= c e^{-\rho_m(x,y)\sigma_m(\lambda)l_m(\lambda) - \rho_h(x,y)\sigma_h(\lambda)l_h(\lambda)} E(x, y, \lambda) s_i(\lambda) d\lambda, \quad (11.8)$$

where $s_i(\lambda)$ ($i = R, G, B$) is the spectral sensitivity of the digital camera. If we assume two conditions: (1) the spectral sensitivity is a narrow band and can be approximated by the delta function as $s_i(\lambda) = \delta(\lambda_i)$ [4]. (2) The skin is illuminated by a single color of illuminant. Then the spectral radiance of illuminant is separable as $E(x, y, \lambda) = p(x, y) E(\lambda)$, and we can obtain the following equation from (11.8):

$$v_i(x, y) = c e^{-\rho_m(x,y)\sigma_m(\lambda)l_m(\lambda) - \rho_h(x,y)\sigma_h(\lambda)l_h(\lambda)} p(x, y) \bar{E}(\lambda_i). \quad (11.9)$$

Since the skin spectral reflectance curve maintains a high correlation in the range of spectral sensitivity, the sensitivity approximation by the delta function is valid in conventional imaging devices. The assumption of a single illuminant color is sometimes unacceptable around a window, where room light and sunlight illuminate the face from different directions. In our technique described here, we could not consider this situation. However, we can actively illuminate the face with a flashlight, and the flashlight will be dominant on the face because of its high radiance.

If we take the logarithm of (11.9) yields following equation with vector and matrix formulation:

$$\mathbf{v}^{\log}(x, y) = -\rho_m(x, y)\boldsymbol{\sigma}_m - \rho_h(x, y)\boldsymbol{\sigma}_h + p^{\log}(x, y)\mathbf{1} + \mathbf{e}^{\log}, \quad (11.10)$$

where,

$$\mathbf{v}^{\log}(x, y) = [\log(v_R(x, y)), \log(v_G(x, y)), \log(v_B(x, y))]^t,$$

$$\boldsymbol{\sigma}_m = [\sigma_m(\lambda_R)l_m(\lambda_R), \sigma_m(\lambda_G)l_m(\lambda_G), \sigma_m(\lambda_B)l_m(\lambda_B)]^t,$$
$$\boldsymbol{\sigma}_h = [\sigma_h(\lambda_R)l_h(\lambda_R), \sigma_h(\lambda_G)l_h(\lambda_G), \sigma_h(\lambda_B)l_h(\lambda_B)]^t,$$
$$\mathbf{1} = [1,1,1]^t,$$
$$\mathbf{e}^{\log} = [\log(E(\lambda_R)), \log(E(\lambda_G)), \log(E(\lambda_B))]^t,$$
$$p^{\log}(x,y) = \log(p(x,y)) + \log(c),$$

are used to simplify the notation. By this equation, the observed signals $\mathbf{v}^{\log}$ can be represented by the weighted linear combination of the three vectors $\boldsymbol{\sigma}_m$, $\boldsymbol{\sigma}_h$, $\mathbf{1}$ with the bias vector $\mathbf{e}^{\log}$, as illustrated in Fig. 11.4. It is also indicated by (11.10) that optical scattering in the skin is simply modeled linearly, and the inverse scattering technique becomes a simple inverse matrix operation.

Here we express the imaging model in (11.10) in detail. The vectors $\boldsymbol{\sigma}_m$, $\boldsymbol{\sigma}_h$ are relative absorbance vectors for the melanin and hemoglobin components, respectively. They are not affected by the illuminant, but they may be affected by the depth of the epidermis and dermis layers, since the mean path length changes with depth. In the color vector space in Fig. 11.4, the strength of the shading on surface $p^{\log}(x,y)$ is directed into the vector $\mathbf{1}$ at any position, and the color of the illuminant is a constant vector $\mathbf{e}^{\log}$ at any position. These two details show the illuminant-independent property of the proposed analysis. These features are used later to remove the shading on the face.

If the irradiance onto the surface is a constant $p^{\log}(x,y) = $ constant in all pixels of the analyzed skin area, we can apply the independent component analysis to the 2D plane spanned by the hemoglobin and melanin vectors $\boldsymbol{\sigma}_m$, $\boldsymbol{\sigma}_h$, as written in earlier section and shown in Fig. 11.5a. The 2D plane can be

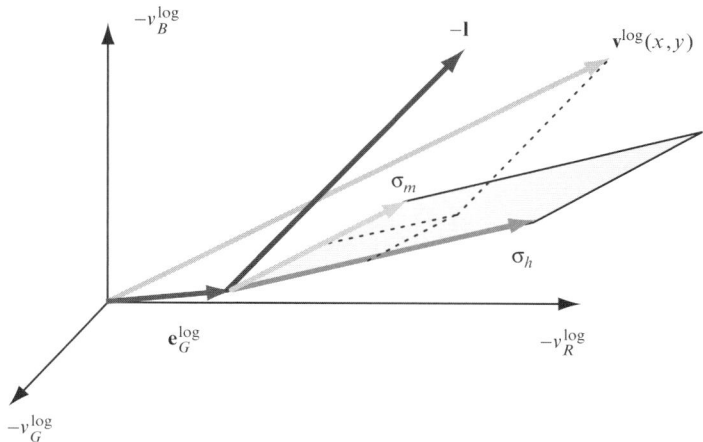

Fig. 11.4. Observed signal, represented by the weighted linear combination of three vectors with a bias vector

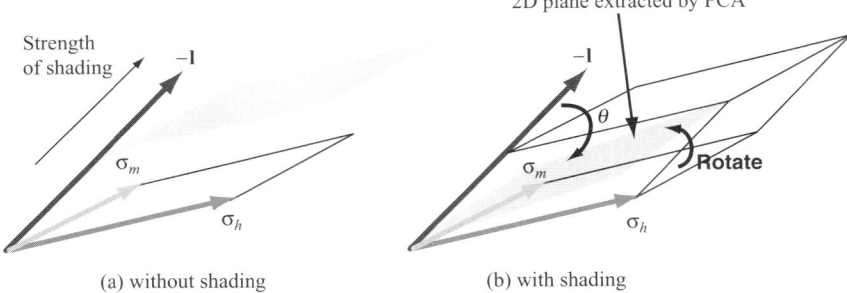

Fig. 11.5. Effect of shading in color vector space

obtained by principal component analysis (PCA), as performed by Tsumura et al. [21].

11.3.2 Finding the Skin Color Plane in the Face and Projection Technique for Shading Removal

In practical, shading on the face is caused by directional light, and results in a wrong estimate for the density of chromophore on the shading area. Since the skin texture of color is homogeneous in the local area, the strength of shading is added to each value of skin color in 3D configuration. If we apply the principal component analysis to the skin color that is influenced by shading in the color vector space, the resultant 2D plane will be rotated from an accurate plane to the direction of vector 1, as shown in Fig. 11.5b.

The procedure to find the accurate skin color plane from the shaded facial image is described in [23]. By using the procedure described in [23], we can find the appropriate 2D skin color plane spanned by the absorbance vectors $\boldsymbol{\sigma}_\mathrm{m}$, $\boldsymbol{\sigma}_\mathrm{h}$. Since the strength of shading is directed to the vector **1** for any device and illuminant, the observed skin color vector can be decomposed as

$$\mathbf{v}^{\log} = [\boldsymbol{\sigma}_\mathrm{m}, \boldsymbol{\sigma}_\mathrm{h}, \mathbf{1}][w_\mathrm{m}, w_\mathrm{h}, w_\mathrm{p}]^\mathrm{t} + \mathbf{e}^{\log}, \tag{11.11}$$

where

$$[w_\mathrm{m}, w_\mathrm{h}, w_\mathrm{p}]^\mathrm{t} = [\boldsymbol{\sigma}_\mathrm{m}, \boldsymbol{\sigma}_\mathrm{h}, \mathbf{1}]^{-1}(\mathbf{v}^{\log} - \mathbf{e}^{\log}).$$

The bias vector $\mathbf{e}^{\log}$ is unknown. Therefore, if we assume that the smallest value of each chromophore in the skin image is zero, then $\mathbf{e}^{\log}$ is calculated by $e_i^{\log} = \min_{x,y}(v_i^{\log}(x,y))$ for each band of color. Based on the above decomposition, the shading term $w_\mathrm{p}\mathbf{1}$ is removed as follows:

$$\mathbf{v}^{\log}_{\mathrm{projection}} = [\boldsymbol{\sigma}_\mathrm{m}, \boldsymbol{\sigma}_\mathrm{h}, \mathbf{0}][\boldsymbol{\sigma}_\mathrm{m}, \boldsymbol{\sigma}_\mathrm{h}, \mathbf{1}]^{-1}(\mathbf{v}^{\log} - \mathbf{e}^{\log}) + \mathbf{e}^{\log}. \tag{11.12}$$

This process is shown in Fig. 11.6 in color vector space. Figure 11.7 shows the results of the shading removal. Fig. 11.7a, d is the original image of skin, and

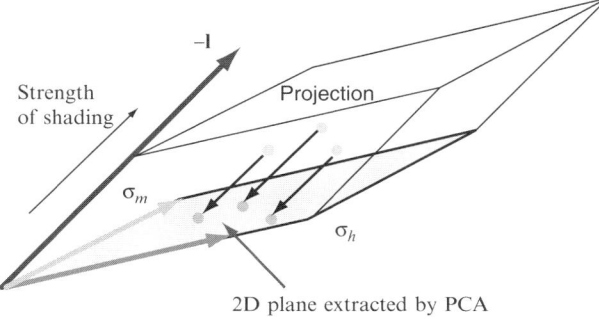

Fig. 11.6. Projection onto the skin color plane for shading removal

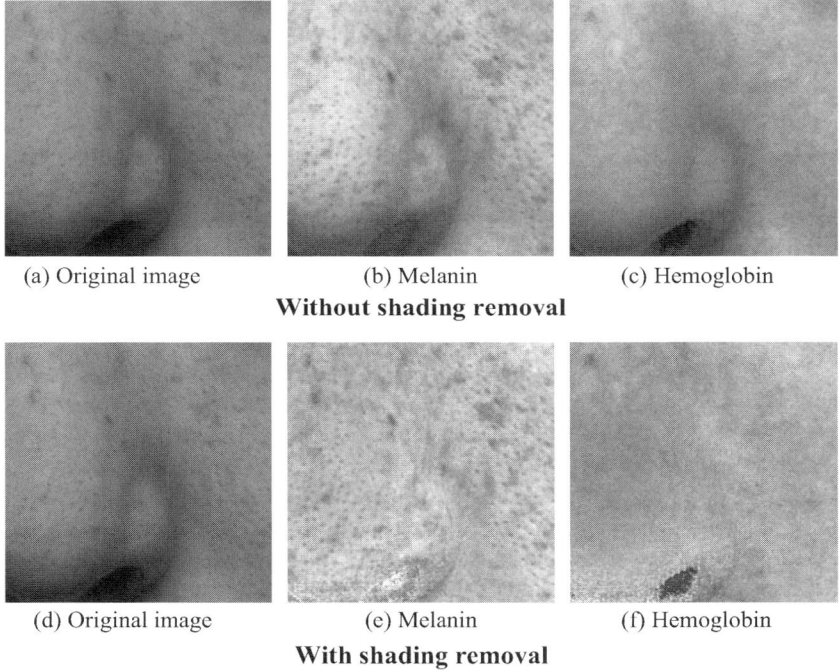

Fig. 11.7. Results of independent component analysis

Fig. 11.7b, c is the extracted melanin and hemoglobin chromophores, respectively, without shading removal. Note that the shading caused by the nose produces a wrong density estimation for chromophore and you can still recognize the nose in Fig. 11.7b, c. Figure 11.7e, f shows melanin and hemoglobin chromophores obtained using shading removal. Thus, in contrast to Fig. 11.7c, e, the shading caused by the nose cannot be observed. From these results, we conclude that the shading removal significantly improves the quality of the analysis.

11.4 Validation of the Analysis

In this section, it is confirmed that the physiological validity of the proposed analysis by practical experiments. The arm of a subject is irradiated by UV-B for the melanin component, and methyl nicotinate ($1\,\mathrm{mg\,ml^{-1}}$ solution) is applied to another arm for the hemoglobin component. The methyl nicotinate is known to increase the hemoglobin. An image of the arm, where UV-B (1.5 Minimum Erythema Dose) was irradiated in local rectangular areas, was taken after two weeks by digital camera (Nikon D1, 2,000 by 1,312 pixels). An image of the arm, where methyl nicotinate was also applied in local round areas, was also taken by digital camera after 30 min of application. Figure 11.8b shows the square patterns caused by melanin, which indicate the biological independent component analysis. Figure 11.8a–c shows the original skin image and the images of the densities for the melanin and hemoglobin components, respectively. On the other hand, Fig. 11.8d–f shows the original skin image for methyl nicotinate and the images of the densities for the melanin and hemoglobin components, respectively. Figure 11.8f also shows

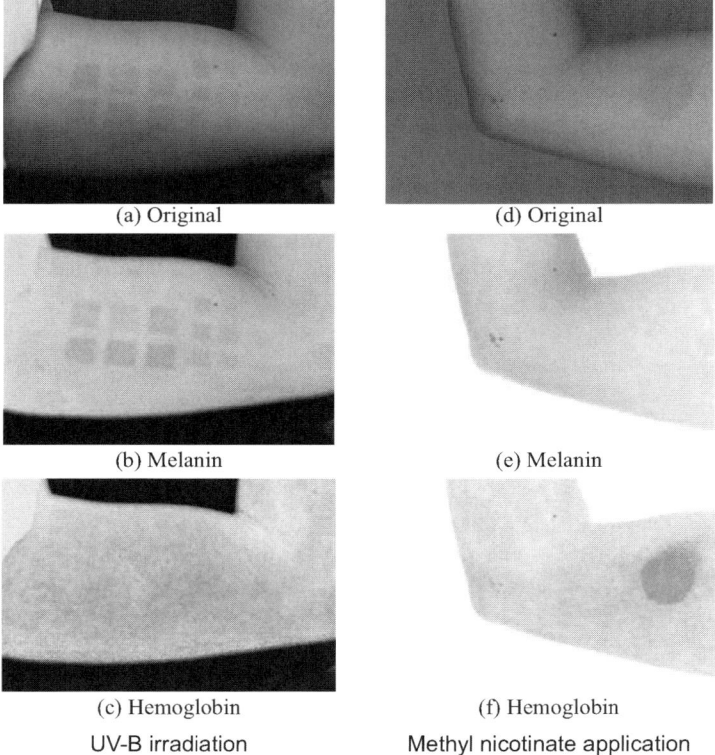

(a) Original (d) Original

(b) Melanin (e) Melanin

(c) Hemoglobin (f) Hemoglobin
UV-B irradiation Methyl nicotinate application

Fig. 11.8. UV-B irradiation and application of methyl nicotinate

the round patterns, which indicate the biological response of hemoglobin to methyl nicotinate. In contrast, the patterns did not appear in the hemoglobin components in Fig. 11.8c and melanin components in Fig. 11.8e. Note that shading in the lower part of the arm is removed, and thus the density of chromophore is extracted effectively on the entire area of the arm. These results are valid physiologically, and show the effectiveness of the proposed method of skin color image analysis. It is also indirectly shown that the approximation for the imaging model also valid in our applications.

11.5 Image-Based Skin Color and Texture Analysis/Synthesis

For the realistic synthesis of skin color and texture, we should change the amount of chromophore spatially with nonlinear processing. In this section, the skin image of a 50-year-old woman was captured, analyzed, processed, and by inversely transforming the analysis we could realistically synthesize the skin color and texture of a 50-year-old woman to that of a 20-year-old woman [23].

At first, we prepared the skin images of the 50- and 20-year-old women. The women are a mother and her daughter; they were paired so that we could suppress all other impressions except for the color and texture of chromophore. Figures 11.9a, c shows the original skin images for mother and daughter, respectively. The extracted melanin images are analyzed by the pyramid-based technique [5]. In this chapter, a Laplacian pyramid is used, since there were no unique orientations in the distribution of chromophore. The analyzed histograms in each level of the pyramid for the 50-year-old woman are matched to the histograms obtained from the analysis of images for the 20-year-old woman (this process is described in detail in later section). Figure 11.10a shows the original melanin texture of a 50-year-old woman, and

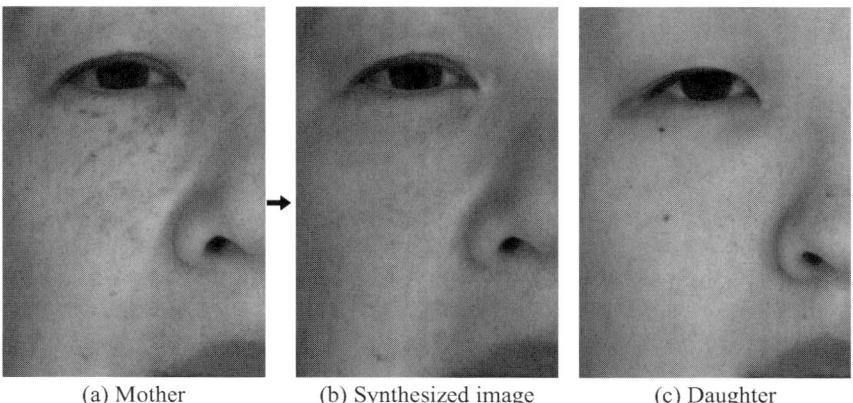

(a) Mother   (b) Synthesized image   (c) Daughter

Fig. 11.9. Skin color and texture synthesis for simulating cosmetic application

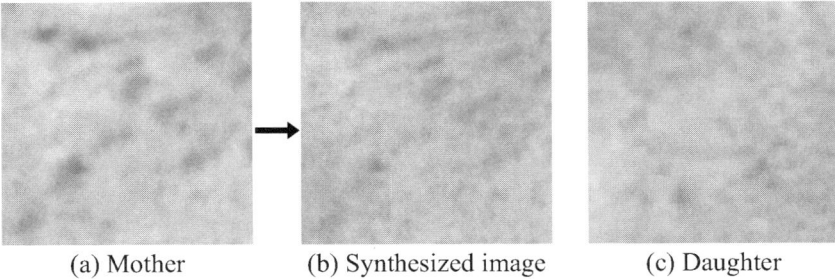

(a) Mother (b) Synthesized image (c) Daughter

Fig. 11.10. Skin texture synthesis for melanin component

Fig. 11.10b shows the synthesized melanin texture obtained from the proposed analysis and process for the original melanin texture of the 50-year-old woman, and Fig. 11.10c shows the melanin texture of the 20-year-old woman used as the target texture. These figures show that the original melanin texture of the 50-year-old woman is changed to the melanin texture of 20-year-old woman maintaining the original distribution of chromophores. Figure 11.9b shows the image synthesized from the proposed analysis and process of original image of the 50-year-old woman in Fig. 11.9a. This figure shows very realistic changes of color and texture in the skin images. From the skin image in Fig. 11.9b, we can conclude that the proposed technique can be used as a very realistic E-cosmetic function.

11.6 Data-Driven Physiologically Based Skin Texture Control

It is difficult to control the appearance continuously by the method we described above, because, although it requires a target texture for histogram matching, it is difficult to collect a variety of melanin texture images due to the limitation in the speed of melanin formation. Therefore, in this section, the database of melanin texture is used to control the skin appearance [22]. Figure 11.11 shows an overview of the process for controlling the skin melanin texture. The input texture is decomposed by the Laplacian pyramid method [3] into layers with different spatial frequencies. The histograms of the layers are analyzed, and feature vector **f** is extracted to represent the histograms. The obtained feature vector is shifted in the principal space of the melanin texture to control the appearance of the texture. This space is constructed by 123 samples of melanin texture from our own database. Figure 11.12 shows part of the database used in this chapter. The shifted feature vector **f'** is inversely transformed into histograms of the layers. The pixel values at each layer of the input texture are transformed by a lookup table created for matching the original histograms and the histograms synthesized by the shifted feature vector **f'**. The transformed layers make up the synthesized texture. By controlling

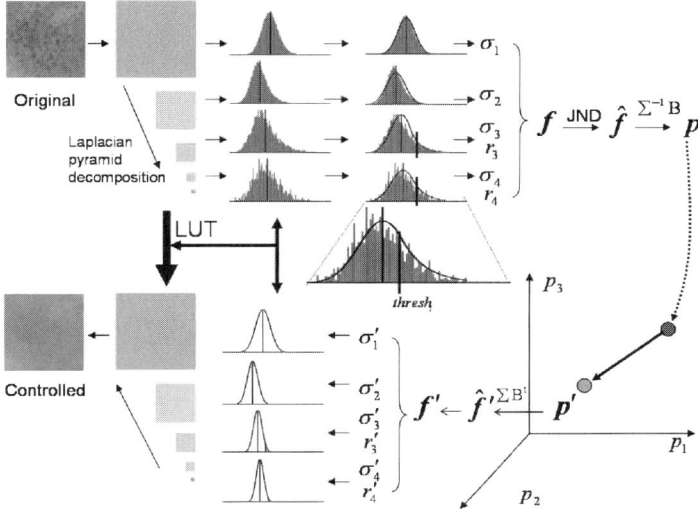

Fig. 11.11. Process of controlling the skin melanin texture

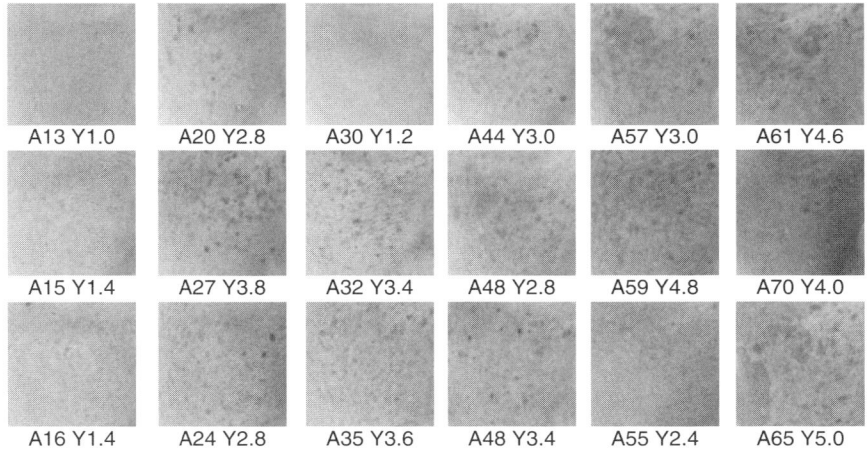

Fig. 11.12. Skin melanin texture database (18 of 123 samples): The values written below each texture are yellow unevenness A and age Y

the feature vector continuously in the principal space, the texture can be animated continuously according to the data-driven, physiologically based skin texture control.

The texture is decomposed into five layers, since five layers are enough to represent the appearance of the skin in our case (The resolution is 6 pixels per mm). The fifth layer is a low-pass component of the texture, and this is not modeled and controlled. The first to fourth layers are modeled and controlled as follows. A normalized histogram $h_i(v)$ is calculated for each layer. The index

i is the number of the layer and v is the value of the normalized components. Since the sum of the normalized values in the histogram is the pixel number, the normalized histogram can be defined as

$$\int_{-\infty}^{\infty} h_i(v)\mathrm{d}v = 1. \tag{11.13}$$

Figure 11.13 shows the histogram at each layer for four examples of textures. It can be seen that the histogram in the first and second layers can be modeled by Gaussian distribution with a mean value of 0. Since the histogram is normalized in (11.13), only the standard deviation σ_i of the histogram is used as the model parameter.

$$\sigma_i = \left[\int_{-\infty}^{\infty} \{h_i(v)\}^2 \mathrm{d}v\right]^{1/2}, \quad i = 1 \quad or \quad 2. \tag{11.14}$$

We can also see that the histogram in the third and fourth layers is modeled not only by Gaussian distribution on the high-density region. The high density region is an important part for expressing the characteristic of the melanin stain. We modeled the histogram using a combination of the Gaussian distribution and exponential distribution in the high-density region. The parameter of the Gaussian distribution is extracted using only the left side of the histogram, as follows:

$$\sigma_i = \frac{1}{2r_{left}}\left[2\int_{-\infty}^{0} \{h_i(v)\}^2 \mathrm{d}v\right]^{1/2}, \quad i = 3 \quad or \quad 4, \tag{11.15}$$

where

$$r_{left} = \int_{-\infty}^{0} h_i(v)\mathrm{d}v.$$

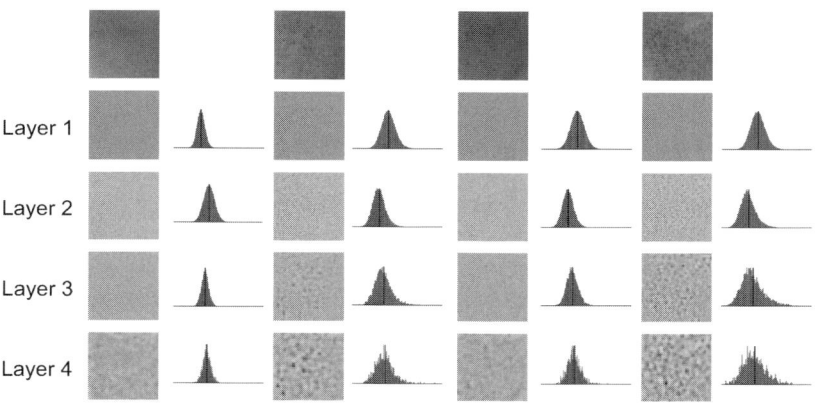

Fig. 11.13. Various types of skin melanin texture and their histograms in layers of the Laplacian pyramid decomposition

We set a threshold value thresh_i empirically to decide the characteristic high-density region ($v >$ thresh_i). The high-density region is modeled by $a_i \exp u_i(v - thresh\_i)$, where a_i is the value of the modeled Gaussian distribution at $v =$ thresh_i. Let us define the function $N(v : m, \sigma)$ as the normalized Gaussian distribution of mean value m and standard deviation σ. Using the residual sum

$$r_i = 1 - \int_{-\infty}^{\text{thresh}\_i} 2r_{\text{left}} N(v : 0, \sigma_i) dv \tag{11.16}$$

in the high-density region ($v >$ thresh_i) after modeling the Gaussian distribution, the attenuation coefficient is obtained by solving the following equation:

$$\int_{\text{thresh}\_i}^{\infty} a_i \exp\{u_i(v - \text{thresh}\_i)\} dv = r_i. \tag{11.17}$$

The modeled Gaussian distribution in the range of $v \leq$ thresh_i and the exponential distribution in the range of $v >$ thresh_i are reconstructed using σ_i in (11.15) and residual sum r_i in (11.16). Now, we have six model parameters, and we define the feature vector as $\mathbf{f} = [\sigma_1, \sigma_2, \sigma_3, \sigma_4, r_3, r_4]^t$. The histograms of the first to fourth layers are well approximated by this feature vector to express the texture.

The principal space is necessary to control the texture in physiologically plausible changes. Therefore, the principal space is constructed based on the texture database. Figure 11.12 shows examples of melanin texture in the texture database. It is also necessary to consider the human sensitivity to the change of texture for each feature, since unobserved physiologically plausible changes will not mean anything to changes in the appearance of the skin. For this purpose, we normalized each component of the feature vector by using the degree of human sensitivity to each component. The degree of human sensitivity is alternatively obtained by using the reciprocal of the just noticeable difference (JND); thus, the normalized feature vector is calculated as

$$\hat{\mathbf{f}} = \begin{bmatrix} 1/\text{JND}\_\sigma_1 & 0 & \cdots & 0 \\ 0 & \ddots & & \vdots \\ \vdots & & \ddots & 0 \\ 0 & \cdots & 0 & 1/JND\_r_2 \end{bmatrix} \mathbf{f} \tag{11.18}$$

by using the just noticeable difference JND_σ_i or JND_r_i for each feature. The JND value is obtained by subjective evaluation of the textures, which have a gradually changing parameter. We apply principal component analysis to the samples of normalized feature vector $\hat{\mathbf{f}}$ in the database to extract the physiologically plausible space of changes and obtain the first, second and third normalized principal vectors $\mathbf{b}_1, \mathbf{b}_2, \mathbf{b}_3$ and components p_1, p_2, p_3.

The components are also normalized by the square root of the eigenvalue for each sample, as follows.

$$\mathbf{p} = \Sigma^{-1} \mathbf{B}(\hat{\mathbf{f}} - \bar{\hat{\mathbf{f}}}), \quad (11.19)$$

where $\mathbf{p} = [p_1, p_2, p_3]^t$, $\mathbf{B} = [\mathbf{b}_1, \mathbf{b}_2, \mathbf{b}_3]^t$, Σ is the diagonal matrix whose components are the square roots of the eigenvalues, and $\hat{\mathbf{f}}$ is the mean vector in the samples. The cumulative proportion for three components is 0.988, therefore, three components are enough to represent the change of texture.

Figure 11.14a, b shows the distribution of the normalized principal components in the principal space. Controlling the feature vector in the principal space is based on pairs of the texture and texture attribute in the database. In this chapter, yellow-unevenness A and age Y are used as attributes of the texture. These values are written below each texture image in Fig. 11.12, and the each plot is related to (a) yellow-unevenness or (b) age. The values of yellow-unevenness are obtained by subjective evaluation by experts in the cosmetics development field, and are commonly used in that field. Based on the pair of normalized principal component $\mathbf{p}_i$ and attribute a_i of the ith sample in N samples, $(\mathbf{p}_1, a_1), (\mathbf{p}_2, a_2), (\mathbf{p}_3, a_3), \ldots, (\mathbf{p}_N, a_N)$ the principal direction $\boldsymbol{\rho}$ of change for that attribute is obtained by linear multiple regression analysis as follows:

$$\begin{bmatrix} \rho \\ b \end{bmatrix} = [\mathbf{P} \ \mathbf{1}] \left([\mathbf{P} \ \mathbf{1}][\mathbf{P} \ \mathbf{1}]^t\right)^{-1} \mathbf{a}, \quad (11.20)$$

where $\mathbf{P} = [\mathbf{p}_1^t, \ldots, \mathbf{p}_N^t]^t$, $\mathbf{a} = [a_1, \ldots, a_N]^t$, $\mathbf{1} = [1, \ldots, 1]^t$, b is the bias value obtained at the same time. The principal direction $\boldsymbol{\rho}$ is drawn in Fig. 11.14a, b for yellow-unevenness and age, respectively. The directions are similar to each other, and the correlation coefficients are 0.72 and 0.48 for yellow-unevenness and age, respectively. By using this principal direction, we can control the feature vector continuously in the principal space as follows:

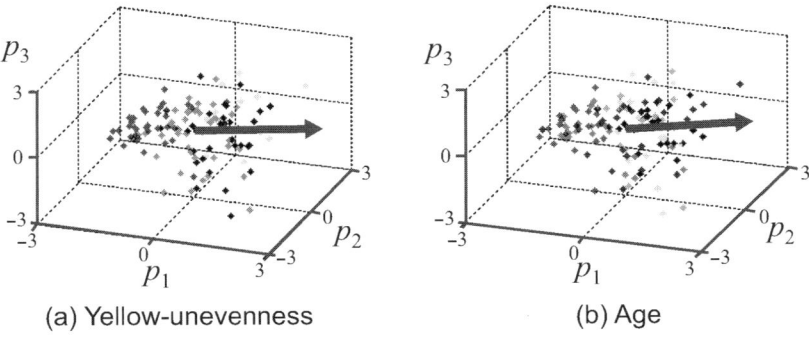

Fig. 11.14. Plots of the principal space of 123 skin melanin texture and the control direction for yellow-unevenness (**a**) and aging (**b**)

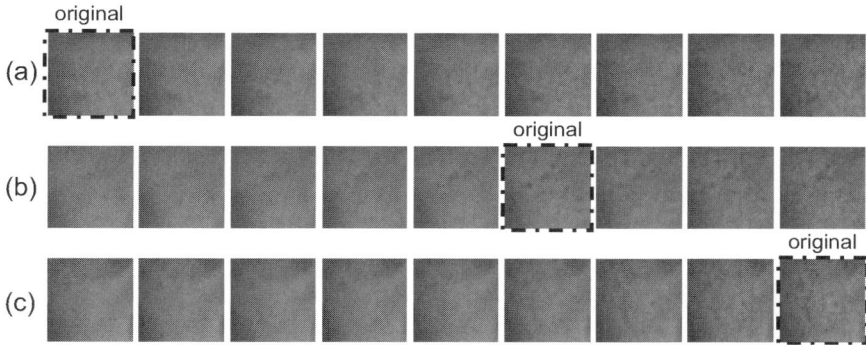

Fig. 11.15. Results of the skin appearance change for three original images by data-driven physiologically based melanin texture control

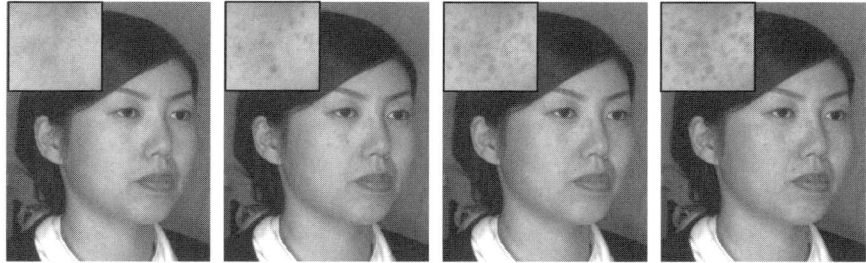

Fig. 11.16. Continuous melanin texture controls based on the data-driven physiological model from the skin texture database

$$\mathbf{p}' = \mathbf{p} + s\boldsymbol{\rho}, \tag{11.21}$$

$$\hat{\mathbf{f}}' = \mathbf{\Sigma}\mathbf{B}^t\mathbf{p}'_i + \bar{\hat{\mathbf{f}}}, \tag{11.22}$$

where s is the tag for controlling the skin texture.

Figure 11.15 shows examples of the resulting continuous syntheses for three original images with different degrees of stain. The realistic continuous change of skin texture can be seen in the series of images. Figure 11.16 also shows the continuous change of facial appearance by the same technique. The second image from the left is the original facial image and texture. It is known that the skin image taken by ultraviolet (UV) light can predict the change of the skin in aging. Figure 11.17a–c shows the original image, UV image, and controlled image using the above method, respectively. It can be seen that the change of texture is predicted well in the proposed method; therefore, it can be said that we can confirm the physiological validity of the proposed melanin texture control.

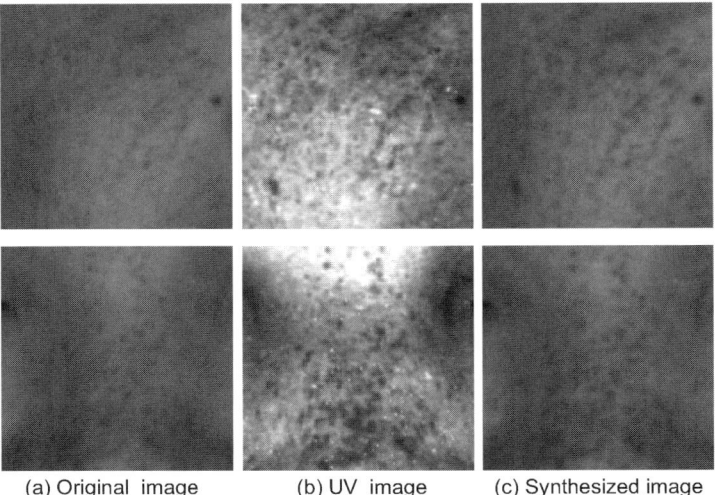

(a) Original image (b) UV image (c) Synthesized image

Fig. 11.17. Comparison of the synthesized image with the UV image, which can be used to predict the change of melanin texture in aging

11.7 Conclusion and Discussion

With the incorporation of the pyramid-based texture synthesis/analysis technique, we could synthesize the realistic change of texture and color image for skin, based on a database for physiological matching.

For the simple analysis of skin image without ill-conditioned problems, we approximated the imaging process as a simple weighted combination of vectors. Our results show the approximation was valid in our application. The bias vector $\mathbf{e}^{\log}$ is obtained from the assumption that the smallest value of each chromophore in the skin image is zero. Actually, this is not true, but this does not affect the analysis of skin chromophore density because the density obtained is a relative value. However, in the synthesis of skin color and texture using the target skin image, the difference of the bias vector causes a change of the mean skin color. Therefore, we matched the histograms between the target image and the synthesized image in optical density space to compensate for this difference of bias color. For scattering in the skin, we performed a linear approximation, based on the modified Lambert–Beer law. By using this linear model and independent component analysis, we could easily perform inverse optical scattering without a priori information of the absorption coefficient and mean path length, which is directed by the scattering coefficient. The spatial distribution of scattering cannot be represented by our method because we modeled the imaging process at each pixel independently. It is necessary to consider the dependence of pixels for synthesizing the change of skin translucency.

In the technique used in this paper, surface reflection is removed by polarization filters, and the surface reflection is added to the synthesized images to obtain the final image. To obtain the surface reflection accurately, we changed the orientation of the polarization filter in front of the camera [15]. The inverse lighting technique is performed in the single image, but the final synthesized image is obtained from the two images produced with the change of orientation of the polarization filter in front of the camera. This is not practical for commercially available digital cameras. Since we have shading components in the proposed method, we will be able to approximately reconstruct the shape from the shading (see [2] as an example). Based on the reconstructed shape, the surface reflection component could be calculated for arbitrary illuminants. In this chapter, we just modified the melanin and hemoglobin components in the synthesis. Further functional synthesis will be performed by changing the components of surface reflection or shading.

Acknowledgments

We thank K. Kamimura, K. Sato, H. Shimizu, K. Takase, S. Okaguchi, R. Usuba at Chiba University for the huge amount of work for their theses under our supervision, Prof. H. Haneishi at Chiba University for his valuable comments. We also thank M. Shiraishi, N. Okiyama, S. Akazaki, K. Hori, M. Shiraishi, H. Nabeshima at Kao Corporation for their help and valuable comments, and Prof. D. Mandic, Prof. T. Tanaka and reviewers for revising this chapter so well. This research is mainly supported by PRESTO, Japan Science and Technology Agency, and a JSPS Grants-in-Aid for Scientific Research (16760031).

References

1. Blanz, V., Vetter, T.: A morphable model for the synthesis of 3d faces. In: Proceedings of SIGGRAPH '99, pp. 187–194 (1999)
2. Brooks, M.J.: Shape from Shading. MIT Press, Cambridge, MA (1989)
3. Burt, P.J., Adelson, E.H.: A multiresolution spline with application to image mosaics. ACM Transactions on Graphics **2**(4), 217–236 (1983)
4. Drew, M.S., Chen, C., Hordley, S.D., Finlayson, G.D.: Sensor transforms for invariant image enhancement. In: Color Imaging Conference, pp. 325–330 (2002)
5. Heeger, D.J., Bergen, J.R.: Pyramid-based texture analysis/synthesis. In: Proceedings of SIGGRAPH '95, pp. 229–238 (1995)
6. Hiraoka, M., Firbank, M., Essenpreis, M., Cope, M., Arridge, S.R., van der Zee, P., Delpy, D.T.: A Monte Carlo investigation of optical pathlength in inhomogeneous tissue and its application to near-infrared spectroscopy. Physics in Medicine and Biology **38**, 1859–1876 (1993)
7. Hunt, R.W.G.: The Reproduction of Colour. Fountain Press (1995)
8. Hyvärinen, A., Karhunen, J., Oja, E.: Independent Component Analysis. Wiley, New York (2001)

9. Hyvärinen, A., Oja, E.: A fast fixed-point algorithm for independent component analysis. Neural Computation **9**(7), 1483–1492 (1997)
10. ICC (International Color Consortium): Specification ICC.1:1998-09, File Format for Color Profiles (1998)
11. IEC 61966-2-1 (1999-10): Multimedia systems and equipment – Colour measurement and management – Part 2-1: Colour management – Default RGB colour space – sRGB (1999)
12. Marschner, S.R.: Inverse lighting for photography. Ph.D. thesis (1998)
13. Marschner, S.R., Greenberg, D.P.: Inverse lighting for photography. In: Proceedings of the Fifth Color Imaging Conference, Society for Imaging Science and Technology (1997)
14. Numata, K., Ri, K., Kira, K.: "E-Make"; A real-time HD skin-make-up machine. Technical Report 76 (1999)
15. Ojima, N., Minami, T., M, K.: Transmittance measurement of cosmetic layer applied on skin by using image processing. In: Proceedings of The 3rd Scientific Conference of the Asian Societies of Cosmetic Scientists, 114 (1997)
16. Ojima, N., Yohida, T., Osanai, O., Akazaki, S.: Image synthesis of cosmetic applied skin based on optical properties of foundation layers. In: Proceedings of International Congress of Imaging Science, pp. 467–468 (2002)
17. Ramamoorthi, R., Hanrahan, P.: A signal-processing framework for inverse rendering. In: Proceedings of SIGGRAPH '01, pp. 117–128 (2001)
18. Sato, Y., Ikeuchi, K.: Temporal-color space analysis of reflection. Journal of Optical Society of America A **11**(11), 2990–3002 (1994)
19. Shafer, S.A.: Using color to separate reflection components. Color research and applications **10**(4), 210–218 (1985)
20. Tominaga, S., Wandell, B.A.: Standard surface-reflectance model and illuminant estimation. Journal of the Optical Society of America A **6**(4), 576–584 (1989)
21. Tsumura, N., Haneishi, H., Miyake, Y.: Independent-component analysis of skin color image. Journal of the Optical Society of America A **16**, 2169–2176 (1999)
22. Tsumura, N., Nakaguchi, T., Ojima, N., Takase, K., Okaguchi, S., Hori, K., Miyake, Y.: Image-based control of skin melanin texture. Applied Optics **45**(25), 6626–6633 (2006)
23. Tsumura, N., Ojima, N., Sato, K., Shiraishi, M., Shimizu, H., Nabeshima, H., Akazaki, S., Hori, K., Miyake, Y.: Image-based skin color and texture analysis/synthesis by extracting hemoglobin and melanin information in the skin. In: Proceedings of SIGGRAPH '03, pp. 770–779 (2003)

12

ICA for Fusion of Brain Imaging Data

Vince D. Calhoun and Tülay Adalı

Many studies are currently collecting multiple types of imaging data and information from the same participants. Each imaging method reports on a limited domain and is likely to provide some common information and some unique information. This motivates the need for a joint analysis of these data. Most commonly, each type of image is analyzed independently and then perhaps overlaid to demonstrate its relationship with other data types (e.g., structural and functional images). A second approach, called *data fusion*, utilizes multiple image-types together in order to take advantage of the "cross" information. In the former approach, any cross information is "thrown" away, hence such an approach, for example, would not detect a change in functional magnetic resonance imaging (fMRI) activation maps that are associated with a change in the brain structure while the second approach would be expected to detect such changes.

12.1 Introduction

In Fig. 12.1, we present a simple motivating example as a problem that the main approach in this chapter addresses. Consider a brain that has a fiber bundle (providing input to a distant location) that is underdeveloped to varying degrees in a group of participants. Suppose that the neuronal activity of a downstream region is decreased relative to the amount of axonal maldevelopment. Traditional analysis would not detect a relationship between these sites – though the fractional anisotropy (FA) data, which provides information about structural connectivity, would reveal slight decreases in the compromised region, fMRI amplitude in the connected site would have a low average and high variability. On the other hand, a coupled approach would reveal an association between the FA and the fMRI activation by leveraging the associations between the FA and fMRI values, thus providing an important piece of evidence about brain connectivity in this group of participants.

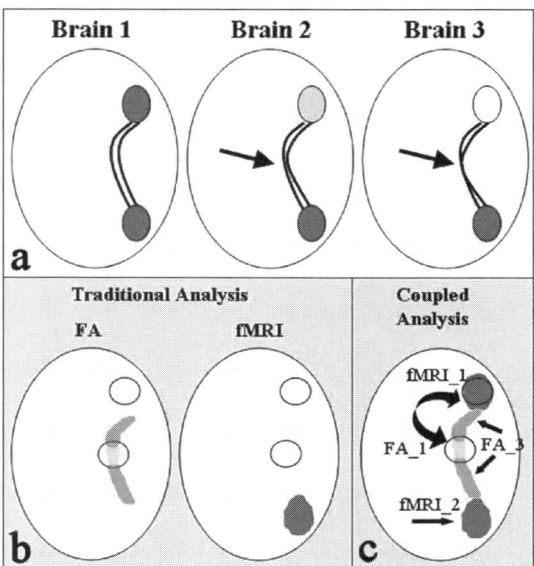

Fig. 12.1. Motivating example in which a fiber bundle providing input to a distant location is underdeveloped to varying degrees in three participants and thus fMRI activity is decreased (fMRI activity at the input is unaffected)

These would be detected in a single component map, i.e., regions FA_1 and fMRI_1, indicated in the figure, would be revealed in a single joint component, whereas the main effect of FA and fMRI (fMRI_2 and FA_3) would be detected in additional component maps.

In this chapter, we first review the basic approaches for fusing information from multiple medical imaging data types. Next, we present a feature-based fusion approach that provides a general framework for fusing information from multiple data types, such as multitask fMRI data, and fMRI and event-related potential (ERP) data. The extracted features for each data type are fused using a data-driven analysis technique, independent component analysis (ICA), which has proved quite fruitful for medical image analysis. The fusion framework we present thus enables the discovery of relationships among data types for given samples, for example, at the group level, to study variations between patients and controls. In the following sections, after a background of some of the common fusion approaches, we introduce the feature-based fusion framework. In particular, we discuss the nature of features and their computation. Next, we present several examples showing the fusion of data from different modalities. The final section discusses the importance of selecting the important features for the joint analysis.

12.2 An Overview of Different Approaches for Fusion

Many studies are currently collecting multiple types of imaging data from the same participants. Each imaging method reports on a limited domain and typically provides both common and unique information about the problem in question. Approaches for combining or fusing data in brain imaging can be conceptualized as having a place on an analytic spectrum with meta-analysis (highly distilled data) to examine convergent evidence at one end and large-scale computational modeling (highly detailed theoretical modeling) at the other end [27]. In between are methods that attempt to perform a direct data fusion [26].

Current approaches for combining different types of imaging information for the most part elect to constrain one type with another type of information – as in electroencephalography (EEG) [16, 22] or diffusion tensor imaging (DTI) [34, 46] being constrained by fMRI or structural MRI (sMRI) data. While these are powerful techniques, a limitation is that they impose potentially unrealistic assumptions upon the EEG or DTI data, which are fundamentally of a different nature than the fMRI data. An alternative approach – which we call *data integration* [1, 47] – is to analyze each data type separately and overlay them – thereby not allowing for any interaction between the data types. For example, a data integration approach would not detect a change in fMRI activation maps that is related to a change in brain structure. One promising direction is to take an intermediate approach in which the processing of each image type is performed using features extracted from different modalities. These features are then examined for relationships among the data types at the group level (i.e., variations among individuals) and specifically, differences in these variations between patients and controls. This approach allows us to take advantage of the "cross" information among data types [1, 47].

Methods such as structural equation modeling (SEM) or dynamic causal modeling (DCM) [21, 40, 42] can be used to examine the correlational structure between regions activated by different tasks or between functional and structural variables [49]. Such approaches are useful for model testing; however, these approaches do not provide an examination of the full set of brain voxels, nor do they allow testing of unknown connections. Alternatively, one could choose to examine correlation (and potentially extend to nonlinear relationship through the use of other criteria such as mutual information) between all points of the data. This approach has been applied to examine functional connectivity in fMRI by computing a 6D matrix of correlations [53]. Such computations are straightforward; however, the drawback is that they are high in dimensionality and hence potentially difficult to interpret. A natural set of tools for avoiding the disadvantages of the above techniques include those that transform data matrices into a smaller set of modes or components. Such approaches include those based upon singular value decomposition (SVD) [39] as well as more recently, independent component analysis

(ICA) [41]. An advantage of ICA over variance-based approaches like SVD or principal component analysis (PCA) is the use of higher-order statistics to reveal hidden structure [28]. We have recently done work showing the value of combining multitask fMRI data [12], fMRI and sMRI data [8], and fMRI and ERP data [13]. One important aspect of the approach is that it allows for the possibility that a change in a certain location in one modality is associated with a change in a *different* location in another modality (or, in the case of ERP one is associating time in ERP with space in fMRI) as we demonstrate with a number of examples in this chapter.

12.3 A Brief Description of Imaging Modalities and Feature Generation

FMRI measures the hemodynamic response related to neural activity in the brain dynamically. SMRI provides information about the tissue types of the brain – gray matter (GM), white matter (WM), cerebrospinal fluid (CSF). Another useful measure of brain function is EEG, which measures brain electrical activity with a higher temporal resolution than fMRI (and lower spatial resolution).

12.3.1 Functional Magnetic Resonance Imaging

fMRI data provide a measure of brain function on a millimeter spatial scale and a sub-second temporal scale. There are a considerable number of available fMRI processing strategies [35]. Two primary approaches include model-based approaches assuming certain hemodynamic properties and often utilizing the general linear model (GLM) [20], and data-driven approaches; one which has proven particularly fruitful is ICA [11, 41], which does not impose a priori constraints upon the temporal hemodynamic evolution.

A strength of the GLM approach is that it allows one to perform specific hypothesis tests, e.g., "where in the brain do these patterns occur?" In contrast, a strength of ICA is its ability to characterize fMRI activations without an a priori hemodynamic model in an exploratory manner, e.g., "what are the temporal and spatial patterns occurring in the brain?" Both approaches have obvious advantages. We next give a brief discussion of preprocessing, GLM analysis, and ICA analysis.

The main preprocessing stages for fMRI are (1) phase fix (correction), (2) registration, and (3) normalization. Phase correction is necessary because each slice is typically acquired sequentially, rather than acquiring all slices simultaneously [44]. Registration is also required because of subject motion between scans. There are numerous algorithms for estimating and correcting for this motion including those based upon Fourier methods, Taylor approximations, Newton's method [19], and others. The third preprocessing stage, normalization, is necessary to (1) compare brains across different individuals, and

(2) use standardized atlases to identify particular brain regions. There are also many methods for applying spatial normalization including maximum likelihood and Newton's methods as well as localized methods.

The most common analysis approach for fMRI is based upon the general linear model, assuming a specific form for the hemodynamic response. In the simplest case the data are modeled as:

$$\mathbf{y}_m = \sum_{i=1}^{R} \mathbf{x}_i \beta_{mi} + \varepsilon_m \quad (12.1)$$

for R regressors, where $\mathbf{y}_m$, $\mathbf{x}_i$, and ε_m are $K \times 1$ for time points $k = 1, 2, \ldots, K$ at brain locations $m = 1, 2, \ldots, M$. The error is typically modeled as Gaussian, independent and identically distributed, zero-mean, with variance σ_v^2.

A complementary data-driven approach can also be utilized. Independent component analysis is a statistical method used to discover hidden factors (sources or features) from a set of measurements or observed data such that the sources are maximally independent. Typically, it assumes a generative model where observations are assumed to be linear mixtures of independent sources, and unlike PCA that uncorrelates the data, ICA works with higher-order statistics to achieve independence. A typical ICA model assumes that the source signals are not observable, statistically independent, and non-Gaussian, with an unknown, but linear, mixing process. Consider an observed M-dimensional random vector denoted by $\mathbf{x} = [x_1, \ldots, x_M]^{\mathrm{T}}$ which is generated by the ICA model:

$$\mathbf{x} = \mathbf{As}, \quad (12.2)$$

where $\mathbf{s} = [s_1, s_2, \ldots, s_N]^{\mathrm{T}}$ is an N-dimensional vector whose elements are assumed independent sources and $\mathbf{A}_{M \times N}$ is an unknown mixing matrix. Typically, $M \geq N$, so that $\mathbf{A}$ is usually of full rank. The goal of ICA is to estimate an unmixing matrix $\mathbf{W}_{N \times M}$ such that $\mathbf{y}$ (defined in (12.3)) is a good approximation to the "true" sources, $\mathbf{s}$:

$$\mathbf{y} = \mathbf{Wx}. \quad (12.3)$$

Independent component analysis has shown to be useful for fMRI analysis for several reasons. Spatial ICA finds systematically nonoverlapping, temporally coherent brain regions without constraining the temporal domain. The temporal dynamics of many fMRI experiments are difficult to study with fMRI due to the lack of a well-understood brain-activation model. ICA can reveal intersubject and interevent differences in the temporal dynamics. A strength of ICA is its ability to reveal dynamics for which a temporal model is not available [14]. Spatial ICA also works well for fMRI as it is often the case that one is interested in spatially distributed brain networks.

12.3.2 Structural Magnetic Resonance Imaging

Structural MRI analysis is defined as the acquisition and processing of T1-, T2-, and/or proton density-weighted images. Multiple structural images are often collected to enable multispectral segmentation approaches. Both supervised and automated segmentation approaches have been developed for sMRI analysis [6, 51]. The near-exponential pace of data collection [18] has stimulated the development of structural image analysis. Advanced methods include the rapidly growing field of computational anatomy [17, 43]. This field combines new approaches in computer vision, anatomical surface modeling [17], differential geometry [43], and statistical field theory [52] to capture anatomic variation, encode it, and detect group-specific patterns. Other approaches include voxel-based methods [2] and manual region-of-interest approaches. Each technique is optimized to detect specific features, and has its own strengths and limitations.

The primary outcome measure in a structural image may include a measure of a particular structure (e.g., volume or surface area) or a description of the tissue type (e.g., gray matter or white matter). There are many methods for preprocessing sMRI data which may include bias field correction [intensity changes caused by radio frequency (RF) or main magnetic field (B_o) inhomogeneities] [15], spatial linear or nonlinear [23] filtering normalization. MR images are typically segmented using a tissue classifier producing images showing the spatial distribution of gray matter, white matter, and CSF. Tissue classifiers may be supervised (where a user selects some points representing each tissue class to guide classification) or unsupervised (no user intervention). Bayesian segmentation methods [2, 50] assign each image voxel to a specific class based on its intensity value as well as prior information on the likely spatial distribution of each tissue in the image. The classification step may be preceded by digital filtering to reduce intensity inhomogeneities due to fluctuations and susceptibility artifacts in the scanner magnetic field. In expectation-maximization (EM) techniques, RF correction and tissue classification steps are combined, using one to help estimate the other in an iterative sequence [51].

12.3.3 Diffusion Tensor Imaging

Diffusion MRI is a technique that measures the extent of water diffusion along any desired direction in each voxel [7]. Such measurements have revealed that diffusion of brain water has strong directionality (anisotropy) attributed to the presence of axons and/or myelination. Diffusion of brain water is often confined to a direction parallel to neuronal fibers. If there is a region where fibers align in a direction, diffusion of water may be restricted to a direction perpendicular to the fibers and tend to diffuse parallel to them. The properties of such water diffusion can be expressed mathematically as a "3×3" [4]. The tensor can be further conceptualized and visualized as an ellipsoid, the three main axes of which describe an orthogonal coordinate system. This ellipsoid can be characterized by six parameters; diffusion constants along the longest, middle, and shortest axes (λ_1, λ_2, and λ_3) and the directions of those axes. Once the diffusion ellipsoid is fully characterized at each pixel of the brain images, local fiber structure can be deduced. For example, if $\lambda_1 \gg \lambda_2 > \lambda_3$ (diffusion is anisotropic), it suggests the existence of dense and aligned fibers within each pixel, whereas isotropic diffusion ($\lambda_1 \sim \lambda_2 \sim \lambda_3$) suggests sparse or not aligned fibers. When diffusion is anisotropic, the direction of λ_1 tells the direction of the fibers. The degree of anisotropy can be quantified using these parameters obtained from diffusion MRI measurements [3, 5].

Among the metrics for quantifying diffusion anisotropy, fractional anisotropy (FA) is considered to be the most robust [45].

$$\text{FA} = \frac{\sqrt{3}}{\sqrt{2}} \frac{\sqrt{(\lambda_1 - \lambda)^2 + (\lambda_2 - \lambda)^2 + (\lambda_3 - \lambda)^2}}{\sqrt{\lambda_1^2 + \lambda_2^2 + \lambda_3^2}}, \text{ where } \lambda = (\lambda_1 + \lambda_2 + \lambda_3)/3.$$

Within the constraints of in-plane resolution, some regions of white matter normally have very high FA, and this probably represents architectural differences in fiber tract organization at the intra-voxel level, i.e., intact fibers crossing within a voxel. Many pathologic processes that cause changes at the micro-structural level, such as demyelination and corruption of microtubules, are likely to cause a significant measurable decrease in FA due to the diminished intra-voxel fiber incoherence.

12.3.4 Electroencephalogram

Electroencephalography is a technique that measures brain function by recording and analyzing the scalp electrical activity generated by brain structures. It is a noninvasive procedure that can be applied repeatedly in patients, normal adults, and children with virtually no risks or limitations. Local current flows are produced when brain cells (pyramidal neurons) are activated. It is believed that contributions are made by large synchronous populations although it is not clear if small populations also make a contribution. The electrical signals are then amplified, digitized, and stored.

Event-related potentials (ERPs) are small voltage fluctuations resulting from evoked neural activity. These electrical changes are extracted from scalp recordings by computer averaging epochs (recording periods) of EEG time-locked to repeated occurrences of sensory, cognitive, or motor events. The spontaneous background EEG fluctuations, which are typically random relative to when the stimuli occurred, are averaged out, leaving the event-related brain potentials. These electrical signals reflect only that activity which is consistently associated with the stimulus processing in a time-locked way. The ERP thus reflects, with high temporal resolution, the patterns of neuronal activity evoked by a stimulus.

Due to their high temporal resolution, ERPs provide unique and important timing information about brain processing and are an ideal methodology for studying the timing aspects of both normal and abnormal cognitive processes. More recently, ICA has been used to take advantage of EEG activity that may be averaged out by computing an ERP [31]. Magnetoencephalography (MEG) is a complementary technique that senses the magnetic field produced by synchronously firing neurons. The MEG system is much more expensive, requiring superconducting sensors, but as the magnetic field is not attenuated by the scalp and skin, allows for improved spatial localization compared to EEG.

12.4 Brain Imaging Feature Generation

Often it is useful to use existing analysis approaches to derive a lower dimensional feature from the imaging data. These features can then be analyzed to integrate or fuse the information across multiple modalities. The data types on which we focus in this chapter are: fMRI, sMRI [including T1- and T2-weighted scans], and EEG. Processing strategies for each of these data types have been developed over a number of years. Each data type, after being preprocessed as above, is reduced into a *feature*, which contributes an input vector from each modality for each subject and each task to the joint ICA framework we introduce in this chapter. A feature is a sub data set extracted from one type of data, related to a selected brain activity or structure. A summary of some features is provided in Table 12.1.

12.5 Feature-Based Fusion Framework Using ICA

Independent component analysis has demonstrated considerable promise for the analysis of fMRI [41], EEG [38], and sMRI data. In this section, we present a data fusion framework utilizing ICA, which we call the joint ICA (jICA).

Table 12.1. Core features for FMRI (Sternberg working memory task [SB], auditory oddball task [AOD]), sMRI, and EEG (AOD task): The fMRI features are computed using the SPM software and represent fMRI activity in response to certain stimuli. The sMRI measures are the result of segmentation of T1-weighted brain images, and the EEG features are time-locked averages to the stimuli

Modality	Core-feature
fMRI SB task	Recognition-related activity [29]
	Encode-related activity [29]
fMRI AOD task	Target-related activity [33]
	Novel-related activity [33]
sMRI	GM concentration [24]
	WM concentration [24]
	CSF concentration [24]
EEG AOD task	Target-related ERP [32]
	Novel-related ERP [32]
DTI	Fractional anisotropy [45]

Note that the ICA approach we described earlier for fMRI data is a first level analysis (i.e., is applied directly to the 4D data without reduction into a feature, and though the basic algorithm is similar, the application details are different from the ICA we propose to utilize at the second level, on the generated features). An amplitude map generated by ICA at the first level would be considered a feature similar to an amplitude map generated by the GLM approach.

Given two sets of data (can be more than two, for simplicity, we first consider two), $\mathbf{X}_F$ and $\mathbf{X}_G$, we concatenate the two data sets side-by-side to form $\mathbf{X}_J$ and write the likelihood as

$$L(\mathbf{W}) = \prod_{n=1}^{N} \prod_{v=1}^{V} p_{J,n}(u_{J,v}), \qquad (12.4)$$

where $u_{J,v}$ denotes the elements of vector $\mathbf{u}_J = \mathbf{W}\mathbf{x}_J$. Here, we use the notation in terms of random variables such that each entry in the vectors $\mathbf{u}_J$ and $\mathbf{x}_J$ correspond to a random variable, which is replaced by the observation for each sample $n = 1, \ldots, N$ as rows of matrices $\mathbf{U}_J$ and $\mathbf{X}_J$. When posed as a maximum likelihood problem, we estimate a *joint* demixing matrix $\mathbf{W}$ such that the likelihood $L(\mathbf{W})$ is maximized.

Let the two data sets $\mathbf{X}_F$ and $\mathbf{X}_G$ have dimensionality $N \times V_1$ and $N \times V_2$, then we have

$$L(\mathbf{W}) = \prod_{n=1}^{N} \left(\prod_{v=1}^{V_1} p_{F,n}(u_{F,v}) \prod_{v=1}^{V_2} p_{G,n}(u_{G,v}) \right). \qquad (12.5)$$

Depending on the data types in question, the above formula can be made more or less flexible.

This formulation assumes that the sources associated with the two data types (F and G) modulate the same way across N samples (usually subjects).

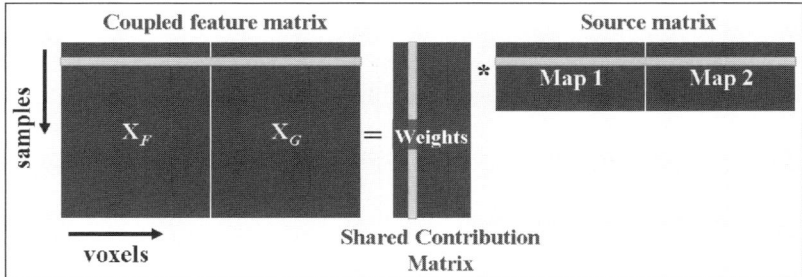

Fig. 12.2. Illustration of model in which loading parameters are shared among features: The feature matrix is organized by placing the features (e.g., SPM map and GM map) from the two modalities side by side. This matrix is then modeled as containing spatially independent joint source images which share common mixing matrix parameters

This is a strong constraint; however, it has a desirable regularization effect to the problem simplifying the estimation problem significantly, which is important especially when dealing with different data types. Also, the framework provides a natural link to two types of data by constraining the contributions to be similar. In addition, it is important to normalize the two data types independently so that they have similar contributions to the estimation and that $V_1 \approx V_2$.

The underlying assumptions for the form given in (12.4) depend on the data types used for F and G. For example, when the two data types belong to the same data type but different tasks, the assumption of $p_J = p_F = p_G$ is more plausible than when dealing different data types. On the other hand, when little is known about the nature of the source distributions in a given problem, imposing a distribution of the same form provides significant advantages yielding meaningful results as we demonstrate with an fMRI-ERP fusion example. In addition, certain nonlinear functions such as the sigmoid function has been noted as providing a robust solution to the ICA problem providing a good match for a number of source distributions, especially when they assume super-Gaussian statistics.

Hence, there are different ways to relax the assumptions made in the formulation above, such as instead of constraining the two types of sources to share the same mixing coefficients, i.e., to have the same modulation across N samples, we can require that the form of modulation across samples for the sources from two data types are correlated but not necessarily the same. We have implemented such an approach, called parallel ICA [36, 37].

12.6 Application of the Fusion Framework

In this section, we show examples of the application of jICA introduced in Sect. 12.5 to real data from multiple tasks/modalities using (1) multitask fMRI data, (2) fMRI/sMRI and (3) ERP/fMRI data. Furthermore, we address the selection of best input features for the jICA data fusion to achieve the best performance, in terms of classifying subjects. At last, we extend our test to analyze the impact of preprocessing steps, by evaluating the performance of jICA on features derived from different preprocessing steps.

In the examples we present, fMRI data were preprocessed using the software package SPM2. SMRI data were segmented into GM, WM, and CSF images using the same program. ICA was used to remove ocular artifacts from the EEG data [30]. The EEG data, then, were filtered with a 20 Hz low pass filter. ERPs were constructed for trials in which participants correctly identified target stimuli, from the midline central position (Cz) because it appeared to be the best single channel to detect both anterior and posterior sources.

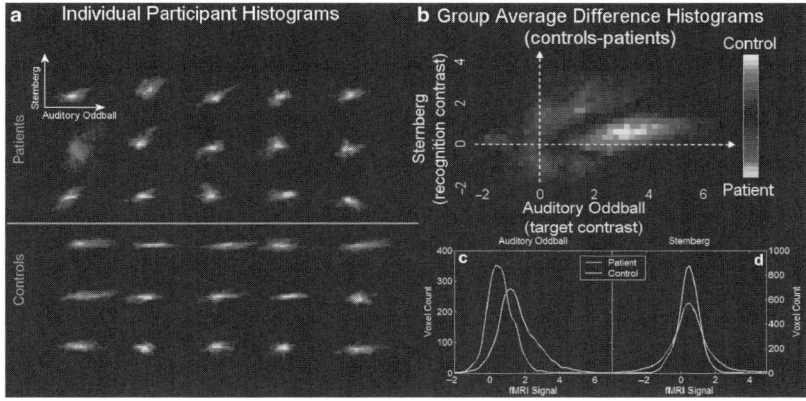

Fig. 12.3. Cross-task 2D histograms for AOD versus SB fMRI activation: Joint 2D histograms for voxels identified in the analysis. Individual (**a**) and group average difference (**b**) histograms [with orange areas (right-hand side of figure) larger in controls and blue areas (left-hand side of figure) larger in patients (see online version for color figures)] are provided along with the marginal histograms for the auditory oddball (SPM contrast image for "targets") (**c**) and Sternberg (SPM contrast image for "recall") (**d**) data

12.6.1 Multitask fMRI

We performed a joint analysis of fMRI data collected from a Sternberg (SB) task and an auditory oddball (AOD) task. Data in each task were collected from 15 controls and 15 patients with schizophrenia. Additional details of the tasks and subjects are provided in [9]. A single joint component was found to discriminate schizophrenia patients and healthy controls. A joint histogram was computed by ranking voxels surviving the threshold for the AOD and SB parts of the joint source in descending order and pairing these two voxel sets. Single subject and group-averaged joint histograms are presented in Fig. 12.3a, b and the marginal histograms for the AOD and SB tasks are presented in Fig. 12.3c, d.

In general, more AOD task voxels were active in the controls and the SB task showed a slight increase standard deviation for the patients. Results also revealed significantly more correlation between the two tasks in the patients ($p < 0.000085$). A possible synthesis of the findings is that patients are activating less, but also activating with a less unique set of regions for these very different tasks, consistent with a generalized cognitive deficit.

12.6.2 Functional Magnetic Resonance Imaging–Structural Functional Magnetic Resonance Imaging

It is also feasible to use jICA to combine structural and functional features. Our approach requires acceptance of the likelihood of gray matter changes being related to functional activation. This is not an unreasonable premise when considering the same set of voxels [48], or even adjacent voxels but as the current study shows, it also requires the acceptance of related gray matter regions and functional regions which are spatially *remote*. Given the functional interconnectedness of widespread neural networks, we suggest that this, also, is a reasonable conception for the relationship between structural and functional changes.

The next example is from a joint-ICA analysis of fMRI data of auditory oddball task and gray matter segmentation data [8]. Auditory oddball target activation maps and segmented gray matter maps were normalized to a study specific template to control for intensity differences in MR images based on scanner, template and population variations.

Results are presented in Fig. 12.4. The AOD part of the joint source is shown in Fig. 12.4a, the GM part of the joint source is shown in Fig. 12.4b, and the ICA loading parameters separated by group and shown in Fig. 12.4c. Only one component demonstrated significantly different loadings in patients and controls (loading for controls was higher than that for patients). Different regions were identified for the fMRI and sMRI data. For display, auditory oddball and gray matter sources were converted to Z-values (i.e., zero mean, unit standard deviation) and thresholded at $|Z| > 3.5$.

The main finding was that the jICA results identified group differences in bilateral parietal and frontal as well as right temporal regions in gray matter associated with bilateral temporal regions activated by the auditory oddball target stimulus. This finding suggests gray matter regions that may serve as a morphological substrate for *changes* in (functional) connectivity. An unexpected corollary to this finding was that, in the regions showing the largest group differences, gray matter concentrations were *increased* in patients versus controls, suggesting that these increases are somehow related to decreases in functional connectivity in the AOD fMRI task.

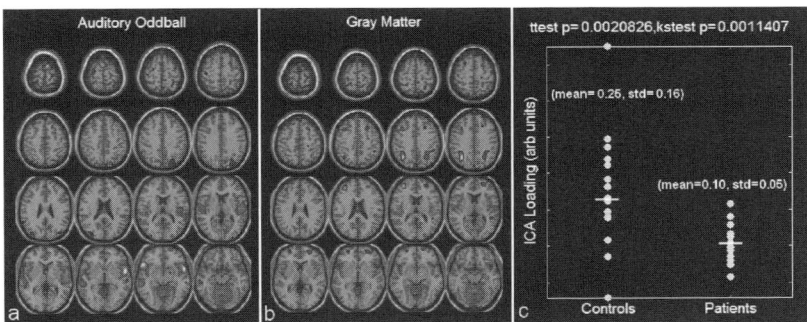

Fig. 12.4. Auditory oddball/gray matter jICA analysis: Only one component demonstrated a significant difference between patients and controls. The joint source map for the auditory oddball (*left*) and gray matter (*middle*) data is presented along with the loading parameters for patients and controls (*far right*)

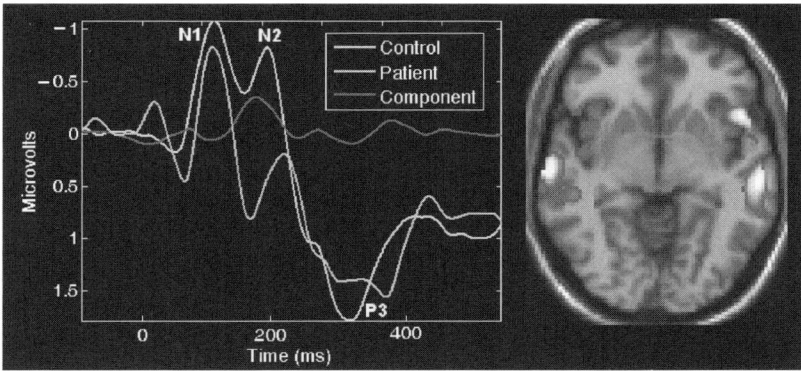

Fig. 12.5. ERP/fMRI jICA: Joint component which showed significantly different loading parameters ($p < 0.0001$) for patients versus controls: (*left*) control (yellow; highest curve) and patient (blue; lowest curve) average ERP plots along with the ERP part of the identified joint component (pink; center curve) (see online version for color figures). (*right*) Thresholded fMRI part of the joint component showing bilateral temporal and frontal lobe regions

12.6.3 Functional Magnetic Resonance Imaging–Event-Related Potential

The feature-based jICA framework can be used for ERP and fMRI data collected from 23 healthy controls and 18 chronic schizophrenia patients, during the performance of the AOD task. Fifteen joint components were estimated from the target-related ERP time courses and fMRI activation maps via the jICA. One joint component was found to distinguish patients and controls using a two-sample t-test ($p < 0.0001$) on patient and control loading parameters. This identified component shows a clear difference in fMRI at bilateral frontotemporal regions implicated in schizophrenia (Fig. 12.5 right) and in ERP at times during the N2/P3 complex (Fig. 12.5 left) which have been previously implicated in patients.

In the same way as for Fig. 12.3 significant voxels/time points were used to generate an ERP versus fMRI histogram for controls (orange) and patients (blue), shown in Fig. 12.6. The controls are clearly showing increases in both fMRI and ERP data. On the right side of Fig. 12.6 is illustrated a simple example motivating the use of a joint analysis of multiple modalities.

12.6.4 Structural Magnetic Resonance Imaging–Diffusion Tensor Imaging

We now present an example showing gray matter and fractional anisotropy maps in a joint ICA analysis in 11 participants. Seven components were estimated and a template of the occipital lobe generated from a previous study

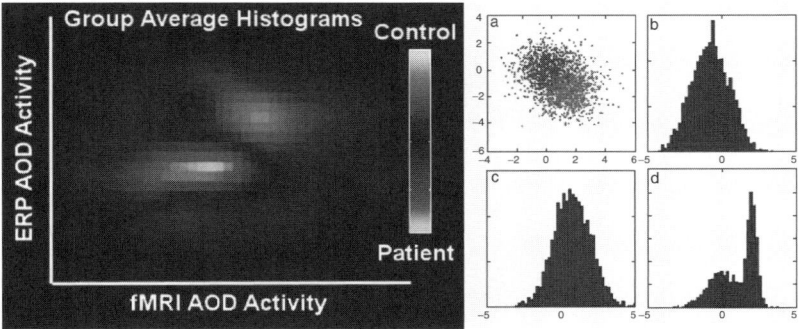

Fig. 12.6. ERP/fMRI Histograms: (*left*) Joint histograms for patients (blue; lower left corner) and controls (orange; upper right corner). (*right*) Simulated data from two Gaussians (**a**) showing a case in which marginal histograms (**b**, **c**) are less able to detect differences between groups whereas the histogram in the direction of maximal separation (**d**) clearly shows the two distributions from patients and controls

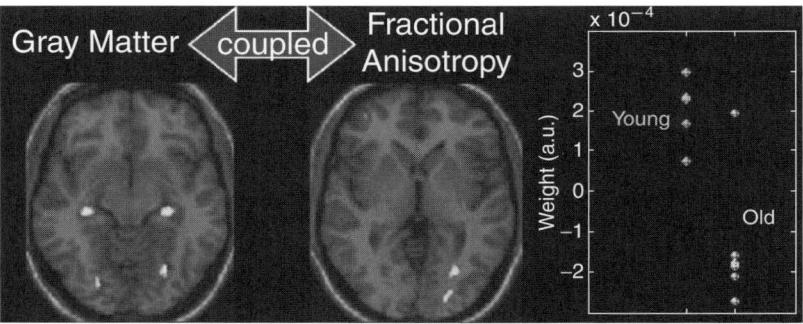

Fig. 12.7. GM and FA Spatial Correspondence (*left*) and corresponding weights (*right*) comparing older and younger participants ($p < 0.002$)

was used to select the FA and joint GM map. A picture of the resulting joint source (for each modality, the slice corresponding to the maximal coupling is displayed) is shown in Fig. 12.7 and demonstrates GM regions in occipital lobe and bilateral lateral geniculate regions are associated with white matter occipital lobe regions consistent with the optic radiations (that is, higher FA in optic radiations is associated with lower GM values in lateral geniculate and visual cortex).

The group was then split into an old (mean age 63 ± 10) and young (mean age 44 ± 10) cohort. The weight parameter calculated from the earlier ICA estimation is plotted in Fig. 12.7 (right) as a function of cohort membership. A highly significant difference was observed with young participants showing higher FA and GM (see) and older participants showing lower FA and GM. This is consistent with loss of gray matter volume (and white matter FA) with age.

Analyzing GM maps and FA maps together can be challenging as the FA images are warped due to the various gradient directions used. In the above analysis, GM and FA are coupled together at an *image* level, but not at a voxel level. Thus misregistration between image-types will not directly affect the results.

12.6.5 Parallel Independent Component Analysis

As discussed in Sect. 12.5, the strong regularization imposed by the jICA framework can be relaxed in a number of ways to allow for more flexibility in the estimation. One such approach we investigated is called parallel independent component analysis (paraICA). As a framework to investigate the integration of data from two imaging modalities, this method is dedicated to identify components of both modalities and connections between them through enhancing intrinsic interrelationships. We have applied this approach to link fMRI/ERP data and also fMRI and genetic data (single nucleotide polymorphism arrays) [36, 37]. Results show that paraICA provides stable results and can identify the linked components with a relatively high accuracy.

The result for fMRI/ERP data is consistent with that found by the joint ICA algorithm [10], where a shared mixing matrix is used for both modalities. The fundamental difference is that paraICA assumes the fMRI and ERP data are mixed in a similar pattern but not identically. ParaICA pays more attention to individually linked components and their connections, while the joint ICA studies intereffects between EEG and fMRI as a whole [10]. It provides a promising way to analyze the detail coupling between hemodynamics and neural activation.

The methods in this section were computed using a new Matlab toolbox called Fusion ICA Toolbox or FIT (http://icatb.sourceforge.net).

12.7 Selection of Joint Components

In some cases, it is important to define criteria for selecting among joint components. For example, when studying two groups, e.g., patients and controls, we may want to determine which combination of joint features is optimal in some sense. We apply our approach to the problem of identifying image-based biomarkers in schizophrenia. A *biomarker* is a characteristic that is objectively measured and evaluated as an indicator of normal biologic processes, pathogenic processes, or pharmacologic responses to a therapeutic intervention. In our proposal, a biomarker would be the joint component resulting from the optimal combination of fMRI, EEG, and sMRI features. We have used two criteria for this, separation of the mixing parameters and separation of the source distributions.

The estimated source distributions for these components are computed separately for each group and the Kullback–Leibler (KL) divergence is computed between the patient (sz) and control (hc) distributions (i.e., $D_{\text{KL}}(p_{sz}(\mathbf{f})$

$||p_{hc}(\mathbf{f}))$, where $\mathbf{f}$ is a multidimensional feature/modality vector). There are several possible divergence measures including J, KL, and Renyi. J divergence is simply a symmetric KL divergence, such that:

$$D_J = (D_{KL}(p,q) + D_{KL}(q,p))/2.$$

Typical reasons for selecting one divergence over another are the ability to express the solution analytically given a particular assumed distributional form or to sensitize the divergence to particular distributional features [e.g., minimizing the Renyi (alpha) divergence with $\alpha = 0.5$ has been shown to be optimal for separating similar feature densities [25]; the limiting case of $\alpha = 1$ results in the KL divergence].

We compute features for (1) AOD target-related fMRI activity (AOD_T), (2) AOD novel-related fMRI activity (AOD_N), (3) SB recognition fMRI activity (SB), and (4) sMRI GM values (GM). All four features were collected on each of 15 controls and 15 patients with schizophrenia.

To evaluate the impact of the choice of divergence measures, we have computed results for several divergence measures. An example of the results is shown in Table 12.2. The number in parenthesis indicates the relative gain

Table 12.2. Several divergence measures for three combinations of fMRI features

	KL	J	Renyi (0.5)
Oddball and Sternberg	0.497 (7.4)	0.480 (5.9)	0.2874 (7.4)
Oddball	0.067 (1.3)	0.081 (1.6)	0.039 (1.5)
Sternberg	0.050	0.050	0.0254

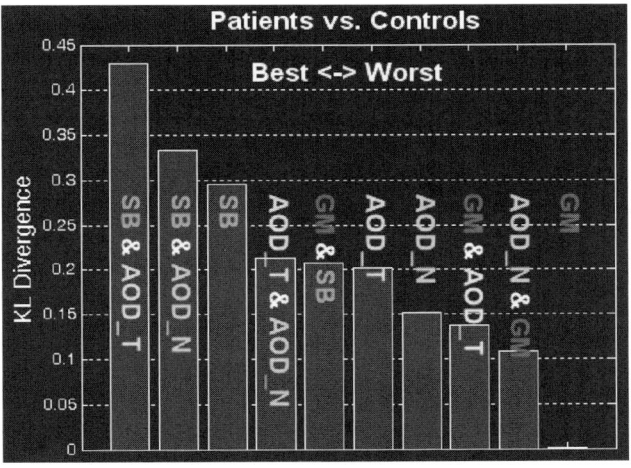

Fig. 12.8. Example of evaluation of KL Divergence for combinations of features comparing two-way fusion with nonfusion. Larger KL Divergence values represent better patient versus control separation

(ratio of current row/previous row) where larger numbers indicate better patient/control separation. We compare the KL and Renyi divergences with alpha = 0.5, as well as their symmetric counterparts.

The results shown in Fig. 12.8 are ranked according to a divergence measure, to determine which combination of features/modalities provides better separation. For example, the result shows us that combining the SB feature with the AOD_N or AOD_T features provides increased separation beyond SB or AOD_T alone. It also suggests that the incorporation of GM tends to decrease the separation. Note that though we have demonstrated above a comparison of patients and controls, this method is a general one, useful for studying a variety of questions. For example, instead of patient versus controls, we could have examined age-related activity only in healthy controls.

12.8 Conclusion

We present a general framework for combining different types of brain imaging data at the group level via features computed from each data type. We also show that by combining modalities in certain ways performance is improved. This approach enables us to take advantages of the strengths and limitations of various modalities in a unified analytic framework and demonstrates that data fusion techniques can be successfully applied to joint brain imaging data to reveal unique information that cannot be evaluated in any one modality. Here, we do not compare jICA with any classification method, since the main purpose of jICA is to discover the components from multiple modalities, which share similar correspondences to subjects.

Acknowledgments

This research was supported in part by grants from the National Institutes of Health under grants 1 R01 EB 000840, R01 EB 005846, and R01 EB 006841 and the National Science Foundation under grant NSF-IIS 0612076.

References

1. Ardnt, C.: Information gained by data fusion. In: SPIE Proceedings (1996)
2. Ashburner, J., Friston, K.J.: Voxel-based morphometry – the methods. NeuroImage **11**, 805–821 (2000)
3. Basser, P.J.: Inferring microstructural features and the physiological state of tissues from diffusion-weighted images. NMR Biomed. **8**, 333–344 (1995)
4. Basser, P.J., Mattiello, J., LeBihan, D.: Estimation of the effective self-diffusion tensor from the NMR spin echo. J. Magn. Reson. B **103**, 247–254 (1994)

5. Basser, P.J., Pierpaoli, C.: Microstructural and physiological features of tissues elucidated by quantitative-diffusion-tensor MRI. J. Magn. Reson. B **111**, 209–219 (1996)
6. Bezdek, J.C., Hall, L.O., Clarke, L.P.: Review of MR image segmentation techniques using pattern recognition. [review]. Med. Phys. **20**, 1033–1048 (1993)
7. Bihan, D.L., Breton, E., Lallemand, D., Grenier, P., Cabanis, E., Laval-Jeantet, M.: MR imaging of intravoxel incoherent motions: Application to diffusion and perfusion in neurologic disorders. Radiology **161**, 401–407 (1986)
8. Calhoun, V.D., Adali, T., Giuliani, N., Pekar, J.J., Pearlson, G.D., Kiehl, K.A.: A method for multimodal analysis of independent source differences in schizophrenia: Combining gray matter structural and auditory oddball functional data. Hum. Brain Map. **27**, 47–62 (2006)
9. Calhoun, V.D., Adali, T., Kiehl, K.A., Astur, R.S., Pekar, J.J., Pearlson, G.D.: A method for multi-task fMRI data fusion applied to schizophrenia. Hum. Brain Map. **27**, 598–610 (2006)
10. Calhoun, V.D., Adali, T., Pearlson, G.D., Kiehl, K.A.: Neuronal chronometry of target detection: Fusion of hemodynamic and event-related potential data. NeuroImage **30**, 544–553 (2006)
11. Calhoun, V.D., Adali, T., Pearlson, G.D., Pekar, J.J.: Spatial and temporal independent component analysis of functional MRI data containing a pair of task-related waveforms. Hum. Brain Map. **13**, 43–53 (2001)
12. Calhoun, V.D., Keihl, K.A., Pearlson, G.D.: A method for multi-task fMRI data fusion applied to schizophrenia. NeuroImage **26** (2005)
13. Calhoun, V.D., Pearlson, G.D., Kiehl, K.A.: Neuronal chronometry of target detection: Fusion of hemodynamic and event-related potential data. NeuroImage **30**, 544–553 (2006)
14. Calhoun, V.D., Pekar, J.J., McGinty, V.B., Adali, T., Watson, T.D., Pearlson, G.D.: Different activation dynamics in multiple neural systems during simulated driving. Hum. Brain Map. **16**, 158–167 (2002)
15. Cohen, M.S., DuBois, R.M., Zeineh, M.M.: Rapid and effective correction of rf inhomogeneity for high field magnetic resonance imaging. Hum. Brain Map. **10**, 204–211 (2000)
16. Dale, A.M., Halgren, E., Lewine, J.D., Buckner, R.L., Paulson, K., Marinkovic, K., Rosen, B.R.: Spatio-temporal localization of cortical word repetition effects in a size judgement task using combined fMRI/MEG. NeuroImage **5**, 592–592 (1997)
17. Fischl, B., Dale, A.M.: Cerebral cortex from magnetic resonance images. Proc. Natl. Acad. Sci. U.S.A **97**, 11050–11055 (2000)
18. Fox, P.T.: The growth of human brain mapping. Hum. Brain Map. **5**, 1–2 (1997)
19. Friston, K., Ashburner, J., Frith, C.D., Poline, J.P., Heather, J.D., Frackowiak, R.S.: Spatial registration and normalization of images. Hum. Brain Map. **2**, 165–189 (1995)
20. Friston, K., Jezzard, P., Turner, R.: Analysis of functional MRI time-series. Hum. Brain Map. **10**, 153–171 (1994)
21. Friston, K.J., Harrison, L., Penny, W.: Dynamic causal modelling. NeuroImage **19**, 1273–1302 (2003)
22. George, J.S., Aine, C.J., Mosher, J.C., Schmidt, D.M., Ranken, D.M., Schlitt, H., Wood, C.C., Lewine, J.D., Sanders, J.A., Belliveau, J.W.: Mapping function in the human brain with magnetoencephalography, anatomical

magnetic resonance imaging, and functional magnetic resonance imaging. J. Clin. Neurophysiol. **12**, 406–431 (1995)
23. Gerig, G., Martin, J., Kikinis, R., Kubler, O., Shenton, M., Jolesz, F.: Unsupervised tissue type segmentation of 3D dual-echo MR head data. IVC **10**, 349–360 (1992)
24. Giuliani, N., Calhoun, V.D., Pearlson, G.D., Francis, A., Buchanan, R.W.: Voxel-based morphometry versus regions of interest: A comparison of two methods for analyzing gray matter disturbances in schizophrenia. Schizophr. Res. **74**, 135–147 (2005)
25. Hero, A., Ma, B., Michel, O., Gorman, J.: Alpha-divergence for classification, indexing and retrieval. Technical Report 328, Dept. EECS, University of Michigan, Ann Arbor (2001)
26. Horwitz, B., Poeppel, D.: How can EEG/MEG and fMRI/PET data be combined? Hum. Brain Map. **17**, 1–3 (2002)
27. Husain, F.T., Nandipati, G., Braun, A.R., Cohen, L.G., Tagamets, M.A., Horwitz, B.: Simulating transcranial magnetic stimulation during PET with a large-scale neural network model of the prefrontal cortex and the visual system. NeuroImage **15**, 58–73 (2002)
28. Hyvärinen, A.: Survey on independent component analysis. Neural Comput. Surveys **2**, 94–128 (1999)
29. Johnson, M.R., Morris, N., Astur, R.S., Calhoun, V.D., Kieh, K.A., Pearlson, G.D.: Schizophrenia and working memory: A closer look at fMRI of the dorsolateral prefrontal cortex during a working memory task. In: Proceedings of CNS (2005)
30. Jung, T.P., Makeig, S., Humphries, C., Lee, T.W., McKeown, M.J., Iragui, V., Sejnowski, T.J.: Removing electroencephalographic artifacts by blind source separation. Psychophysiology **37**, 163–178 (2000)
31. Jung, T.P., Makeig, S., Westerfield, M., Townsend, J., Courchesne, E., Sejnowski, T.J.: Analysis and visualization of single-trial event-related potentials. Hum. Brain Map. **14**, 333–344 (2001)
32. Kiehl, K.A., Smith, A.M., Hare, R.D., Liddle, P.F.: An event-related potential investigation of response inhibition in schizophrenia and psychopathy. Biol. Psychiat. **48**, 210–221 (2000)
33. Kiehl, K.A., Stevens, M., Laurens, K.R., Pearlson, G.D., Calhoun, V.D., Liddle, P.F.: An adaptive reflexive processing model of neurocognitive function: Supporting evidence from a large scale ($n = 100$) fMRI study of an auditory oddball task. NeuroImage **25**, 899–915 (2005)
34. Kim, D.S., Ronen, I., Formisano, E., Kim, K.H., Kim, M., van Zijl, P., Ugurbil, K., Mori, S., Goebel, R.: Simultaneous mapping of functional maps and axonal connectivity in cat visual cortex. In: Proceedings of HBM, Sendai, Japan (2003)
35. Lange, N., Strother, S.C., Anderson, J.R., Nielsen, F.A., Holmes, A.P., Kolenda, T., Savoy, R., Hansen, L.K.: Plurality and resemblance in fmri data analysis. NeuroImage **10**, 282–303 (1999)
36. Liu, J., Calhoun, V.D.: Parallel independent component analysis for multimodal analysis: Application to fMRI and EEG data. In: Proceedings of ISBI (2007)
37. Liu, J., Pearlson, G.D., Windemuth, A., Ruano, G., Perrone-Bizzozero, N.I., Calhoun, V.D.: Combining fMRI and SNP data to investigate connections between brain function and genetics using parallel ICA, Hum. Brain Map., In Press

38. Makeig, S., Jung, T.P., Bell, A.J., Ghahremani, D., Sejnowski, T.J.: Blind separation of auditory event-related brain responses into independent components. Proc. Natl. Acad. Sci. **94**, 10,979–10,984 (1997)
39. McIntosh, A.R., Bookstein, F.L., Haxby, J.V., Grady, C.L.: Spatial pattern analysis of functional brain images using partial least squares. NeuroImage **3**, 143–157 (1996)
40. McIntosh, A.R., Gonzalez-Lima, F.: Structural equation modeling and its application to network analysis in functional brain imaging. Hum. Brain Map. **2**, 2–22 (1994)
41. McKeown, M.J., Makeig, S., Brown, G.G., Jung, T.P., Kindermann, S.S., Bell, A.J., Sejnowski, T.J.: Analysis of fMRI data by blind separation into independent spatial components. Hum. Brain Map. **6**, 160–188 (1998)
42. Mechelli, A., Price, C.J., Noppeney, U., Friston, K.J.: A dynamic causal modeling study on category effects: Bottom-up or top-down mediation? J. Cogn. Neurosci. **15**, 925–934 (2003)
43. Miller, M.I., Trouve, A., Younes, L.: Euler–Lagrange equations of computational anatomy. Annu. Rev. Biomed. Eng **4**, 375–405 (2002)
44. van de Moortele, P.F., Cerf, B., Lobel, E., Paradis, A.L., Faurion, A., Bihan, D.L.: Latencies in fmri time-series: Effect of slice acquisition order and perception. NMR Biomed. **10**, 230–236 (1997)
45. Pajevic, S., Pierpaoli, C.: Color schemes to represent the orientation of anisotropic tissues from diffusion tensor data: Application to white matter fiber tract mapping in the human brain. Magn. Reson. Med. **42**, 526–540 (1999)
46. Ramnani, N., Lee, L., Mechelli, A., Phillips, C., Roebroeck, A., Formisano, E.: Exploring brain connectivity: A new frontier in systems neuroscience. functional brain connectivity. In: Trends Neurosci., vol. 25, pp. 496–497. Dusseldorf, Germany (2002)
47. Savopol, F., Armenakis, C.: Mergine of heterogeneous data for emergency mapping: Data integration or data fusion? In: Proceedings of ISPRS (2002)
48. Thomsen, T., Specht, K., Hammar, A., Nyttingnes, J., Ersland, L., Hugdahl, K.: Brain localization of attentional control in different age groups by combining functional and structural MRI. NeuroImage **22**, 912–919 (2004)
49. Vitacco, D., Brandeis, D., Pascual-Marqui, R., Martin, E.: Correspondence of event-related potential tomography and functional magnetic resonance imaging during language processing. Hum. Brain Map. **17**, 4–12 (2002)
50. Warfield, S.K., Kaus, M., Jolesz, F.A., Kikinis, R.: Adaptive, template moderated, spatially varying statistical classification. Med. Image Anal. **4**, 43–55 (2000)
51. Wells, W.M., Grimson, W.E.L., Kikinis, R., Jolesz, F.A.: Adaptive segmentation of mri data. IEEE Trans. Med. Imag. **20**, 429–442 (1996)
52. Worsley, K.J., Andermann, M., Koulis, T., MacDonald, D., Evans, A.C.: Detecting changes in nonisotropic images. Hum. Brain Map. **8**, 98–101 (1999)
53. Worsley, K.J., Cao, J., Paus, T., Petrides, M., Evans, A.C.: Applications of random field theory to functional connectivity. Hum. Brain Map. **6**, 364–367 (1998)

Part IV

Knowledge Extraction in Brain Science

13

Complex Empirical Mode Decomposition for Multichannel Information Fusion

Danilo P. Mandic, George Souretis, Wai Yie Leong, David Looney, Marc M. Van Hulle, and Toshihisa Tanaka

Information "fusion" via signal "fission" is addressed in the framework of empirical mode decomposition (EMD). In this way, a general nonlinear and non-stationary signal is first decomposed into its oscillatory components (fission); the components of interest are then combined in an ad hoc or automated fashion to provide greater knowledge about a process in hand (fusion). The extension to the field of complex numbers $\mathbb{C}$ is particularly important for the analysis of phase-dependent processes, such as those coming from sensor arrays. This allows us to combine the data driven nature of EMD with the power of complex algebra to model amplitude-phase relationships within multichannel data. The analysis shows that the extensions of EMD to $\mathbb{C}$ are not straightforward and that they critically depend on the criterion for finding local extrema within a complex signal. For rigour, convergence of EMD is addressed within the framework of fixed point theory. Simulation examples on information fusion for brain computer interface (BCI) support the analysis.

13.1 Introduction

Signals obtained by standard data acquisition techniques are real-valued and naturally, the processing of such signals is performed in the field of real numbers $\mathbb{R}$ (or $\mathbb{R}^n$ for multichannel recordings). Notice, however, that a complex representation of a real signal can be both intuitive and useful, since this way both the amplitude and phase relationships between the data streams can be modelled simultaneously. Additionally, several important signal processing areas (telecommunications, sonar, radar, etc., to mention but a few) use complex-valued data structures.

For single channel signal analysis, Fourier spectral theory is the best established tool for the complex (frequency) domain processing of linear and stationary signals. Since real-world signals are typically non-stationary (and nonlinear), for which Fourier analysis is not well suited, we need to resort to time frequency analysis techniques such as the short time Fourier transform

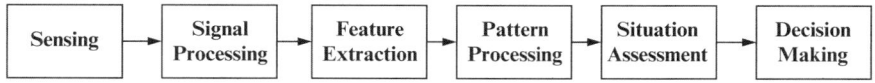

Fig. 13.1. The "waterfall model" of information fusion

(STFT) and wavelet transform (WT). Despite the power of these techniques, they still rely on some kind of projection on a set of pre-defined bases; this makes some areas of their application, particularly when focusing on high frequency content, rather limited [7]. An alternative approach is empirical mode decomposition (EMD) [7], a data driven time–frequency analysis technique that has an enormous appeal for nonlinear, non-stationary signals.

13.1.1 Data Fusion Principles

Knowledge extraction and data fusion are at the very core of modern technology, their principles have been long studied in various areas of information theory, signal processing and computing [5, 12, 16, 17]. A recent overview of data and sensor fusion models can be found in [12], whereas a sequential data fusion model in the field of complex numbers $\mathbb{C}$ is introduced in [11]. One well-established information fusion model is the so-called waterfall model, shown in Fig. 13.1. This framework comprises several stages of processing, starting from the raw data acquisition through to situation assessment and decision making stages. Notice that the algorithms for signal processing, feature extraction, and pattern processing naturally belong to the area of statistical signal processing.

By design, EMD has a natural ability to perform both signal conditioning (denoising) and feature extraction. We further show that EMD provides a unifying framework for both the information *fission* (the phenomenon by which observed information is decomposed into a set its components) and *fusion*, whereby several stages from Fig. 13.1, such as Signal Processing, Feature Extraction and Situation Assessment can be naturally performed by EMD. The Decision Making stage can be achieved based on EMD either by inspection or by post-processing in the form of some machine learning algorithm.

The aim of this work is therefore to provide theoretical and practical justification for the use of both real and complex EMD in knowledge extraction and information fusion. For convenience, this is achieved by an extension of the standard, real-valued, EMD method to the field of complex numbers $\mathbb{C}$. The analysis is general and is supported by an extensive set of simulations on point processes, such as spike trains coming from networks of spiking neurons.

13.2 Empirical Mode Decomposition

Empirical mode decomposition [7] is a technique to adaptively decompose a given signal, by means of a process called the sifting algorithm, into a finite set of oscillatory components. These components, called "intrinsic mode functions" (IMFs), represent the oscillation modes embedded in the data. The IMFs act as a naturally derived set of basis functions for the signal; EMD can thus be seen as an exploratory data analysis technique. In fact, EMD and the Hilbert–Huang transform comprise the so-called "Hilbert Spectral Analysis" [7]; a unique spectral analysis technique employing the concept of instantaneous frequency. In general, the EMD aims at representing an arbitrary signal via a number of IMFs and the residual. More precisely, for a real-valued signal $x[k]$, the EMD performs the mapping

$$x[k] = \sum_{i=1}^{N} c_i[k] + r[k], \qquad (13.1)$$

where the $c_i[k]$, $i = 1, \ldots, N$ denote the set of IMFs and $r[k]$ is the trend within the data (also referred to as the last IMF or residual). By design, an IMF is a function which is characterized by the following two properties: the upper and lower envelope are symmetric; and the number of zero-crossings and the number of extrema are exactly equal or they differ at most by one.

To extract the IMFs from a real-world signal, the *sifting* algorithm, which is described in Table 13.1 is employed. Following the sifting process, the Hilbert transform can be applied to each IMF separately. This way, it is possible to generate analytic signals, having an IMF as the real part and its Hilbert transform as the imaginary part, that is $x + j\mathcal{H}(x)$ where $\mathcal{H}$ is the Hilbert transform operator. Equation (13.1) can therefore be augmented to its analytic form given by

Table 13.1. The EMD algorithm

---

1. Connect the local maxima of the signal with a spline. Let U denote the spline that forms the upper envelope of the signal;
2. Connect the local minima of the signal with a spline. Let L denote the spline that forms the lower envelope of the signal;
3. Subtract the mean envelope $m = \frac{U+L}{2}$ from the signal to obtain a proto-IMF;
4. Repeat Steps 1, 2 and 3 above until the resulting signal is a proper IMF (as described above). The IMF requirements are checked indirectly by evaluating a stoppage criterion, originally proposed as:

$$\sum_{k=0}^{T} |h_{n-1}[k] - h_n[k]|^2 \, h_{n-1}^2[k] \leq SD \qquad (13.2)$$

where $h_n[k]$ and $h_{n-1}[k]$ represent two successive sifting iterates. The SD value is usually set to 0.2–0.3;
5. After finding an IMF, this same IMF is subtracted from the signal. The residual is regarded as new data and fed back to Step 1 of the algorithm;
6. The algorithm is completed when the residual of Step 5 is a monotonic function. The last residual is considered to be the trend.

$$X(t) = \sum_{i=1}^{n} a_i(t) e^{j\theta_i(t)}, \tag{13.3}$$

where the trend $r(t)$ is purposely omitted, due to its overwhelming power and lack of oscillatory behaviour. Observe from (13.3), that now the time-dependent amplitude $a_i(t)$ can be extracted directly and that we can also make use of the phase function $\theta_i(t)$. Furthermore, the quantity $f_i(t) = \frac{d\theta_i}{dt}$ represents the instantaneous frequency [2]; this way by plotting the amplitude $a_i(t)$ versus time t and frequency $f_i(t)$, we obtain a time–frequency–amplitude representation of the entire signal called the Hilbert Spectrum. It is this combination of the concept of instantaneous frequency and EMD that makes the framework so powerful as a signal decomposition tool.

To illustrate the operation of EMD, consider a signal which consists of two added sine waves with different frequencies, shown in Fig. 13.2 (left) (the original signal is in the first row, followed by the corresponding IMFs C1–C2, for convenience the last IMF is denoted by R3), and its Hilbert spectrum (Fig. 13.2 (right)). Observe the "fission" process performed by EMD, whereby the original signal is split into a set of IMFs (C1–C2) and residual (R3). The frequencies of the sine waves composing the original signal are clearly visible in the time–frequency spectrum (Fig. 13.2 (right)).

Another example of the usefulness of EMD in both the "Signal Processing" and "Feature Extraction" module from the "waterfall" information fusion model from Fig. 13.1 is the so-called blood volume estimation [14]. In this application, a sensor in human heart records both the blood flow (near sine wave) and the superimposed electrocardiogram (ECG). The goal is to extract pure blood volume signal, since this information is crucially important in heart surgery. This is also a very difficult signal processing problem since the ECG and blood volume are coupled and standard signal processing techniques (both supervised and blind) are bound to fail. By applying EMD to this problem, we were able to both denoise the useful blood volume signal (remove the ECG

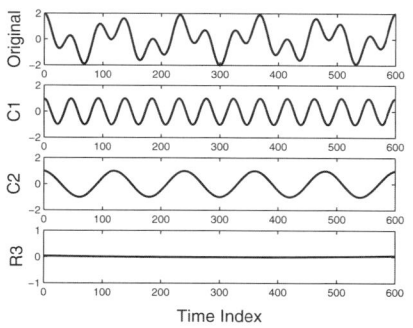 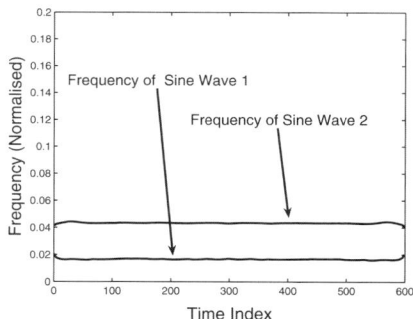

Fig. 13.2. EMD of two concatenated sine waves. (*Left*) Intrinsic mode functions and (*right*) Hilbert spectrum

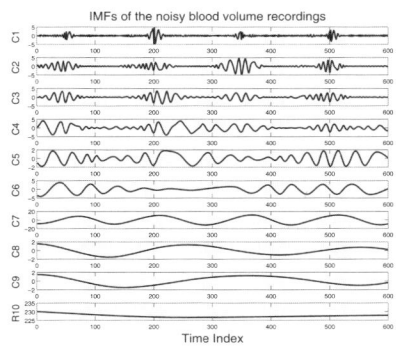
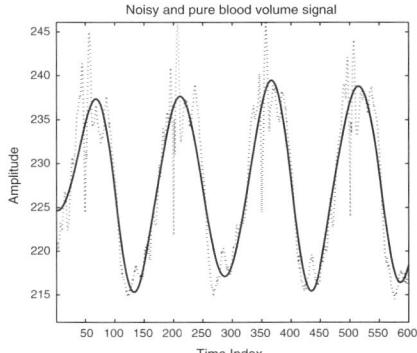

Fig. 13.3. EMD for *signal processing* (denoising) and *feature extraction* from coupled noisy blood volume recordings. (*Left*) Intrinsic mode functions; (*right*) The original noisy blood volume signal (*dotted line*) and the extracted pure blood volume constructed by fusion of IMFs C7–C9 (*solid line*)

artefact) and to extract (by visual inspection) the IMFs which represents blood volume (IMFs C7–C9), as illustrated in Fig. 13.3. This demonstrates the "fusion" via "fission" abilities of EMD.

13.3 Ensemble Empirical Mode Decomposition

A drawback of EMD is the appearance of disparate scales across *different* IMF components, known as mode mixing. This is often the result of signal intermittency and can leave the IMF components devoid of physical meaning. One answer to this problem is a noise-assisted data analysis (NADA) method called Ensemble EMD (EEMD) [6]. The EEMD defines the true IMF components as the mean of an ensemble of trials, each consisting of the signal corrupted by additive white noise of finite variance. The principle of the EEMD is simple: the added white noise populates the whole time–frequency space uniformly, facilitating a natural separation of the frequency scales that reduces the occurrence of mode mixing.

By an ensemble average, the effect of the added white noise with standard deviation ε decreases as

$$\varepsilon_n = \frac{\varepsilon}{\sqrt{N}} \quad \Leftrightarrow \quad ln\,\varepsilon_n + \frac{\varepsilon}{2}\ln N = 0, \tag{13.4}$$

where N is the number of ensemble members, and ε_n is defined as the final standard deviation of for each ensemble member (a difference between the input signal and the sum of corresponding IMFs).

To illustrate the performance of EEMD, we consider a set of highly noisy recordings of brain neuronal activity obtained from a cortex implant (trains

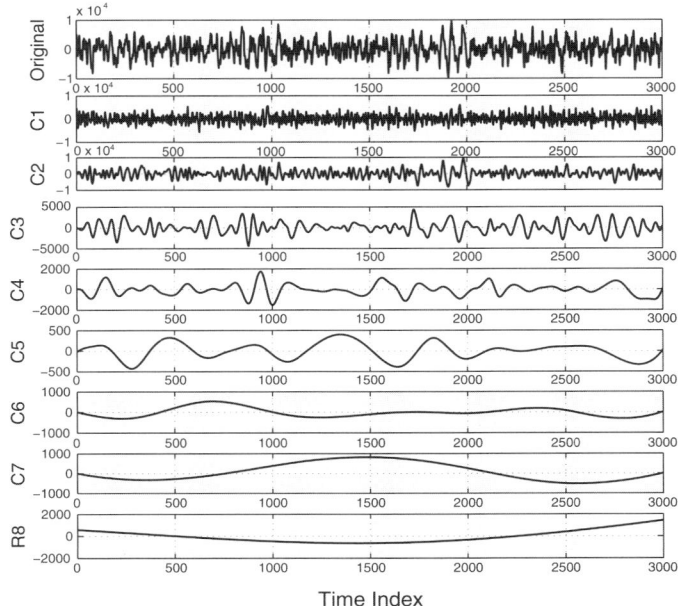

Fig. 13.4. Ensemble empirical mode decomposition of recorded neuronal spikes

of spiking neurons). The data set was obtained within the European Project Neuroprobes (www.neuroprobes.org); the data were sampled at 44 kHz, and the Probe NP-PA-09 was used for which the impedance was 28.8 $k\Omega$. The neuronal spikes were produced from the stimulation of the tongue of a monkey; although the spikes of neuronal activity were recorded, they could not be detected due to a very high background noise level. Figure 13.4 shows the recorded signals and the IMFs (C1–R8) produced by the EEMD algorithm; notice that spiky signals were detected in IMF C4 (due to the analogue amplifiers these spikes are somewhat integrated). The IMF of interest was then compared with the originally recorded noisy signal in Fig. 13.5, clearly indicating the ability of EEMD to extract the signal of interest.

To further demonstrate the knowledge extraction abilities of EMD with real data, we consider its application in the removal of unwanted ocular artefacts (caused by eye activity of the subject) from electroencephalogram (EEG) data. The unwanted artefacts are a major obstacle for real-time brain computer interface (BCI) applications, since they occupy the low frequency range of EEG (0–13 Hz) and cannot be removed by conventional filtering methods due to overlapping of their respective spectra in a non-stationary and nonlinear environment. To equip EMD to perform the *Decision Making* stage of the information fusion model (Fig. 13.1), in the postprocessing stage we employed an adaptive filtering type machine learning algorithm which automatically combined the relevant IMFs to produce the signal with the artefact removed. This is verified on the EEG data obtained from the Fp1 electrode (placed

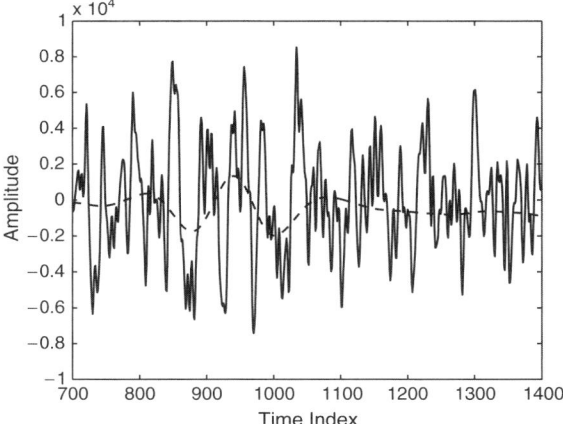

Fig. 13.5. The main component of the extracted brain neuronal recording, that is, IMF C4 from Fig. 13.4 (*thick dotted line*) compared with the original noisy signal (*solid line*)

close to the forehead); the original signal and several of the extracted IMFs are plotted in Fig. 13.6. Note the inherent "fission", that is, the original signal was decomposed into a number of its oscillating modes. An artefact-free reconstruction of the EEG signal (see Fig. 13.7) was achieved by fusion of the relevant IMFs, that is the IMFs that correspond to instantaneous frequencies outside of the range dominated by the ocular artefacts [9]. The fusion process employed a block-based approximation [18], based on minimisation of the mean square error.

We now investigate the singular problem of detecting the ocular peaks as they occur in the EEG signal. Consider again Fig. 13.6 and the IMFs obtained in regions contaminated by an artefact. Since there is no discernable change in the frequencies of the IMFs surrounding artefact events, standard Hilbert–Huang analysis will fail to detect the presence of the relatively significant change in the signal dynamics, a shortcoming of standard EMD signal analysis. After a visual inspection of the IMFs, it is clear that several of the IMF peaks coincide at the time instant of the artefact. In other words, the IMFs are in phase when the signal peak occurs. This example demonstrates that even in the case of real-valued data, a rigorous analysis of phase (joint complex valued amplitude and phase) activity is necessary for complete analysis.

13.4 Extending EMD to the Complex Domain

Complex data are a bivariate quantity with a mutual dependence between the real and imaginary components. Additionally, arithmetic operations on $\mathbb{C}$ form a unique algebra. Although an attractive solution might appear to

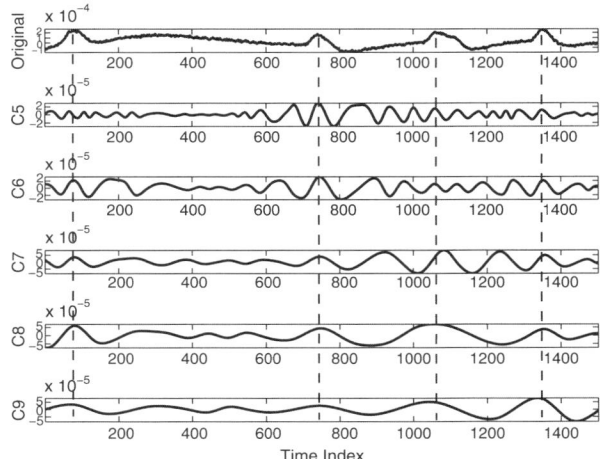

Fig. 13.6. The EEG signal, with blink artefacts at time instants 80, 740, 1,070 and 1,350, is displayed at the top of the figure. Additional plots show some of the extracted IMFs. Note there is no discernable change in frequency surrounding peaks (denoted on the plots by the *dashed line*), but that the IMF components at the event of spike are in phase

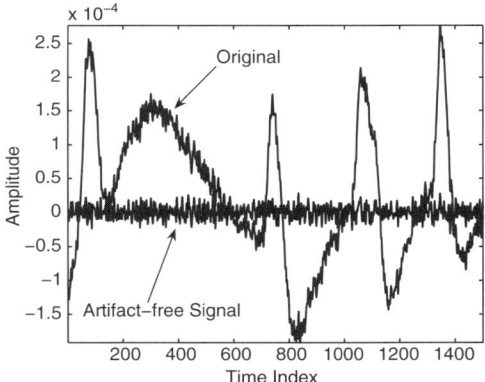

Fig. 13.7. Automated fusion of "relevant" IMFs to reconstruct artefact-free EEG signal (*solid line*)

be the application of real EMD to both the real and imaginary components separately, this mapping onto two independent univariate quantities will crucially ignore any mutual information that existed between the original components (phase). The same can be said, for example, of the separate application of EMD to the amplitude and phase functions of a complex data set. Furthermore, a critical aspect of EMD is that the IMFs have a physical interpretation.

We shall now examine the effectiveness of our two recently introduced extensions of EMD to the complex domain in the proposed framework. Each of the algorithms strive to retain capabilities and properties of real-valued EMD. The first method, termed "Complex Empirical Mode Decomposition", is based on the direct use of the properties of the Hilbert transform [15]. The second method, "Rotation Invariant Empirical Mode Decomposition", is a generic extension of the real EMD [1].

13.4.1 Complex Empirical Mode Decomposition

The first method to "complexify" EMD was introduced in 2007, termed Complex Empirical Mode Decomposition [15]. The method is rigorously derived, based on the inherent relationship between the positive and negative frequency components of a complex signal and the properties of the Hilbert transform. The idea behind this approach is rather intuitive: first note that a complex signal has a two-sided, asymmetric spectrum. The complex signal can therefore be converted into a sum of analytic signals by a straightforward filtering operation that extracts the opposite sides of the spectrum. Direct analysis in $\mathbb{C}$ can subsequently be achieved by applying standard EMD to both the positive and negative frequency parts of the signal. Given a complex signal $x[k]$, real-valued components corresponding to the positive and negative sides of the spectrum can be extracted as

$$x_+[k] = \mathcal{R}\mathcal{F}^{-1}\left\{X(e^{j\omega}) \cdot H(e^{j\omega})\right\},$$
$$x_-[k] = \mathcal{R}\mathcal{F}^{-1}\left\{X(e^{j\omega}) \cdot H(e^{-j\omega})\right\} \quad (13.5)$$

where $\mathcal{F}^{-1}$ is the inverse Fourier transform, $\mathcal{R}$ is an operator that extracts the real part of a signal, and $H(e^{j\omega})$ is an ideal filter with the transfer function

$$H(e^{j\omega}) = \begin{cases} 1, & \omega > 0 \\ 0, & \omega < 0 \end{cases}.$$

Given that $x_+[k]$ and $x_-[k]$ are real-valued, standard EMD can be applied to obtain a set of IMFs for each analytic signal. This can be expressed as

$$x_+[k] = \sum_{i=1}^{N_+} x_i[k] + r_+[k],$$
$$x_-[k] = \sum_{i=-N_-}^{-1} x_i[k] + r_-[k], \quad (13.6)$$

where $\{x_i[k]\}_{i=1}^{N_+}$ and $\{x_i[k]\}_{i=-N_-}^{i=1}$ denote sets of IMFs corresponding, respectively, to $x_+[k]$ and $x_-[k]$, whereas $r_+[k]$ and $r_-[k]$ are the respective residuals. The original complex signal can be reconstructed by

$$x[k] = (x_+[k] + \jmath\mathcal{H}[x_+[k]]) + (x_-[k] + \jmath\mathcal{H}[x_-[k]])^*, \qquad (13.7)$$

where $\mathcal{H}$ is the Hilbert transform operator. To conclude the derivation, a single set of complex IMFs corresponding to the complex signal $x[k]$ is given by

$$x[k] = \sum_{i=N_-,i\neq 0}^{N_+} y_i[k] + r[k], \qquad (13.8)$$

where $r[k]$ is the residual and $y_i[k]$ is defined by

$$y_i[k] = \begin{cases} (x_+[k] + \jmath\mathcal{H}[x_+[k]]), & i = 1, \ldots, N_+, \\ (x_-[k] + \jmath\mathcal{H}[x_-[k]])^*, & i = -N_-, \ldots, -1 \end{cases}.$$

To illustrate this technique, consider a problem involving real-world data of several streams: neuronal spike modelling in BCI. The task is to detect any synchronisation that occurs between spiking neuron time series. For our simulations, two spiking neuron time series (see Fig. 13.8) were generated using the tool described in [13]. The simulation time was set to 0.2 s and the sampling frequency to 10 kHz. We note that there are synchronised spikes at time instants 750, 1,350 and $1,800 \times 10^{-1}$ ms.

Although the two time series ($x_1[k]$ and $x_2[k]$) are real-valued, a complex signal, $z[k] = x_1[k] + \jmath x_2[k]$, is constructed with the signals representing the real and imaginary parts, respectively. The IMFs and the Hilbert spectra corresponding to the positive and the negative frequency parts of the signal $z[k]$ are shown, respectively, in Figs. 13.9 and 13.10. Observe that complex EMD has failed to reveal any synchronised events between the data streams. To its advantage though, it has a straightforward, intuitive mathematical derivation. As an example of its mathematical robustness, it preserves the dyadic filter property of standard EMD. In other words, the algorithm acts as dyadic filter bank when processing complex noise [15].

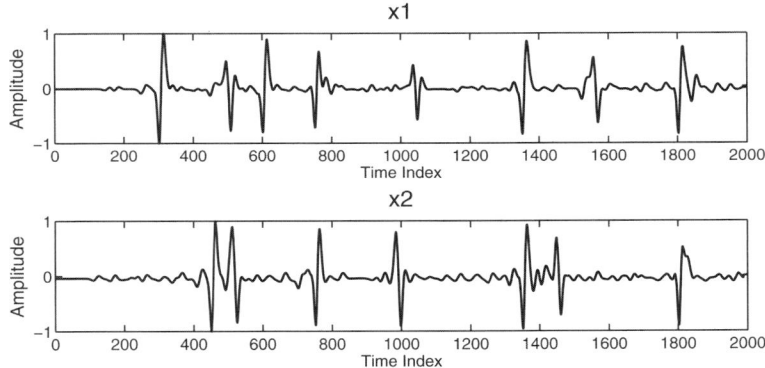

Fig. 13.8. Point processes x_1 and x_2, note the synchronised spikes at time instants 750, 1,350, and $1,800 \times 10^{-1}$ ms

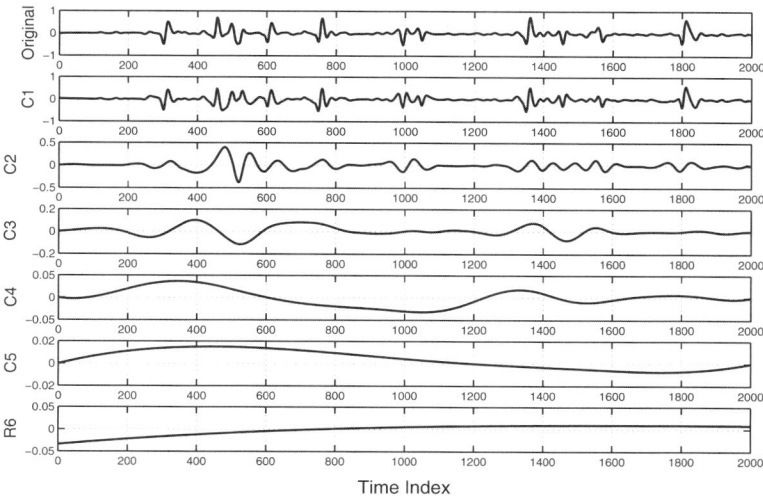

Fig. 13.9. IMFs corresponding to the positive frequencies of the signal $z[k] = x_1[k] + \jmath x_2[k]$

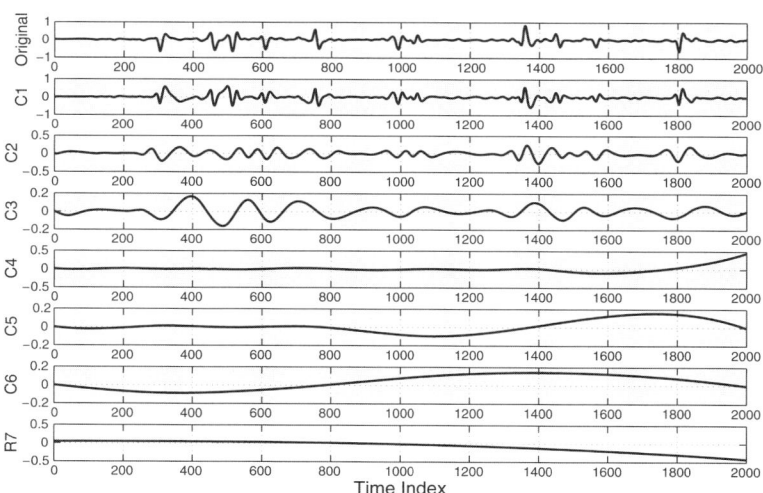

Fig. 13.10. The IMFs corresponding to the negative frequencies of the signal $z[k] = x_1[k] + \jmath x_2[k]$, note that compared to Fig. 13.10 there are a different number of IMFs

It is in fact the precise mathematical derivation that deprives the resulting IMFs of their physical connection with the original data set. This is demonstrated by observing Figs. 13.9 and 13.10 and noting that the number of IMFs obtained from the positive and negative part of the complex signal $z[k]$ are different. Another disadvantage is the ambiguity at zero frequency, a problem inherited from the properties of the Hilbert transform. Furthermore, the method cannot be extended to higher dimensions due to the limitation of representing a signal by its positive and negative frequencies. It should be finally noted that the mean should be removed before applying complex EMD as the Hilbert transform ignores constants.

$$\mathcal{H}\{x[k] + c\} = \mathcal{H}\{x[k]\}, \tag{13.9}$$

where $\mathcal{H}$ is the Hilbert transform operator.

13.4.2 Rotation Invariant Empirical Mode Decomposition

Rotation invariant EMD, introduced in [1], proposes a way of extending EMD theory so that it operates fully in $\mathbb{C}$. This is achieved through the use of complex splines. Unlike Complex EMD, by design this method generates an equal number of IMFs for the real and imaginary parts; these can be given physical meaning thus retaining an important property of real-valued EMD. A critical aspect of the derivation of EMD in $\mathbb{C}$ is the definition of an extremum. Several possible approaches are suggested in [1] such as the extrema of the modulus and the locus where the angle of the first-order derivative (with respect to time) changes sign; this way it can be assumed that a local maxima will be followed by a local minima (and vice versa). This definition is equivalent to the extrema of the imaginary part, that is, for a complex signal $Z(t)$ (for convenience we here use a continuous time index t)

$$\angle \dot{Z}(t) = 0 \Rightarrow \angle \{\dot{x}(t) + \jmath \cdot \dot{y}(t)\} = 0,$$
$$\Rightarrow tan^{-1}\frac{\dot{y}(t)}{\dot{x}(t)} = 0 \Rightarrow \dot{y}(t) = 0. \tag{13.10}$$

To illustrate the rotation invariant EMD, consider a set of wind[1] speed and direction measurements, which have been made complex, as illustrated in Fig. 13.11. The problem of the choice of the criterion for the extrema of the complex function and one IMF generated by the rotation invariant EMD are shown in Fig. 13.12. Clearly, the choice of criterion for finding the extrema of a complex signal is not unique, the extracted complex IMFs, however, do possess physical meaning; consider IMF 6 from Fig. 13.12 which reflects the level of detail for a given time-scale.

13.4.3 Complex EMD as Knowledge Extraction Tool for Brain Prosthetics

Consider again the complex signal $z[k] = x_1[k] + \jmath x_2[k]$ from Sect. 13.4.2, constructed from the two real-valued spiking neuron time series. The task is to detect the event of spike synchronisation, that is when the two neurons fire

---

[1] Publicly available from *http://mesonet.agron.iastate.edu/request/awos/1min.php*

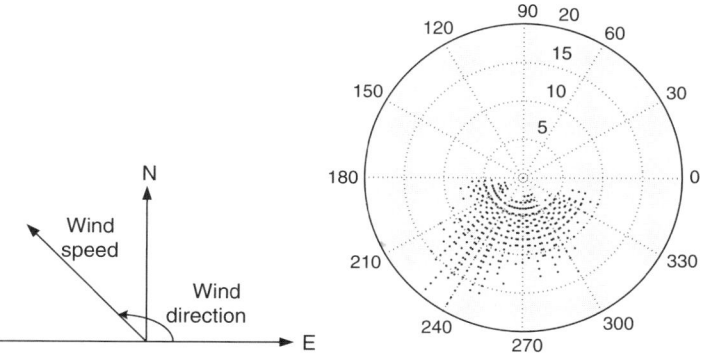

Fig. 13.11. Wind as a complex vector field. (left) A complex vector of wind speed v and direction d (ve^{jd}) and (right) the wind lattice [4], a set of complex valued wind data

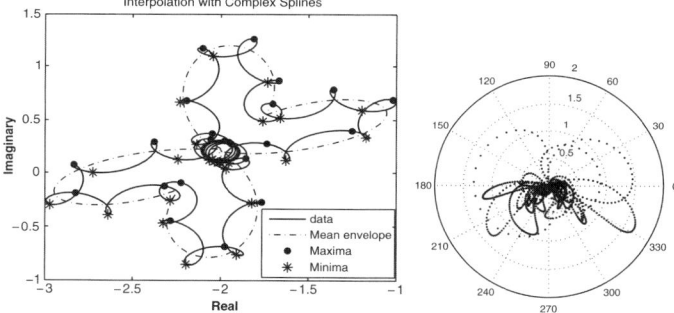

Fig. 13.12. Wind field analysed by the rotation invariant complex EMD. (left) The original complex signal and the extrema and (right) IMF 6 extracted from the wind data

simultaneously. Rotation invariant EMD allows us to achieve this, due to the equal number of IMFs for the real and imaginary part of $z[k]$, as shown in Figs. 13.13 and 13.14, showing, respectively, IMFs obtained using the "extrema of the modulus" and "zero-crossings of the angle of the first derivative" criteria. By design, two spiking events will be synchronised if the phase of $z[k]$ is $\frac{\pi}{4}$. Since the spikes are high-frequency events, most of the relevant information is contained in the first few IMFs. To illustrate the detection of synchronised spikes, consider the first IMF obtained by the "modulus" criterion (top of Fig. 13.13). The magnitude and phase of this first IMF is given in Fig. 13.15. Note that large peaks in the magnitude of the first IMF do not exclusively indicate synchronisation of the spikes, however, from the phase obtained we can easily distinguish between arbitrary maxima events and synchronisation events. From Fig. 13.15, as desired, all synchronised spiking events occur at the same phase of $\frac{\pi}{4}$ (denoted by the iso-phase line).

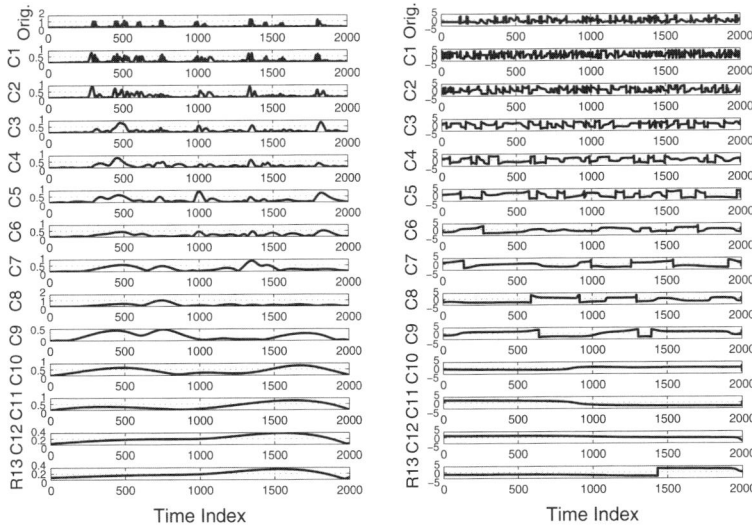

Fig. 13.13. Magnitude and phase of IMFs determined by rotation invariant EMD. Criterion: extrema of the modulus

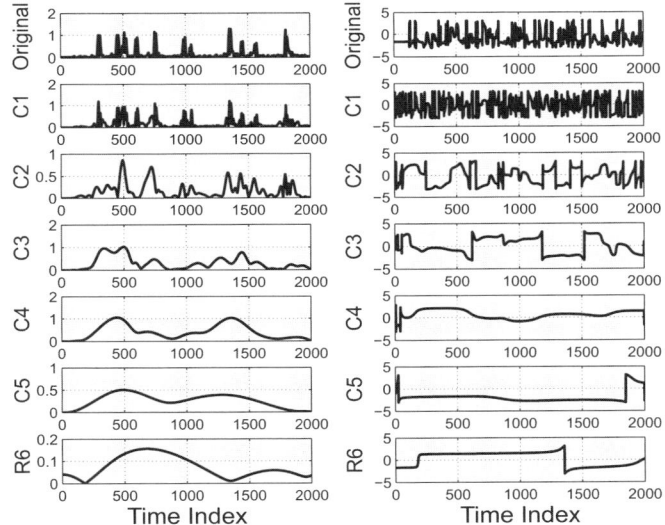

Fig. 13.14. Magnitude and phase of IMFs determined by rotation invariant EMD. Criterion: zero crossings of the angle of the first derivative. This criterion is equivalent to finding the extrema of the imaginary part; the corresponding IMFs are identical

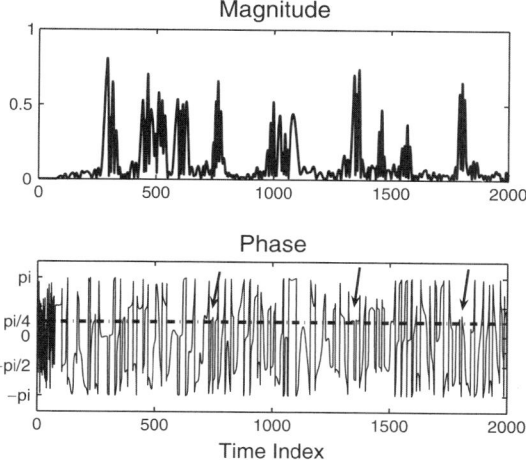

Fig. 13.15. The magnitude and phase of the first IMF extracted with the modulus criterion can be used to detect spike synchronisation

13.5 Empirical Mode Decomposition as a Fixed Point Iteration

The sifting process within EMD can be viewed as iterative application of a nonlinear operator T on a signal. The action of the operator can be described by:

$$T[h_k] = h_{k+1} = h_k - m_k(h_k) \qquad (13.11)$$

where h_k denotes the result of the kth iteration and m_k is the kth local mean signal (depending on h_k). From their definition, IMFs have a zero local mean, therefore, for an extracted IMF, (13.11) becomes:

$$T[h_k] = h_k \qquad (13.12)$$

indicating that IMFs can be viewed as fixed points of the iteration $h_{k+1} = T[h_k]$.

Theoretically, the iteration (13.11) could converge in only one step. However, since the true local mean is unknown, a spline approximation is used. Indeed, the existence of hidden scales, can result in the appearance of new extrema after every round of sifting. For these reasons, sifting is carried out several times, until the stoppage requirement (e.g., (13.2)) is satisfied. Practically, sifting to the extreme would remove all the amplitude modulation and would result in purely frequency-modulated components (IMFs). This is not desired because this drains the physical meaning from the IMFs [7].

To summarise, from the series of approximations (see Table 13.1):

$$h_1, h_2, \ldots, h_k, \ldots, h_\infty \qquad (13.13)$$

EMD picks up at each iteration a term that retains enough variation according to the predefined stoppage criterion. A slight modification of the stoppage criterion to a symmetric one such as:

$$d_1(h_k, h_{k-1}) = \sum_{t=0}^{T} \frac{|h_{k-1}(t) - h_k(t)|^2}{T} \tag{13.14}$$

or

$$d_2(h_k, h_{k-1}) = \sum_{t=0}^{T} \frac{|h_{k-1}(t) - h_k(t)|^2}{|h_{k-1}(t) + h_k(t)|^2} \tag{13.15}$$

would immediately turn the stoppage criterion into a metric, as desired (by the contraction mapping theorem, CMT) [10]. In combination with an analytic expression for the mean value of the signal such as the one proposed in [3], this would lead to the verification of the Lipschitz continuity of the series (13.13). The Lipschitz continuity can be used to determine:

1. The existence of IMFs, via convergence of (13.13), through application of the contraction mapping theorem [8] or
2. Uniqueness and conditions on the class of signals for which the series converges
3. The speed of convergence

13.6 Discussion and Conclusions

The concept of multichannel information fusion via the inherent fission abilities of standard empirical mode decomposition (EMD) has been introduced. This is achieved by starting from standard real EMD and extending it to the fusion of data in higher dimensions. It has been shown that even for real-valued data, their processing in the field of complex numbers $\mathbb{C}$ offers great advantages, for instance, when the phase information is important. Both the real and complex EMD have been shown to provide a unified approach to information fusion via fission, naturally performing data processing (denoising) and feature extraction. In addition, it has been illustrated that EMD also provides the decision stage within the information fusion framework. This can be achieved either by inspection or by intelligent fusion of intrinsic mode functions (IMFs) through a suitable machine learning algorithm. The IMFs obtained from the rotation invariant EMD have been shown to have physical meaning, which greatly enhances the range of applications of this technique. A rigorous theoretical framework for the analysis of EMD using the fixed point theory (FPT) methodology has also been addressed.

The simulations have been conducted on real-world problems in brain computer interface (BCI), including eye blink artefact removal from electroencephalograms (EEG), and spike detection and synchronisation for brain prosthetics based on an array of implanted electrodes in the surface of the brain cortex.

Acknowledgements

The authors acknowledge support from the NeuroProbes project, 6th framework programme of the European Commission (IST-2004-027017).

References

1. Altaf, M.U.B., Gautama, T., Tanaka, T., Mandic, D.P.: Rotation invariant complex empirical mode decomposition. In: Proceedings of the International Conference on Acoustics, Speech and Signal Processing (ICASSP), vol. 3, pp. 1009–1012 (2007)
2. Cohen, L.: Instantaneous anything. In: Proceedings of the IEEE International Conference on Acoustics, Speech and Signal Processing (ICASSP), vol. 5, pp. 105–108 (1993)
3. Delechelle, E., Lemoine, J., Niang, O.: Empirical mode decomposition: An analytical approach for sifting process. IEEE Signal Processing Letters **12**, 764–767 (2005)
4. Goh, S.L., Chen, M., Popovic, D.H., Aihara, K., Obradovic, D., Mandic, D.P.: Complex-valued forecasting of wind profile. Renewable Energy **31**(1733–1750), tba (2006)
5. Hall, D.L., Llinas, J.: An introduction to multisensor data fusion. Proceedings of the IEEE **85**(1), 6–23 (1997)
6. Hu, J., Yang, Y.D., Kihara, D.: EMD: an ensemble algorithm for discovering regulatory motifs in DNA sequences. BMC Bioinformatics **7**, 342 (2006)
7. Huang, N.E., Shen, Z., Long, S.R., Wu, M.L., Shih, H.H., Quanan, Z., Yen, N.C., Tung, C.C., Liu, H.H.: The empirical mode decomposition and the Hilbert spectrum for nonlinear and non-stationary time series analysis. Proceedings of the Royal Society A **454**, 903–995 (1998)
8. Leibovic, K.N.: Contraction mapping with application to control processes. Journal of Electronics and Control pp. 81–95 (1963)
9. Looney, D., Li, L., Rutkowski, T.M., Mandic, D.P., Cichocki, A.: Ocular artifacts removal from eeg using emd. In: Proceedings of the 1st International Conference on Cognitive Neurodynamics (ICCN) (2007)
10. Mandic, D.P., Chambers, J.A.: Recurrent Neural Networks for Prediction: Architectures, Learning Algorithms and Stability. Wiley (2001)
11. Mandic, D.P., Goh, S.L., Aihara, K.: Sequential data fusion via vector spaces: Fusion of heterogeneous data in the complex domain. International Journal of VLSI Signal Processing Systems **48**(1–2), 98–108 (2007)
12. Mandic, D.P., Obradovic, D., Kuh, A., Adali, T., Trutschell, U., Golz, M., Wilde, P.D., Barria, J., Constantinides, A., Chambers, J.: Data fusion for modern engineering applications: An overview. In: Proceedings of the IEEE International Conference on Artificial Neural Networks (ICANN'05), pp. 715–721 (2005)
13. Smith, L.S., Mtetwa, N.: A tool for synthesizing spike trains with realistic interference. Journal of Neuroscience Methods **159**(1), 170–180 (2006)
14. Souretis, G., Mandic, D.P., Griselli, M., Tanaka, T., Hulle, M.M.V.: Blood volume signal analysis with empirical mode decomposition. In: Proceedings of the 15th International DSP Conference, pp. 147–150 (2007)

15. Tanaka, T., Mandic, D.P.: Complex empirical mode decomposition. IEEE Signal Processing Letters **14**(2), 101–104 (Feb 2007)
16. Wald, L.: Some terms of reference in data fusion. IEEE Transactions on Geosciences and Remote Sensing **37**(3), 1190–1193 (1999)
17. Waltz, E., Llinas, J.: Multisensor Data Fusion. Artech House (1990)
18. Weng, B., Barner, K.E.: Optimal and bidirectional optimal empirical mode decomposition. In: Proceedings of the International Conference on Acoustics, Speech and Signal Processing (ICASSP), vol. 3, pp. 1501–1504 (2007)

14

Information Fusion for Perceptual Feedback: A Brain Activity Sonification Approach

Tomasz M. Rutkowski, Andrzej Cichocki, and Danilo P. Mandic

When analysing multichannel processes, it is often convenient to use some sort of "visualisation" to help understand and interpret spatio-temporal dependencies between the channels, and to perform input variable selection. This is particularly advantageous when the levels of noise are high, the "active" channel changes its spatial location with time, and also for spatio-temporal processes where several channels contain meaningful information, such as in the case of electroencephalogram (EEG)-based brain activity monitoring. To provide insight into the dynamics of brain electrical responses, spatial sonification of multichannel EEG is performed, whereby the information from active channels is fused into music-like audio. Owing to its "data fusion via fission" mode of operation, empirical mode decomposition (EMD) is employed as a time–frequency analyser, and the brain responses to visual stimuli are sonified to provide audio feedback. Such perceptual feedback has enormous potential in multimodal brain computer and brain machine interfaces (BCI/BMI).

14.1 Introduction

Brain computer and brain machine interfaces (BCI/BMI) are typically based on the monitoring of brain electrical activity by means of the electroencephalogram (EEG). Owing to its non-invasive nature, the EEG-based BCI/BMI are envisaged to be at the core of future "intelligent" prosthetics, and are particularly suited to the needs of the handicapped. The development of such devises, however, is time consuming, since brain responses to the same stimulus depend on the user's attention level and vary across users. Given that BCI/BMI are developed by multidisciplinary teams, it would be very advantageous to have some sort of universally understood representation of brain responses in the design phase, and perceptual feedback in the training phase. This chapter proposes to achieve this by *brain electrical activity sonification*, that is, by converting the time–frequency features of multichannel EEG into an audible music-like signal, which is easy to interpret. Other industries which would benefit greatly from such a tool include the entertainment and automotive industries, where e.g. the control and navigation in a computer-aided application can be achieved by BCI/BMI. To achieve this, the onset of "planning an action" should recorded from the scalp, and the relevant information should be "decoded" from the multichannel EEG.

Signal processing challenges in the processing of brain electrical responses (arising mostly due to the non-invasive nature of EEG) include the detection, estimation and interpretation of the notoriously noisy multichannel EEG recordings, together with cross-user transparency [10]. The set of extracted EEG features should provide sufficient information to the machine learning algorithms within BCI/BMI (minimum description); this set should also be large enough to allow for generalisation and cross-user differences [11]. An additional challenge comes from the fact that due to the nature of the information processing mechanism within the brain, the set of features that describes cognitive processes is not likely to be static. It should be designed so as to be real-time adaptive to accommodate the temporally non-stationary and spatially time varying number of "active channels".

To help in the design and application of BCI/BMI, we propose to employ auditory feedback, and thus provide "visualisation" of the brain states in the form of a music-like audio, that is, to perform "sonification" of brain electrical activity. Since, for example, the auditory pathway within the brain is not involved in the processing of visual stimuli, brain sonification has the potential to improve the efficiency and response times in visual BCI/BMI. Perhaps the first commercial application of sonification has been in Geiger counters, and sonification has since been considered as an alternative to standard visualisation techniques. The sonification of EEG signals can be therefore defined as a *procedure in which electrical brain activity captured from human scalp is transformed into an auditory representation* [5]. An earlier approach to sonify brain states is our previously introduced single channel EMDsonic [9].

In this chapter, we make use of the characteristics of human auditory perception (such as the temporal and pressure resolution) to provide simultaneous (multichannel) sonification of the recorded EEG. We then analyse the potential of this audio feedback in the representation and understanding of evoked potentials in the steady-state visual evoked potential (SSVEP) setting, and discuss the possibility of enhanced training of the users of BCI/BMI. The feature extraction from brain electrical responses is performed based on empirical mode decomposition (EMD). Some practical issues in the operation of EMD are also addressed, these include the envelope approximation and the fusion of multichannel EMD features. For mode detail on EMD see Chap. 13.

The chapter is organised as follows. EMD is first introduced as a time–frequency analysis technique suitable for the multichannel EEG recordings. The intrinsic mode functions (IMF) within EMD are identified as features which allow to create multiple spatially localised pianorolls. Based on the pianorolls and MIDI interfaces, "musical" scores for every of the spatial channels are obtained. Finally, the proposed sonification approach is illustrated in the SSVEP setting.

14.2 Principles of Brain Sonification

The concept of sonification is easy to accept in the BCI/BMI framework (most humans can differentiate between a large number of musical pieces). It is, however, important to realise that there is no unique way to generate such audio feedback from EEG. Unlike biofeedback, the audio feedback does not impact the subsequent brain responses to visual stimuli, which makes brain sonification a convenient perceptual feedback [2] in multimodal brain computer interfaces.

Our approach, depicted in Fig. 14.1, is based on multichannel information fusion, whereby the time–frequency features across the spatially distributed EEG channels are combined into music-like events. Once EEG signals are captured, they are mapped into the time–frequency domain based on the "data fusion via fission" property of EMD (decomposition into a set of AM–FM basis functions). The insight into the mental states via sonification also allows the

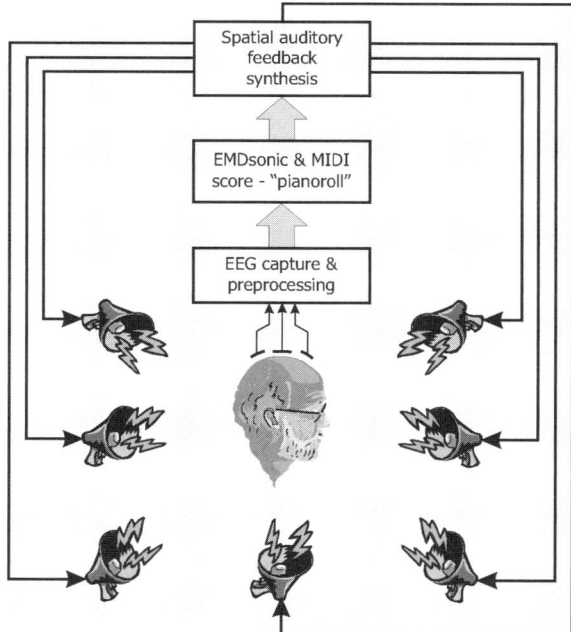

Fig. 14.1. The perceptual feedback realised by EEG sonification in surround sound environments. The mapping of EEG features onto a discrete set of musical variables provides a convenient insight into the dynamics and spatial patterns of brain responses

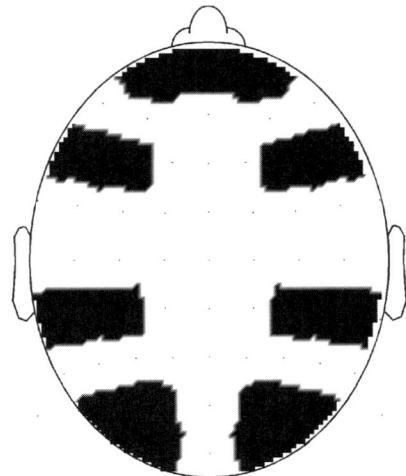

Fig. 14.2. A view of the human head from above. The dark regions represent locations of EEG electrodes. These locations correspond roughly to the positions of the loudspeakers in a surround sound environment

users to learn to modulate, or enhance their own brain responses to certain tasks. In this chapter, we consider the steady-state visual evoked potential (SSVEP) responses [4], since due to their stable behaviour and long response times, the audio feedback can operate in near real-time. Since the additional "audio" modality can be presented to the user by a surround sound system (see Fig. 14.1), for convenience, the EEG electrodes in the SSVEP experiment were located so as to reflect the position of the seven loudspeakers, as shown in Fig. 14.2. This allows us to synchronise the audio feedback with the dynamics of the brain activities at the different spatial locations on the scalp. The practical usefulness of EMD in this context has been illustrated in [9]. The level of detail (richness of information source) obtained by EMD is addressed in Sect. 14.3.1.

14.3 Empirical Mode Decomposition

To decompose the EEG signals into time–frequency components suitable for audio representation, we use empirical mode decomposition (EMD) [3], a procedure which decomposes the signal into a set of so-called intrinsic mode functions (IMF). By design, the IMFs have well defined instantaneous frequencies, and can often have physical meaning. The desired properties of IMFs which make them well suited for the extraction of time–frequency (non-linear dynamics) features from the data include

- Completeness
- Orthogonality

- Locality
- Adaptiveness

To obtain an IMF, it is necessary to remove local riding waves and asymmetries, these are estimated from local envelopes based on the *local* minima and maxima of the signal in hand. There are several approaches to the approximation of signal envelopes; their usefulness in the decomposition of EEG is discussed Sect. 14.3.1.

Since IMFs represent oscillatory modes within a signal, every IMF satisfies the following two conditions:

1. The number of extrema and the number of zero crossings should be equal or differ at most by one.
2. Within every IMF, the mean value of the sum of the envelope produced by connecting the local maxima and the envelope defined by the local minima should be zero.

Every IMF comprises only one mode of oscillation, thus, no complex riding waves are allowed. An IMF is not limited to be a narrow-band signal, as it would be in the traditional Fourier or wavelet decomposition, it can be both amplitude and frequency modulated at once, it can also exhibit non-stationary or non-linear behaviour.

The process of extracting IMFs from a signal is called the *sifting process* [3] and can be summarised as follows:

1. Identify the extrema of the signal waveform $x(t)$
2. Generate "signal envelopes" by connecting local maxima of $x(t)$ by a cubic spline, connect local minima by another cubic spline
3. Determine the local mean, m_i, by averaging the upper and lower signal envelopes
4. Subtract the local mean from the data to obtain:

$$h_i = x(t) - m_i \qquad (14.1)$$

5. Repeat until there are no more IMFs to extract

A "proper" IMF is the first component containing the finest temporal scale in the signal. Within EMD, the residual r_i is generated by subtracting the IMF from the signal, and it contains information about longer time-scales which are further resifted to find additional IMFs.

14.3.1 EEG and EMD: A Match Made in Heaven?

The application of the Hilbert transform to the IMFs and the corresponding Hilbert spectra allow us to represent the temporal and spatial dynamics of the EEG responses in the *amplitude–instantaneous frequency–time* coordinate system. This way, finding an IMF corresponds to the detection, estimation, and enhancement of band-limited components (*fission*) of the original

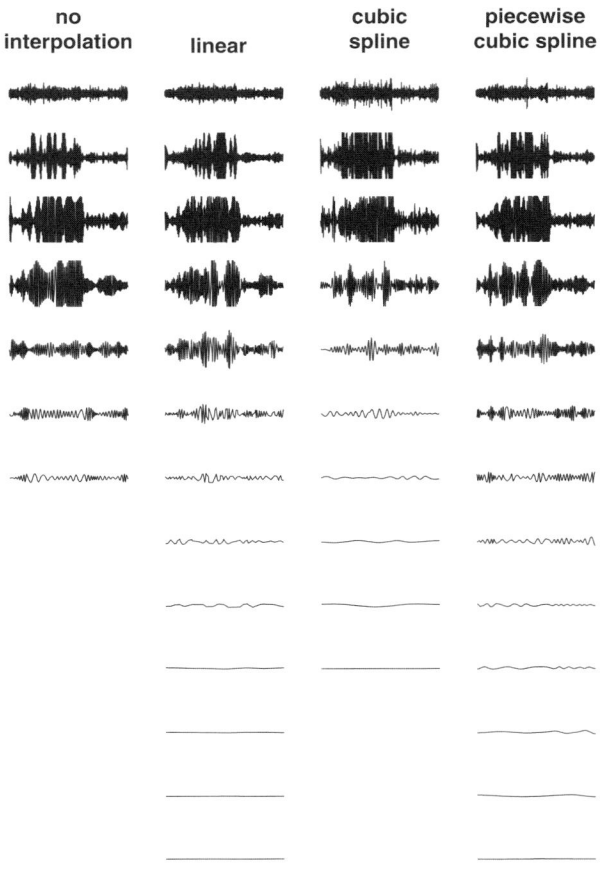

Fig. 14.3. Envelope approximation within EMD for a spatially averaged SSVEP response. Columnwise from *left* to *right*: no interpolation, linear interpolation, cubic spline interpolation, and piecewise cubic spline interpolation. The first column (no interpolation) results in simplest decomposition which still retains enough detail for the sonification

Table 14.1. The index of orthogonality for different envelope estimation strategies

Envelope estimation method	Number of IMFs	IO
No interpolation	7	0.1941
Linear	13	0.1697
Cubic spline	10	0.1118
Piecewise cubic spline	14	0.1103

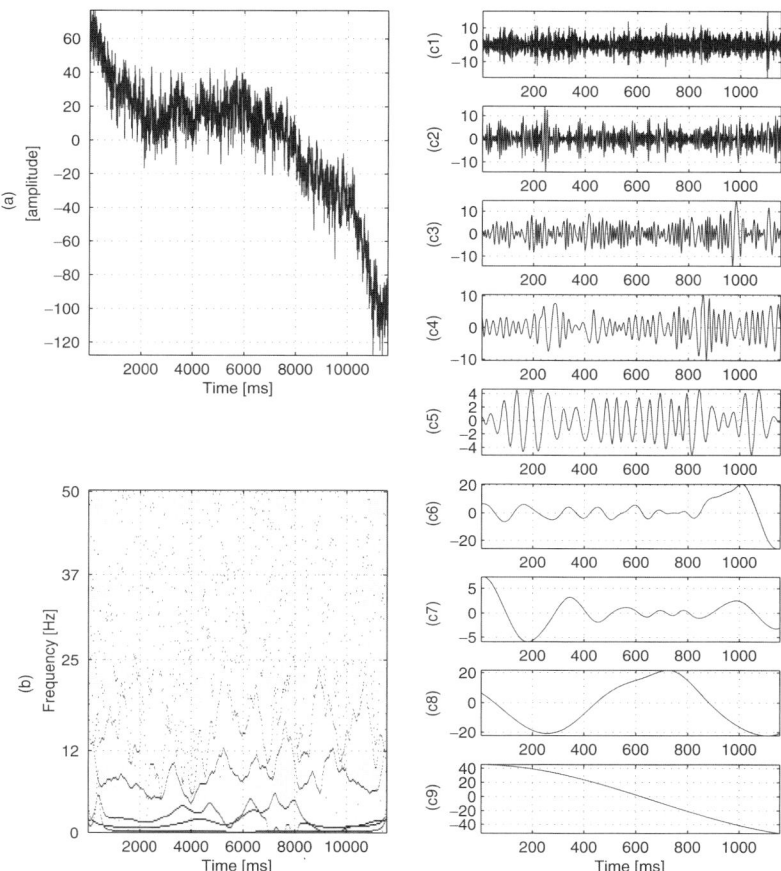

Fig. 14.4. Decomposition of raw EEG using EMD; notice the drift in the signal. Clockwise from the *top left* panel: (**a**) Raw EEG signal recorded from the frontal electrode Fp1. (**c**) The panels (**c1**)–(**c9**) show the nine IMFs extracted sequentially (IMFs are ordered from the highest to the lowest frequency). (**b**) The Hilbert–Huang spectrum ranging from 1 to 50 Hz, produced by [8]

recordings. To sonify the brain activity based on the time–frequency features, such components are then *fused* into single musical events (chords). This also helps in eliminating riding waves from the signal, which ensures no fluctuations of the instantaneous frequency caused by an asymmetric waveform.

Estimation of the signal envelope within EMD is a first and critical step towards finding IMFs, and the subsequent sonification. Since the decomposition is empirical, it was originally proposed to use cubic splines for this purpose [3], although it is clear that the choice of envelope estimation technique has a critical influence on the number and shapes of extracted IMFs. To illustrate the influence of the envelope approximation on the EMD features, we have considered several envelope estimation techniques including: linear

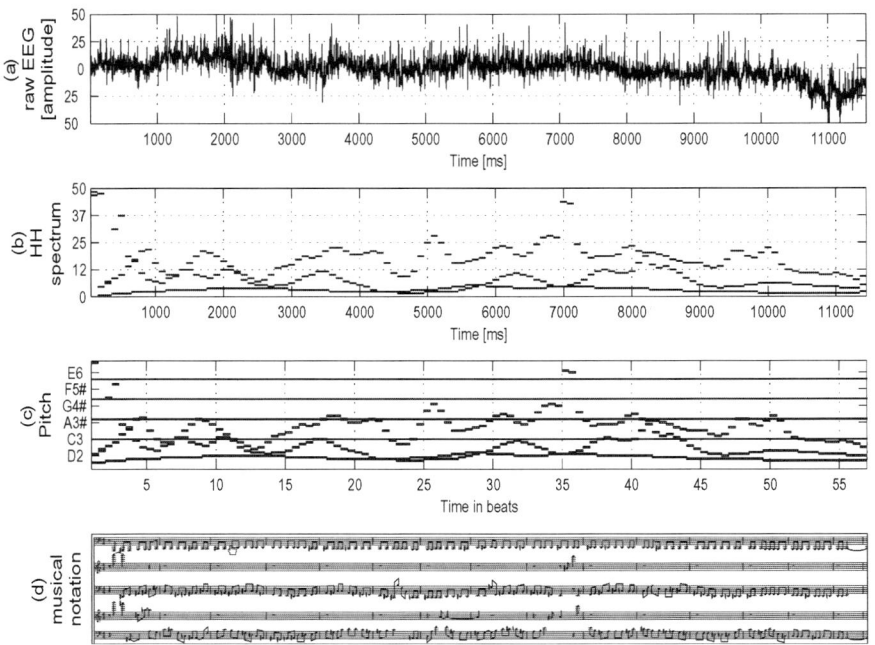

Fig. 14.5. Application of EMD to EEG sonification. (**a**) The original raw EEG signal captured during a steady-state visual evoked paradigm experiment; (**b**) Hilbert–Huang spectrum of the EEG signal; (**c**) "pianoroll" composed from the Hilbert–Huang spectrogram (similarity of the diagrams in panels (**b**) and (**c**) the resolution of the MIDI transformation) and (**d**) the MIDI representation converted into a musical score sheet

connections of the local extrema within the signal, linear interpolation, cubic splines, and piecewise cubic splines. As expected, the more complex the interpolation technique employed, the less distortion in the decomposition. Figure 14.3 illustrates the IMFs obtained for the different interpolation techniques. From the computational complexity point of view, the first column in Fig. 14.3 is the most attractive. It is, however, clear that the degree of complexity of the envelope approximation method corresponds directly to the resolution in the time–frequency plane and to the degree of orthogonality between the IMFs. To evaluate the interpolation techniques from the component orthogonality point of view, we employ the index of orthogonality (IO), given by

$$\text{IO} = \sum_{\forall_{i \neq j}} \frac{\mathbf{c}_i^T \mathbf{c}_j}{\mathbf{x}^T \mathbf{x}}, \qquad (14.2)$$

where vector $\mathbf{x}$ denotes the EEG signal, $\mathbf{c}_j$ is the jth IMF, and $(\cdot)^T$ denotes the vector transpose operator. The values of index IO for the various envelope estimation approaches are shown in Table 14.1. Observe that computationally

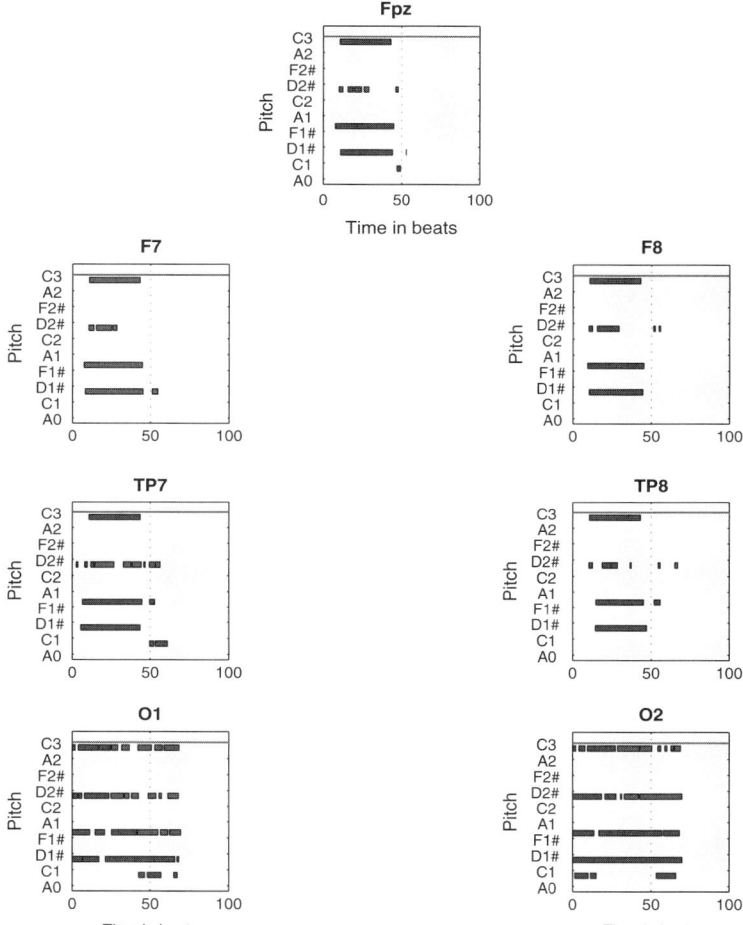

Fig. 14.6. Pianorolls representing musical scores at the seven locations of the EEG electrodes (see Fig. 14.8 for more detail). Longer activity is observed in the occipital area (electrodes O1 and O2)

simpler methods result in "less-orthogonal" components, whereas the number of obtained IMFs do not differ significantly among the methods considered.

14.3.2 Time–Frequency Analysis of EEG and MIDI Representation

An example of EMD applied to a real-world EEG recording is shown in Fig. 14.4. A single channel EEG shown in panel (a) was decomposed into eight oscillatory components (right-hand side column in Fig. 14.4), whose

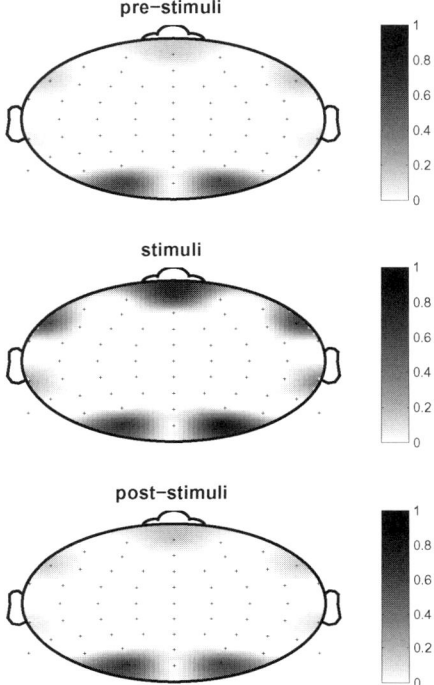

Fig. 14.7. The normalised powers of EEG during the SSVEP experiment. Observe the constant activity at the back of the heads (visual cortex area), since the subjects had their eyes open during the stimulus presentation. During the stimulus the same activity was also detected in the frontal and parietal areas (*middle panel*)

Hilbert spectrum is depicted in panel (b). The very low frequency components correspond to slow amplitude drifts caused by amplifiers or a loosely connected reference electrode, and can be removed. Clearly, the frequency content is rich enough for the sonification to be performed; also observe the time-varying nature of the spectrogram, which reflects the non-stationarity of EEG.

The time–frequency representations of EEG can be transformed into musical scores using standard MIDI procedures [1]. This mapping is illustrated in Fig. 14.5, where a single channel EEG recording from Fig. 14.5a is decomposed using EMD to produce the Hilbert–Huang spectrum shown in Fig. 14.5b. Figure 14.5c illustrates the mapping of the Hilbert–Huang spectrum onto the pianoroll (direct mapping by searching for ridges [7]). The ridge-based tracking of the maximum activities within the narrow-band IMF spectra allows for the locations and duration of the ridges to be transformed into musical scores, as shown in Fig. 14.5d. The pianorolls so obtained from the spatially distributed EEG electrodes are shown in Fig. 14.6, and can be played together using a surround sound system.

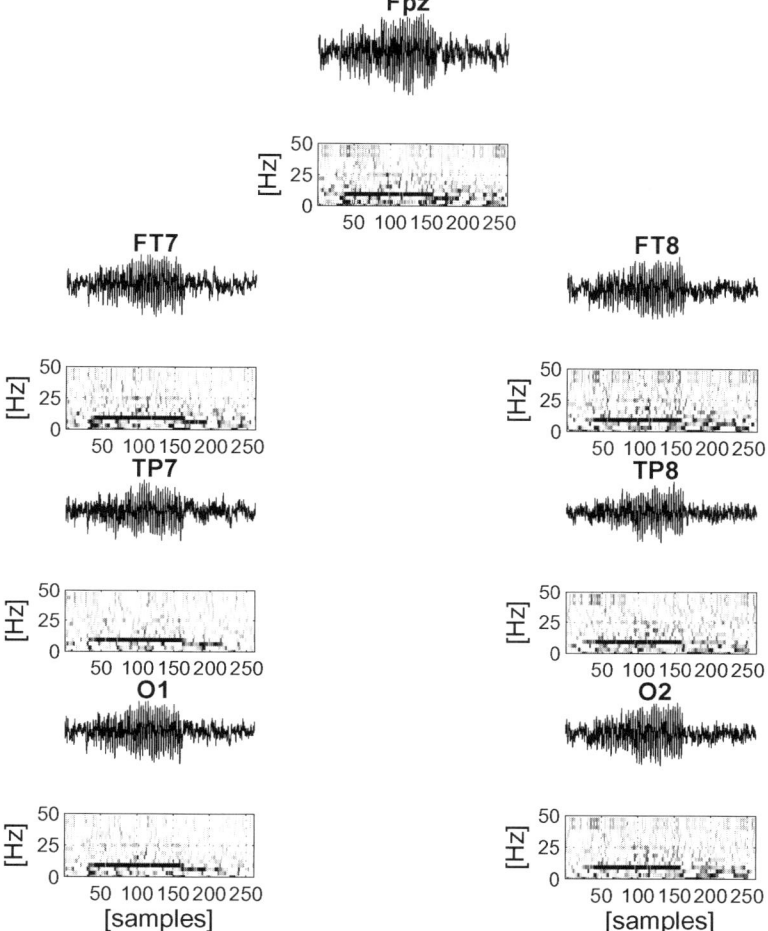

Fig. 14.8. EEG signals captured during steady-state visual stimulation and their Hilbert–Huang spectra showing steady-state frequency responses

14.4 Experiments

The experiments were conducted with the aim of creating auditory feedback in BCI/BMI based on SSVEP. This response is characterised by a very stable (and easy to recognise) response of long duration [9], this facilitates the use of audio feedback for enhancement of the responses. Since the induced brain responses have unique patterns among subjects [6], so too do the pianorolls (as shown in Fig. 14.6), this allows us to incorporate the audio feedback into the visual SSVEP experimental framework, and thus provide multimodal BCI/BMI.

During the experiments, subjects were asked to concentrate on a flashing chessboard displayed on the computer monitor. EMD-based sonification was used to produce an auditory feedback indicating the level of concentration of the subject. The so-produced audio was then played to the subject for every trial, to help accurately to classify the frequencies of the flashing stimuli. Different levels of activity (spatial patterns) recorded at time instances: (1) before the presentation of the visual stimuli (pre-stimuli); (2) during the visual stimulus (flashing checkerboard) which results in SSVEPs (stimuli); and (3) after the stimulus (post-stimuli), are shown in Fig. 14.7. The EEG activities were recorded at the locations illustrated in Fig. 14.2, where the darker areas indicate higher power levels.

To further illustrate the enhancement of the SSVEP responses achieved through the sonification, Fig. 14.8 shows the time series of the SSVEPs (recorded at the locations shown in Fig. 14.7), together with their spectrograms (the corresponding pianorolls are shown in Fig. 14.6). Clearly, by the EMD-based decomposition of EEG, we have obtained the dominant time–frequency components; these can be each mapped onto a convenient frequency range to provide the fusion of such components in the form of a single music event (one chord per time sample).

To summarise, from Fig. 14.7 it can be seen that the activities at the frontal and parietal brain areas exhibit a degree of synchronisation during the stimuli presentation, together with the relatively higher amplitudes. The proposed auditory feedback allows the user to modulate their own brain waves to enhance the desired responses to the visual stimuli, hence performing multimodal BCI/BMI.

14.5 Conclusions

This chapter presents an approach to the sonification of the multichannel recordings of brain electrical activity. This has been achieved based on the "data fusion via fission" property of empirical mode decomposition (EMD), which allows for the fusion of the spatial brain activity in the form of music like audio. This provides an additional modality to enhance the performance of standard (visual) brain computer and brain machine interfaces (BCI/BMI), and is physiologically justified since the visual and audio stimuli are processed in different parts of the cortex. In addition, the EMD-based time–frequency decomposition provides features which are well suited to describe the time varying dynamics of electroencephalogram (EEG) recordings. The Hilbert–Huang spectra of the intrinsic mode functions (IMF) allow for a convenient implementation of music theory to generate so-called pianorolls. The subsequent use of MIDI features allows for the wealth of the music production results to be applied to the sonification of brain recordings. The experiments have been conducted in a neurofeedback situation, and have confirmed the usefulness of the multichannel sonification in multimodal BCI and BMI.

References

1. Eerola, T., Toiviainen, P.: MIR in Matlab: The MIDI toolbox. In: Proceedings of the 5th International Conference on Music Information Retrieval, ISMIR2004, pp. 22–27. Audiovisual Institute, Universitat Pompeu Fabra, Barcelona, Spain (2004)
2. Fahle, M., Poggio, T. (eds.): Perceptual Learning. MIT, Cambridge, MA (2002)
3. Huang, N., Shen, Z., Long, S., Wu, M., Shih, H., Zheng, Q., Yen, N.C., Tung, C., Liu, H.: The empirical mode decomposition and the Hilbert spectrum for nonlinear and non-stationary time series analysis. Proceedings of the Royal Society A: Mathematical, Physical and Engineering Sciences **454**(1971), 903–995 (1998)
4. Kelly, S.P., Lalor, E.C., Finucane, C., McDarby, G., Reilly, R.B.: Visual spatial attention control in an independent brain–computer interface. IEEE Transactions on Biomedical Engineering **52**(9), 1588–1596 (2005)
5. Miranda, E., Brouse, A.: Interfacing the brain directly with musical systems: On developing systems for making music with brain signals. Leonardo **38**(4), 331–336 (2005)
6. Palaniappan, R., Mandic, D.P.: Biometric from the brain electrical activity: A machine learning approach. IEEE Transactions on Pattern Analysis and Machine Intelligence **29**(4), 738–742 (2007)
7. Przybyszewski, A., Rutkowski, T.: Processing of the incomplete representation of the visual world. In: Proceedings of the First Warsaw International Seminar on Intelligent Systems, WISIS'04. Warsaw, Poland (2004)
8. Rilling, G., Flandrin, P., Goncalves, P.: On empirical mode decomposition and its algorithms. In: Proceedings of IEEE-EURASIP Workshop on Nonlinear Signal and Image Processing, NSIP-03. IEEE (2003)
9. Rutkowski, T.M., Vialatte, F., Cichocki, A., Mandic, D., Barros, A.K.: Knowledge-Based Intelligent Information and Engineering Systems, *Lecture Notes in Artificial Intelligence*, vol. 4253, chap. Auditory Feedback for Brain Computer Interface Management – An EEG Data Sonification Approach, pp. 1232–1239. Springer, Berlin, Heidelberg, New York (2006)
10. Wolpaw, J.R., Birbaumer, N., McFarland, D.J., Pfurtscheller, G., Vaughan, T.M.: Brain–computer interfaces for communication and control. Clinical Neurophysiology **113**, 767–791 (2002)
11. Wolpaw, J.R., McFarland, D.J.: Control of a two-dimensional movement signal by a noninvasive brain–computer interface in humans. Proceedings of National Academy of Sciences of the United States America **101**(51), 17849–17854 (2004)

15
Advanced EEG Signal Processing in Brain Death Diagnosis

Jianting Cao and Zhe Chen

In this chapter, we present several electroencephalography (EEG) signal processing and statistical analysis methods for the purpose of clinical diagnosis of brain death, in which an EEG-based preliminary examination system was developed during the standard clinical procedure. Specifically, given the real-life recorded EEG signals, a robust principal factor analysis (PFA) associated with independent component analysis (ICA) approach is applied to reduce the power of additive noise and to further separate the brain waves and interference signals. We also propose a few frequency-based and complexity-based statistics for quantitative EEG analysis with an aim to evaluate the statistical significance differences between the coma patients and quasi-brain-death patients. Based on feature selection and classification, the system may yield a binary decision from the classifier with regard to the patient's status. Our empirical data analysis has shown some promising directions for real-time EEG analysis in clinical practice.

15.1 Introduction

In clinical practice, brain death is referred to the complete, irreversible and permanent loss of all brain and brainstem functions. Nowadays, EEG has been widely used in many countries to evaluate the absence of cerebral cortex function for determining brain death [14, 28]. Since EEG recordings might be corrupted by some artifacts or various sources of interfering noise, extracting informative features from noisy EEG signals and evaluating their significance is crucial in the process of brain death diagnosis [5, 12, 13, 18].

This chapter focuses on applying signal processing techniques to clinical EEG data recorded from 35 patients for the purpose of EEG-based preliminary examination and diagnosis of brain death. The preliminary examination procedure will be introduced in the standard brain death diagnosis procedure with a safe EEG examination and analysis procedure.

For each individual patient's EEG recording session, a robust PFA and ICA method is applied to separate the independent source components, followed by Fourier and time–frequency analysis [6, 7]. The recorded EEG signals are stored to a database in which the patients are categorized into two groups such as the deep coma group and quasi-brain-death (by using the term "quasi-", we mean the diagnosis is not yet ultimate and might subject to further medical tests) group based on the primary clinical diagnosis and EEG data analysis results. To evaluate the quantitative differences between two groups of patients, the relative power ratio and four complexity measures are proposed for quantitative EEG analysis [12, 13]. It is our belief that these systematic and quantitative studies of EEG measurements would be invaluable in neurology and medicine.

15.2 Background and EEG Recordings

15.2.1 Diagnosis of Brain Death

The brain death is defined as the cessation and irreversibility of all brain and brainstem functions. Based on this definition, the basic clinical criterion has been established in most countries. For example, the Japanese criterion (Takeuchi criterion) includes a few major items for brain death diagnosis:

- *Coma test.* Motor responses of the limbs to painful stimuli.
- *Pupil test.* Pupils' response to light and pupils dilatation.
- *Brainstem reflexes test.* E.g., gag, coughing, corneal reflexes, etc.
- *Apnea test.* Patient's capability to breath spontaneously.
- *EEG confirmatory test.* No electrical activity occurs above $2\,\mu V$.

In the standard process of brain death diagnosis, it often involves certain risks and takes a long time. For example, to determine the patient's spontaneous respiration, removing temporarily the respiratory machine is necessary during the apnea test [20]. Moreover, in the EEG confirmatory test, to observe electrical activities above $2\,\mu V$ at the sensitivity of $2\,\mu V\,mm^{-1}$, the recordings should last for 30 min, and the test will be repeated again after 6 h. Therefore, to reduce the risk and to save the precious time for the medical care in clinical practice, it is desirable to develop a practical yet safe and reliable tool in the diagnosis of brain death.

15.2.2 EEG Preliminary Examination and Diagnosis System

We introduced an EEG preliminary examination procedure into the standard brain death diagnosis process (Fig. 15.1a). As seen from the flowchart, after the coma test, pupils test, and brainstem reflexes test conducted for a patient, an EEG preliminary examination and diagnosis system comes in at the bedside of patient. The purpose of the EEG examination is to explore advanced signal

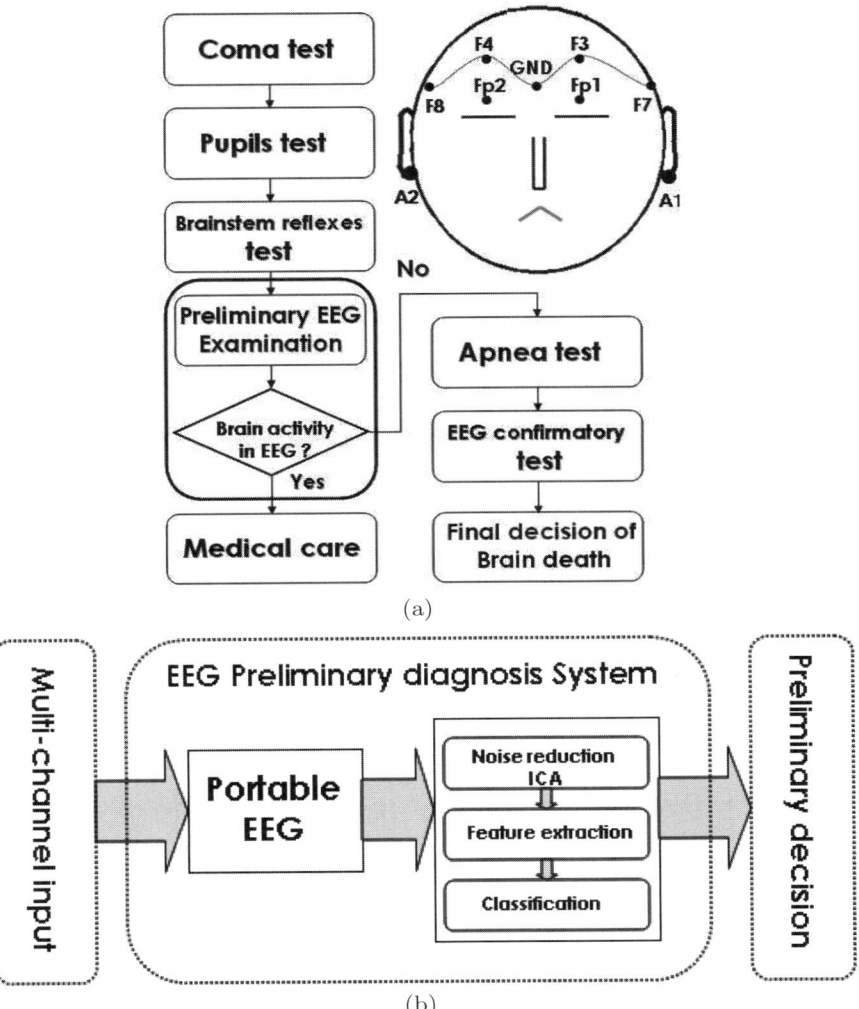

Fig. 15.1. EEG preliminary examination system: (**a**) EEG preliminary examination procedure and the electrode layout and (**b**) the preliminary diagnosis system

processing tools to discover whether any brain wave activity occurs in the patient's brain. If the decision is positive (i.e., indicating the presence of brain activities), it suggests to sidestep the further tests of brain death diagnosis, and go directly to spend more time on the patient's medical care. On the contrary, if the result of EEG preliminary examination is negative, the apnea test and EEG confirmatory test will be executed afterwards as in the standard brain death diagnosis procedure.

It is worth noting that the EEG preliminary examination is *not* a substitute of the standard process of brain death diagnosis. Instead, it is our belief that if the EEG pre-diagnosis is reliable and its results are significant, it would provide a simple and risk-free diagnosis tool in the intensive care unit (ICU) of the hospital without jeopardizing the patient's life.

The EEG preliminary diagnosis system (Fig. 15.1b) includes a portable EEG measurement device and the EEG-oriented signal processing tools. The EEG measurement part consists of a small number of electrodes, a portable amplifier, and a portable computer associated with data acquisition tools. The signal processing tools include several routines of noise reduction, source separation, feature extraction, as well as statistical tests.

In the EEG pre-diagnosis system, the recorded EEG signals are first stored in the portable computer. For the individual patient's EEG data session, a PFA+ICA approach is applied to reduce the noise and separate the source components, followed by the Fourier and time–frequency analysis for visual inspection. For the selected EEG data, several complexity measure-based features are extracted for the multichannel time series. In the quantitative EEG (qEEG) analysis, statistical tests are applied to evaluate the differences between two groups of patients (deep coma group vs. quasi-brain-death group). The database and results can also be incrementally updated based on the increasingly recorded EEG data session whenever new recordings from a patient are available. The essential goal of our preliminary brain death diagnosis system is to obtain a preliminary yet fast decision for the brain death diagnosis based on the real-time EEG recordings.

15.2.3 EEG Recordings

In the present study, the EEG experimental protocols were executed in the Shanghai Huashan Hospital in affiliation with the Fudan University (Shanghai, China). The EEG data were directly recorded at the bedside of patient in the ICU of the hospital, where the level of environmental noise was relatively high since many medical machines are switched on (e.g., respiratory machine). The EEG recording instruments was a portable NEUROSCAN ESI-32 amplifier associated with a laptop computer. Either DC or AC power can be adopted during the EEG measurement.

During EEG recording, a total of nine electrodes were placed on the forehead of the patient lying on the bed, which mainly cover the non-hairy or least hairy area of the scalp. Specifically, six channels are placed at Fp1, Fp2, F3, F4, F7, and F8, and two electrodes that connect the two ears are used as the reference, namely $(A1 + A2)/2$; an additional channel, GND, serves as the ground (see Fig. 15.1a for an illustration). The EEG sampling rate was 1,000 Hz, and the resistances of electrodes were set under $8\,k\Omega$. Instead of using the EEG cap in the standard EEG recording, we used individual electrodes with high-chloride electrolyte gel (suitable for using DC power) in all data recording sessions.

It should be noted that we use this simple and practical EEG measurement technique (especially the electrode layout) to avoid minimum interference with the regular medical treatment since all patients were in supine positions. In comparison with the standard EEG recording setup in a relatively quiet and shielded room, the EEG recordings in the ICU might be contaminated by various environmental noise. However, this disadvantage can be somewhat alleviated by applying advanced signal processing techniques that we will describe below.

With the permission of the patients' families, a total of 35 patients have been examined by using EEG from June 2004 to March 2006, with age range from 17 to 85 years. All patients were examined by the coma test, pupils test, and brainstem reflexes test by two physicians. The patients were classified into a deep-coma group (19 patients) and a quasi-brain-death group (16 patients in 18 cases) before the EEG recording (two patients belong to both coma and quasi-brain-death cases depending on the specific recording date). Note that the examinations and diagnoses from two expert physicians are independent and the clinical opinions at this stage is not final – that is exactly why the advanced EEG signal processing tools might come in to play a confirmatory role. Because the health conditions of patients varied, each patient might have a different number of recorded data sessions at the same or different day. Finally, a total of 64 sessions' recordings from 35 patients were stored in the computer.

Notably, during the EEG recordings all patients appeared in deep coma, and were lying down in the bed with eyes closed. Consequently, no ocular or muscle artifact was observed. However, the heartbeat rhythm can be observed from specific patients. In addition, when AC power was used, interference from power noise might appear at several harmonic frequencies (e.g., 50, 100 Hz), which can be easily eliminated by applying a notch filter to the EEG signal in the data preprocessing stage.

15.3 EEG Signal Processing

Independent component analysis (ICA) is a powerful signal processing tool for blindly separating statistically independent source signals. Various standard ICA methods [4, 8, 9, 16, 21] have been widely used for physiological data analysis [19]. In this section, we present a robust data analysis procedure for decomposing the EEG raw signals recorded from the comatose patients during the EEG preliminary examination. Our proposed procedure has two steps. In the first step, a robust noise reduction and prewhitening technique is presented. In the second step, a robust score function derived from the parameterized *student t-distribution model* [6, 7] is applied to decompose the mixtures of sub-Gaussian and super-Gaussian source signals.

15.3.1 A Model of EEG Signal Analysis

The signal model for the EEG measurements can be formulated by

$$\mathbf{x}(t) = \mathbf{A}\mathbf{s}(t) + \boldsymbol{\xi}(t), \quad t = 1, 2, \ldots, \tag{15.1}$$

where $\mathbf{x}(t) = [x_1(t), \ldots, x_m(t)]^{\mathrm{T}}$ represent the measured electrode signals at time t. The source signals (including brain activities, interferences, etc.) are represented by the vector $\mathbf{s}(t) = [s_1(t), \ldots, s_n(t)]^{\mathrm{T}}$, and $\boldsymbol{\xi}(t) = [\xi_1(t), \ldots, \xi_m(t)]^{\mathrm{T}}$ represents the additive sensor noise at each electrode. Since the source components are overlapped, and propagated rapidly to the electrodes, the element of the mixing matrix $\mathbf{A} \in \mathbb{R}^{m \times n} = (a_{ij})$ can be roughly viewed as an attenuation coefficient during propagation between the ith sensor and jth source.

In the model (15.1), the sources $\mathbf{s}$, the number n, the additive noise $\boldsymbol{\xi}$, and matrix $\mathbf{A}$ are all unknown, and only the sensor signals $\mathbf{x}$ are accessible. It is assumed that the source components $\mathbf{s}$ are mutually independent, as well as being statistically independent of the noise components $\boldsymbol{\xi}$. Each source component s_i contributes to at least two sensors, and a noise component ξ_i contributes at most to only one sensor. Moreover, the noise components $\boldsymbol{\xi}$ themselves are assumed to be mutually uncorrelated.

There are two types of noise components that need to be reduced or discarded in the EEG data analysis. The first type is the additive sensor noise at the individual sensors. The standard ICA approaches usually fail to reduce this type of noise especially when the noise level is high. Therefore, we will first apply a robust prewhitening technique in the preprocessing stage to reduce the power of additive noise. The second type of noise is hidden in the source components, and this type of interference can be discarded after the source separation.

15.3.2 A Robust Prewhitening Method for Noise Reduction

In this section, we first describe the standard *principal component analysis* (PCA) approach for prewhitening. Next, we show that this standard PCA approach can be extended to *principal factor analysis* (PFA) for reducing high-level noise.

Let us rewrite (15.1) in a matrix form as follows:

$$\mathbf{X}_{(m \times N)} = \mathbf{A}_{(m \times n)} \mathbf{S}_{(n \times N)} + \boldsymbol{\Xi}_{(m \times N)}, \tag{15.2}$$

where N denotes observed data sample points. Without loss of generality, the signal $\mathbf{S}$ and the noise $\boldsymbol{\Xi}$ are both assumed to have zero mean. When the sample size N is sufficiently large, the covariance matrix of the observed signal $\mathbf{X}$ is written as

$$\boldsymbol{\Sigma} = \mathbf{A}\mathbf{A}^{\mathrm{T}} + \boldsymbol{\Psi}, \tag{15.3}$$

where $\boldsymbol{\Sigma} = \mathbf{X}\mathbf{X}^T/N$, and $\boldsymbol{\Psi} = \boldsymbol{\Xi}\boldsymbol{\Xi}^T/N$ is a diagonal matrix. Since the sources are mutually independent and all uncorrelated with the additive noise, we obtain $\mathbf{S}\mathbf{S}^T/N \to \mathbf{I}$ and $\mathbf{S}\boldsymbol{\Xi}^T/N \to \mathbf{0}$. For convenience, we further assume that $\mathbf{X}$ has been normalized by $\sqrt{N}$ so that the covariance matrix can be written as $\mathbf{C} = \mathbf{X}\mathbf{X}^T$.

For the averaged-trial EEG data, the noise variance $\boldsymbol{\Psi}$ is relatively small. A cost function for fitting the model to the data is to make $\|\mathbf{C} - \mathbf{A}\mathbf{A}^T\|_F$ as small as possible. It is well known that the principal components can be found by employing the *eigenvalue decomposition* (EVD). That is, the solution of $\mathbf{A}\mathbf{A}^T$ for seeking n principal components can be obtained by

$$\hat{\mathbf{A}}\hat{\mathbf{A}}^T = \mathbf{U}_n \boldsymbol{\Lambda}_n \mathbf{U}_n^T, \tag{15.4}$$

where $\boldsymbol{\Lambda}_n$ is a diagonal matrix that contains the largest n eigenvalues of matrix $\mathbf{C}$, and the column vectors of $\mathbf{U}_n$ are the corresponding eigenvectors. In light of (15.4), one possible solution for $\hat{\mathbf{A}}$ is

$$\hat{\mathbf{A}} = \mathbf{U}_n \boldsymbol{\Lambda}_n^{1/2}, \tag{15.5}$$

and then the whitened components can be obtained by $\mathbf{z} = \boldsymbol{\Lambda}_n^{-1/2} \mathbf{U}_n^T \mathbf{x}$. Note that the covariance matrix is $\mathbb{E}\{\mathbf{z}\mathbf{z}^T\} = \boldsymbol{\Lambda}_n$ in that the elements of $\mathbf{z}$ are uncorrelated. Applying this standard prewhitening technique to the averaged-trial EEG data analysis, a successful result has been reported [19].

For the single-trial EEG data, the signal-to-noise ratio (SNR) is typically poor and much lower compared to the averaging-trial data. This means that the diagonal elements of $\boldsymbol{\Psi}$ cannot be ignored in the model. In this case, we can fit $\mathbf{A}\mathbf{A}^T$ to $\mathbf{C} - \boldsymbol{\Psi}$ by EVD. That is, choosing the columns of $\mathbf{A}$ as eigenvectors of $\mathbf{C} - \boldsymbol{\Psi}$ corresponding to the largest n eigenvalues so that the sum of the squares in each column is identical to the corresponding eigenvalue.

In the case when the noise variance $\boldsymbol{\Psi}$ is unknown, we may use an iterative method to estimate from the data [7]. To do that, we employ the following cost function:

$$L(\mathbf{A}, \boldsymbol{\Psi}) = \text{tr}[\mathbf{A}\mathbf{A}^T - (\mathbf{C} - \boldsymbol{\Psi})]^2 \tag{15.6}$$

and minimize it by setting $\frac{\partial L(\mathbf{A}, \boldsymbol{\Psi})}{\partial \boldsymbol{\Psi}} = 0$, whereby the estimate of $\boldsymbol{\Psi}$

$$\hat{\boldsymbol{\Psi}} = diag\{\mathbf{C} - \hat{\mathbf{A}}\hat{\mathbf{A}}^T\} \tag{15.7}$$

is obtained. The estimate $\hat{\mathbf{A}}$ can be obtained in the same way as in (15.5).

Both the matrix $\mathbf{A}$ and the diagonal elements of $\boldsymbol{\Psi}$ have to be jointly estimated from the data. The estimate $\hat{\mathbf{A}}$ is obtained by the standard PCA. The estimate $\hat{\boldsymbol{\Psi}}$ is obtained by the so-called *unweighted least squares* method as in factor analysis. Once the estimates $\hat{\mathbf{A}}$ and $\hat{\boldsymbol{\Psi}}$ converge to stationary points, we can compute the score matrix with the *Bartlett method* as

$$\mathbf{Q} = [\hat{\mathbf{A}}^T \hat{\boldsymbol{\Psi}}^{-1} \hat{\mathbf{A}}]^{-1} \hat{\mathbf{A}}^T \hat{\boldsymbol{\Psi}}^{-1}. \tag{15.8}$$

Using the above result, the whitened signals can be obtained by $\mathbf{z} = \mathbf{Q}\mathbf{x}$. Note that the covariance matrix is $\mathbb{E}\{\mathbf{z}\mathbf{z}^T\} = \mathbf{\Lambda}_n + \mathbf{C}\mathbf{\Psi}\mathbf{C}^T$, which implies that the subspaces of the source signals are orthogonal.

The robust prewhitening approach plays the same role of decorrelation as in the standard PCA, but the noise variance $\mathbf{\Psi}$ is taken into account. The difference between them is that the standard PCA tries to fit both diagonal and off-diagonal elements of $\mathbf{C}$, whereas PFA tries to fit off-diagonal elements of $\mathbf{C}$ only. Therefore, the robust approach enables us to reduce a high-level noise that is very important in the single-trial neurophysiological signal analysis.

As a simple demonstration of the robustness of PFA for noise reduction, we present a source decomposition example by applying the standard PCA and PFA methods to some high-level noisy signals recorded from a 64-channel magnetoencephalographic (MEG) system. The data recording was based on the phantom experiment in which a sine wave (13 Hz) source is known in advance. The MEG was used here since EEG was more difficult to simulate using the phantom. Based on single-trial MEG recordings, we show one representative single-trial data session and present the comparative results between PCA and PFA in Fig. 15.2. As seen from the result, the PFA is capable of extracting the sine wave while the PCA is not. Since the standard PCA fits

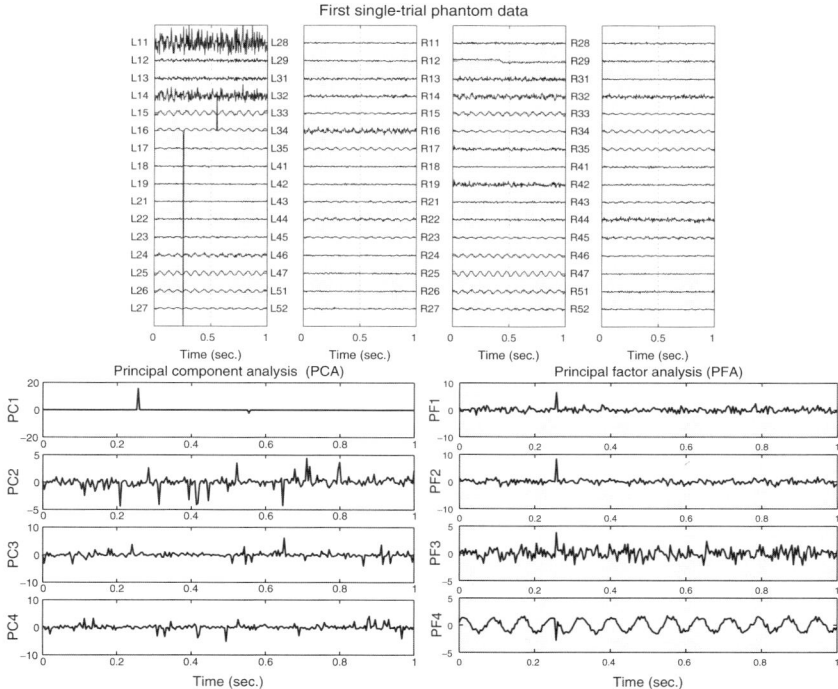

Fig. 15.2. Comparison of PFA vs. PCA for the single-trial MEG phantom data

both diagonal and off-diagonal elements of $\mathbf{C}$, and the principal components are determined by accounting for the maximum variance of the observed variables. If the power of additive noise is much greater than that of the signal, then the extracted principal component will be noise. However, by subtracting the noise variance $\boldsymbol{\Psi}$ from $\mathbf{C}$ as $\mathbf{C} - \boldsymbol{\Psi}$ as in our robust prewhitening approach, the sources are determined by the intercorrelation of the variables.

15.3.3 Independent Component Analysis

After prewhitening of the noisy observations, the transformed signals $\mathbf{z} = \mathbf{Q}\mathbf{x}$ are obtained, where the power of noise, and mutual correlation of signals have been reduced. However, it is insufficient to obtain the independent components, since an orthogonal matrix generally contains additional degrees of freedom. To obtain independent source components, the independent sources $\mathbf{y} \in \mathbb{R}^n$ can then be separated from a linear transformation as

$$\mathbf{y}(t) = \boldsymbol{W}\mathbf{z}(t), \tag{15.9}$$

where $\boldsymbol{W} \in \mathbb{R}^{n \times n}$ is called the demixing matrix, which can be estimated using specific ICA algorithms [4, 8, 9, 16, 21]. In this section, we introduce the natural gradient-based ICA algorithm to estimate the matrix $\boldsymbol{W}$.

The Kullback–Leibler (KL) divergence is a contrast function that measures the mutual information among the output signals y_i by computing the divergence between the joint probability density function (pdf) $p_y(\mathbf{y})$ and the marginal pdfs $p_i(y_i)$:

$$D(\mathbf{y}|\boldsymbol{W}) = \int p_y(\mathbf{y}) \log \frac{p_y(\mathbf{y})}{\prod_{i=1}^n p_i(y_i)} d\mathbf{y}. \tag{15.10}$$

In (15.10), $D(\mathbf{y}|\boldsymbol{W}) = 0$ if and only if the independence condition $p_y(\mathbf{y}) = \prod_{i=1}^n p_i(y_i)$ holds.

The natural gradient-based ICA algorithm has been developed by Amari et al. [4] and also independently developed by Cardoso, which was termed the relative gradient [8]. It has been proved that the natural gradient greatly improves the learning efficiency in blind source separation [2]. Applying the natural gradient to minimize the KL divergence (15.10) yields the following learning rule for $\boldsymbol{W}$ [4]

$$\Delta \boldsymbol{W}(t) = \eta \left[\mathbf{I} - \varphi(\mathbf{y}(t))\mathbf{y}^{\mathrm{T}}(t) \right] \boldsymbol{W}(t). \tag{15.11}$$

When prewhitening has been performed and the demixing matrix $\boldsymbol{W}$ is restricted to be orthogonal, the learning rule can be extended to the following form as [3]:

$$\Delta \boldsymbol{W}(t) = -\eta \left[\varphi(\mathbf{y}(t))\mathbf{y}^{\mathrm{T}}(t) - \mathbf{y}(t)\varphi^{\mathrm{T}}(\mathbf{y}(t)) \right] \boldsymbol{W}(t). \tag{15.12}$$

When the prewhitening is included in the separation process, the above learning rule becomes the *equivariant adaptive source separation via independence* (EASI) algorithm [8]

$$\Delta \mathbf{W}(t) = \eta \left[\mathbf{I} - \mathbf{y}(t)\mathbf{y}^\mathrm{T}(t) - \varphi(\mathbf{y}(t))\mathbf{y}^\mathrm{T}(t) + \mathbf{y}(t)\varphi^\mathrm{T}(\mathbf{y}(t)) \right] \mathbf{W}(t). \quad (15.13)$$

In (15.11)–(15.13), $\eta > 0$ is a learning-rate parameter, and $\varphi(\cdot)$ is the vector of score function whose optimal individual components are given by

$$\varphi_i(y_i) = -\frac{\mathrm{d}}{\mathrm{d}y_i} \log p_i(y_i) = -\frac{\dot{p}_i(y_i)}{p_i(y_i)}, \quad (15.14)$$

where $\dot{p}_i(y_i) = \mathrm{d}p_i(y_i)/\mathrm{d}y_i$.

The natural gradient-based algorithms described in (15.11)–(15.13) rely on the appropriate choice of score functions $\varphi_i(\cdot)$ that further depend on the pdf $p_i(y_i)$ of the sources, which are usually not available. Therefore, one practical approach is to employ a hypothetical pdf model. Several algorithms have been developed for separating the mixtures of sub- and super-Gaussian sources (e.g., [17]). Here, we will use a parametric method based on the student t-distribution model. We will not delve into the details of the theoretical justification but show some advantages of the derived algorithm [7]: (1) the score function derived by the parameterized t-distribution density model is able to separate the mixtures of sub-Gaussian and super-Gaussian sources by combining the t-distribution density with a family of light-tailed distribution (sub-Gaussian) density models; (2) the score function is determined by the value of the kurtosis that corresponds to the source distribution; and (3) the algorithm is stable and is robust to the mis-estimation of parameter (kurtosis) as well as the outlier samples. The proposed score functions have the following form

$$\varphi_i(y_i) = \alpha \lambda_\alpha \, \mathrm{sgn}(y_i) |\lambda_\alpha y_i|^{\alpha - 1}, \quad \kappa_\alpha = \hat{\kappa}_i \leq 0, \quad (15.15)$$

$$\varphi_i(y_i) = \frac{(1+\beta) y_i}{y_i^2 + \frac{\beta}{\lambda_\beta^2}}, \quad \kappa_\beta = \hat{\kappa}_i > 0, \quad (15.16)$$

where α and β are two parameters obtained from the estimate of the kurtosis $\hat{\kappa}_i(y_i)$, and $\lambda_\alpha, \lambda_\beta$ are two scaling constants.

The implementation of our ICA algorithm is summarized as follows:

- Calculate the output $\mathbf{y}$ by using (15.9) with the given transformed data $\mathbf{z}$ and an initial value of $\mathbf{W}$.
- Calculate the kurtosis as $\hat{\kappa}_i = \hat{m}_4/\hat{m}_2^2 - 3$ where the second- and fourth-order moments are estimated by using $\hat{m}_j(t) = [1 - \lambda]\hat{m}_j(t-1) + \lambda y_i^j(t), (j = 2, 4)$, where λ is a small positive coefficient.
- Construct two look-up tables for the kurtosis values $\kappa_\alpha = \frac{\Gamma(\frac{5}{\alpha})\Gamma(\frac{1}{\alpha})}{\Gamma^2(\frac{3}{\alpha})} - 3$ and $\kappa_\beta = \frac{3\Gamma(\frac{\beta-4}{2})\Gamma(\frac{\beta}{2})}{\Gamma^2(\frac{\beta-2}{2})} - 3$ in advance, and seek the closest α or β from the table according to the value of $\hat{\kappa}_i$.

- Calculate the scaling constants as $\lambda_\alpha = \left[\frac{\Gamma(\frac{3}{\alpha})}{m_2 \Gamma(\frac{1}{\alpha})}\right]^{1/2}$ or $\lambda_\beta = \left[\frac{\beta \Gamma(\frac{\beta-2}{2})}{2 m_2 \Gamma(\frac{\beta}{2})}\right]^{1/2}$ using the estimate $\hat{m}_2$ and α or β.
- Calculate the score function $\varphi(\cdot)$ by using (15.15) or (15.16) and update W by using (15.11), or (15.12), or (15.13).

In addition, depending on the quality of signal measurements, we can employ the so-called multiple ICA estimation approach to bootstrap the performance. Namely, instead of using one estimated demixing matrix, we may employ the product of several estimated demixing matrices as $\mathbf{W} = \prod_{l=1}^{\ell} \mathbf{W}_l \in \mathbb{R}^{n \times n}$, where $\mathbf{W}_l$ is computed by using a specific ICA algorithm at stage l ($l = 1, \ldots, \ell$). For example, we could first employ the second-order statistics-based algorithm [21] or the fourth-order statistics-based algorithm [9] to obtain a demixing matrix estimate $\mathbf{W}_1$ at the first stage, then further employ our robust ICA algorithm to obtain another estimate $\mathbf{W}_2$. In our experiences, such a "boosting" procedure could overcome the ill-conditioning problem of $\mathbf{A}$ and could improve the performance of source decomposition in many practical cases.

15.3.4 Fourier Analysis and Time–Frequency Analysis

Upon separating the independent components, we employ the standard Fourier spectrum analysis to estimate the power spectra of individual independent source components. Notably, the amplitudes of the separated components as well as their power spectra have no quantitative physical unit meaning, since the outputs of the ICA algorithm all have scaling indeterminacy (see Fig. 15.3). By visual inspection of the power spectra, we can empirically determine whether the components may contain EEG brain waves. For a closer examination, we also resort to the time–frequency analysis tool, such as the *Wigner–Ville distribution*, to visualize the ongoing temporal signals in a time–frequency plane. Compared to the one-dimensional power spectrum, the two-dimensional time–frequency map may clearly reveal the time-varying spectral information of the specific signal of interest. See Fig. 15.3 for an illustrated example.

15.4 EEG Preliminary Examination with ICA

Because the EEG measures the "smeared" ensemble activities of synchronous firings from millions of neurons in the brain, it indicates the specific activity in which the brain is engaged and provides some hints about the consciousness or comatose status. Provided the brain is not completely dead, it is highly likely that some brain waves might be extracted from the EEG measurements.

In our experimental analysis, we are particularly interested in the delta waves (1–4 Hz), upper theta (6–8 Hz) and alpha (8–12 Hz) waves. This is

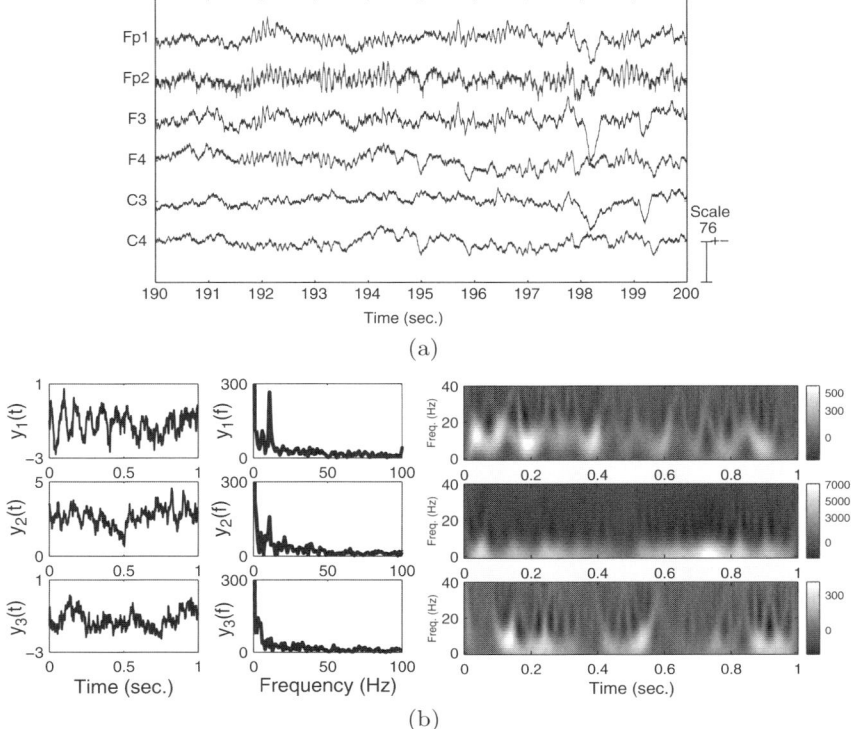

Fig. 15.3. The ICA result for a patient with extracted EEG brain waves: (**a**) the recorded EEG signals and (**b**) an extracted alpha-wave component in the time and frequency domains. The power spectra in (**b**) are shown with arbitrary unit (subject to scale uncertainty), and the time–frequency maps are visualized with logarithm transformation of the power

mainly because the delta waves occur in deep sleep and in some abnormal processes; the theta waves are strong during internal focus, meditation, and spiritual awareness, and they relate to the subconscious status that reflects the state between wakefulness and sleep; the alpha waves are responsible for mental coordination, self-control of relaxation, which are believed to bridge the conscious to subconscious state [22].

In the following, we present two representative case studies of coma and quasi-brain-death patients. The patients in both cases were initially in a deep-coma state and then approached to different outcomes.

15.4.1 Extracted EEG Brain Activity from Comatose Patients

We first demonstrate our result with an example of a 18-year-old male patient. The patient had a primary cerebral disease and was admitted to the hospital on May 20, 2004. After one month hospitalization, the patient lost his consciousness and remained in a deep-coma state. His pupils were dilated, and the respiratory machine was used. On June 22, the patient was completely unresponsive to external visual, auditory, and tactile stimuli and was incapable of any communication. The symptom of the patient was very similar to a brain death case. On the same day, the EEG examination was applied at the bedside of ICU. The recorded EEG data in the first session lasted 5 min (for a total three sessions). As an example, a small time window of 10-s EEG is shown in Fig. 15.3a. By applying the developed signal processing method described in Sect. 15.3 to the recorded signals, we obtained the results in time, frequency, and time–frequency domains shown in Fig. 15.3b. As observed from the figure, a typical α-wave component (12.5 Hz) was extracted. In this case, the number of source was assumed to be three. This is because we hope to extract relative strong components from the observations. When we assumed the number of the decomposed components is equal to the number of sensors, similar results were obtained. Without loss of generality, we also applied the same procedure to many randomly selected time windows of the recorded EEG signals, and similar results were also observed.

In this example, although the symptom of this patient was very similar to a brain death case, our EEG data analysis indicates that the patient still had physiological brain activities. In fact, the patient came back to consciousness slowly after that day. On August 31, the patient was able to respond to simple questions, and was released from the hospital after several months.

By applying our robust ICA algorithm to the 35 patients' EEG recordings, we succeeded in extracting brain activities from 19 patients (see Table 15.1). In general, it is difficult to observe the high-frequency range brain waves (such as the beta or gamma bands) considering the fact that they are more relevant to the high-level cognitive tasks, which seemed nearly impossible for all comatose patients. However, in our experiments, the β-wave component was found to occur in few patients, this might be due to the effect of drug medication.

15.4.2 The Patients Without EEG Brain Activities

For the remaining 16 patients, we have not been able to extract any brain activities from their recordings; in other words, only noise or interference components were extracted. In some cases, the heartbeat component were extracted.

Here, we present a case study of a quasi-brain-death patient. The patient was a 48-years-old male with a virus encephalitis. On October 18, 2005, the patient was in a deep-coma state, his pupils were dilated but had a very weak visual response, and the respiratory machine was used. The EEG examination was applied to the patient at same day. With our proposed signal processing procedure, a θ-wave component was found since the patient has a weak visual response (see in Table 15.1, No. 7 patient).

Table 15.1. The results of extracted brain waves from 19 patients (δ band: 1–4 Hz, θ band: 4–8 Hz, α band: 8–12 Hz, β band: 13–30 Hz)

No.	Gender	Age	Extracted brain waves
1	Male	18	δ-wave, θ-wave, α-wave
2	Male	40	θ-wave
3	Male	85	θ-wave, α-wave
4	Male	65	θ-wave
5	Female	64	θ-wave
6	Female	23	δ-wave
7	Male	48	θ-wave
8	Female	17	α-wave
9	Male	18	δ-wave, θ-wave
10	Male	66	θ-wave
11	Female	84	δ-wave, θ-wave, α-wave, β-wave
12	Male	79	θ-wave
13	Male	48	θ-wave, α-wave
14	Female	73	θ-wave
15	Male	64	θ-wave, α-wave
16	Male	83	δ-wave, θ-wave
17	Female	67	θ-wave, α-wave
18	Female	82	δ-wave
19	Male	37	β-wave

However, in the next day (October 19), the patient's condition appeared worse and he became completely unresponsive to external visual, auditory, and tactile stimuli. The EEG examination was applied to the same patient at the same day, and the recordings lasted 5 min. As an illustration, we apply ICA to one-second duration of the recorded EEG signals, and the result is shown in Fig. 15.4. In this case, the number of the decomposed components is assumed to be identical to number of sensors, and we extracted five interference components and a heartbeat component, but no brain wave activity was found. Notably, our data analysis result was identical to the later clinical diagnosis (at the time of EEG examination we carefully used the term "quasi-brain-death" because no final conclusion could be reached before further tests were applied according to the rigorous brain death criterion). In fact, the patient's heartbeat stopped after two days.

15.5 Quantitative EEG Analysis with Complexity Measures

Thus far, the signal processing tools (i.e., PFA, ICA) we have discussed earlier are *qualitative*. On the other hand, we also hope to explore *quantitative* measures for EEG analysis; this will be the focus of the current section.

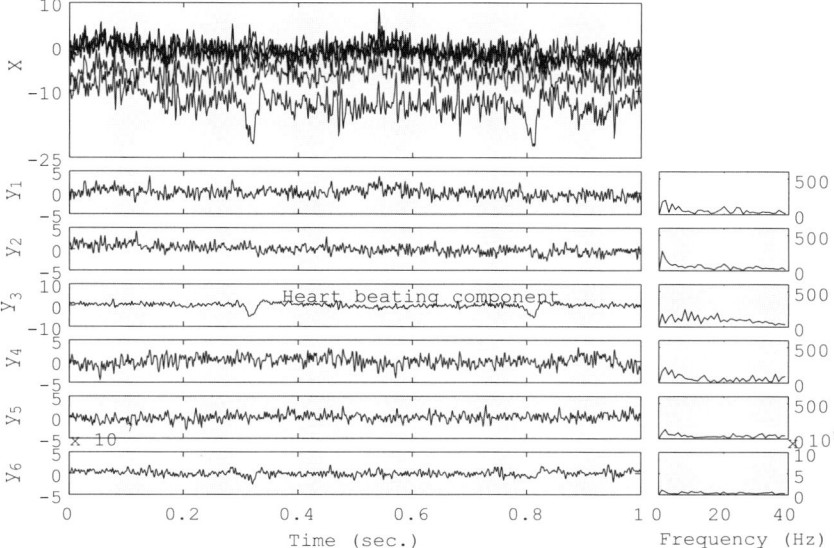

Fig. 15.4. The ICA result for a quasi-brain-death patient without EEG brain activity

After the preliminary clinical diagnosis and signal processing analysis for the EEG recordings, the patients can be categorized into two groups: the deep coma group and the quasi-brain-death group. For evaluating the quantitative differences between two patient groups, the quantitative EEG (qEEG) analysis was conducted [12, 13]. The goal of this procedure is to discover some informative features relevant to the EEG signals that are useful in distinguishing from these two groups (deep coma vs. quasi-brain-death) and to further evaluate their statistical significance (using all available subjects' recordings). In the present study, we propose four statistical measures for qEEG analysis, which are described later in more detail.

15.5.1 The Approximate Entropy

The approximate entropy (ApEn) is a quantity that measures the regularity or predictability of a random signal or time series [24]. To compute the ApEn(m, r) of a time series $\{x(k)\}$, $k = 1, \ldots, N$, the series of vectors of length m, $\mathbf{v}(k) = [x(k), x(k+1), \ldots, x(k+m-1)]^{\mathrm{T}}$ is first constructed from the signal samples $\{x(k)\}$.

Let $D(i, j)$ denote the distance between two vectors $\mathbf{v}(i)$ and $\mathbf{v}(j)$ ($i, j \leq N - m + 1$), which is defined as the maximum difference in the scalar components of $\mathbf{v}(i)$ and $\mathbf{v}(j)$, namely $D(i, j) = \max_{l=1,\ldots,m} |v_l(i) - v_l(j)|$. Then we further compute the metric $N^{m,r}(i)$, which represents the total number of vectors $\mathbf{v}(j)$ whose distance with respect to the generic vector $\mathbf{v}(i)$ is lower than r, namely $D(i,j) \leq r$. The index r is the *tolerance parameter*. Let us now define $C^{m,r}(i)$, the probability to find a vector that differs from $\mathbf{v}(i)$ less than the distance r, as follows:

$$C^{m,r}(i) = \frac{N^{m,r}(i)}{N-m+1}, \quad \phi^{m,r} = \frac{\sum_{i=1}^{N-m+1} \log C^{m,r}(i)}{N-m+1}. \tag{15.17}$$

The ApEn statistic is given by

$$\mathrm{ApEn}(m,r) = \phi^{m,r} - \phi^{m+1,r}. \tag{15.18}$$

Consequently, the ApEn statistic of a time series $\{x(k)\}$ measures the logarithmic likelihood that runs of patterns of length m that are close to each other will remain close in the next incremental comparisons, $m+1$. A greater likelihood of remaining close (i.e., with high regularity) produces a smaller ApEn value, and vice versa. Typically, a signal with low regularity produces a greater ApEn value. The two parameters, m and r, are chosen in advance to compute the ApEn. The typical values $m = 2$ and r between 10% and 25% of the standard deviation of the time series $\{x(k)\}$ are often used in practice [1, 18, 24]. The ApEn statistic is invariant to the scaling of the time series, hence it is irrelevant to the power of the signal.

15.5.2 The Normalized Singular Spectrum Entropy

The *normalized singular spectrum entropy* (NSSE) is a complexity measure arisen from calculating the singular spectrum of a delay-embedded time series [25]. Given an N-point univariate time series $\{x(k)\}$, $k = 1, \ldots, N$, a delay-embedding matrix $\mathbf{X} \in \mathbb{R}^{m \times (N-m+1)}$ (where $N \gg m$) is constructed as follows:

$$\mathbf{X} = \begin{bmatrix} x(1) & x(2) & \cdots & x(N-m+1) \\ x(2) & x(3) & \cdots & x(N-m+2) \\ \vdots & \vdots & \ddots & \vdots \\ x(m) & x(m+1) & \cdots & x(N) \end{bmatrix}.$$

Applying the *singular value decomposition* (SVD) to matrix $\mathbf{X}$, we obtain

$$\mathbf{X} = \mathbf{U}\mathbf{S}\mathbf{V}^\mathrm{T}, \tag{15.19}$$

where $\mathbf{U}$ is an $m \times (N-m+1)$ matrix, and $\mathbf{V}$ is an $(N-m+1) \times (N-m+1)$ square matrix, both of which are orthogonal such that $\mathbf{U}^\mathrm{T}\mathbf{U} = \mathbf{V}^\mathrm{T}\mathbf{V} = \mathbf{I}$. The matrix $\mathbf{S}$ is degenerate and contains a $p \times p$ (where $p = \mathrm{rank}(\mathbf{X})$) diagonal matrix with nonzero singular values $\{\sigma_1, \ldots, \sigma_p\}$ appearing in the diagonal; if $\mathbf{X}$ has full rank, then $p = m$.

Without loss of generality, let us assume that $p = m$, then the *singular spectrum entropy* (SSE) of the matrix $\mathbf{X}$ is defined as [25]

$$\text{SSE}(m) = -\sum_{i=1}^{m} s_i \log s_i, \tag{15.20}$$

where $\{s_i\}$ denotes the normalized singular spectrum

$$s_i = \frac{\sigma_i}{\sum_{j=1}^{m} \sigma_j}. \tag{15.21}$$

Notably, when $s_i = \frac{1}{m}$ for all i, it follows that $\max\{\text{SSE}(m)\} = \log m$. Hence, we can define the normalized SSE (NSSE) by

$$\text{NSSE}(m) = \frac{-\sum_{i=1}^{m} s_i \log s_i}{\log m}, \tag{15.22}$$

such that $0 \leq \text{NSSE}(m) \leq 1$.

It is noteworthy that the SSE is closely related to the *S-estimator* that was proposed in [10]. Specifically, with our notation, the S-estimator can be defined as

$$S = 1 + \frac{\sum_{i=1}^{m} \lambda'_i \log \lambda'_i}{\log m}, \tag{15.23}$$

where $\{\lambda'_i\}$ ($i = 1, \ldots, m$) denotes the *normalized* eigenvalues obtained from diagonalizing $\mathbf{XX}^\mathrm{T}$ (i.e., $\lambda'_i = \lambda_i/\text{tr}(\mathbf{XX}^\mathrm{T}) = \frac{\lambda_i}{\sum_{j=1}^{m} \lambda_j}$, where $\{\lambda_i\}$ denotes the eigenvalues of matrix $\mathbf{XX}^\mathrm{T}$). As known from the relationship between EVD and SVD, it can be inferred that $\lambda_i = \sigma_i^2$ ($i = 1, \ldots, m$, assuming both in decreasing order); hence, we obtain

$$\lambda'_i = \frac{\lambda_i}{\sum_{j=1}^{m} \lambda_j} = \frac{\sigma_i^2}{\sum_{j=1}^{m} \sigma_j^2}, \tag{15.24}$$

which bears some similarity with (15.21).

It is noteworthy that both the SSE (or NSSE) statistic and S-estimator are rotational invariant to the coordinate transformation of the original time series. This is because the singular values or the eigenvalues are invariant to any orthogonal transform. In other words, we can also calculate these statistics in the frequency domain by using the fast Fourier transform (FFT) algorithm.

15.5.3 The C_0 Complexity

The C_0 complexity, originally developed in [11] and later modified in [15], is a complexity measure for characterizing the regularity of a time series.

Let $\{x(k), k = 1, \ldots, N\}$ be a time series with length N, and let $\{X(j), j = 1, \ldots, N\}$ denote its corresponding N-point Fourier transform series; let $G_N = \frac{1}{N} \sum_{j=1}^{N} |X_N(j)|^2$ denote the mean power of the signal, and define

$$X'_N(j) = \begin{cases} X_N(j), & \text{if } |X_N(j)|^2 > G_N, \\ 0, & \text{if } |X_N(j)|^2 \leq G_N. \end{cases} \quad (15.25)$$

Let $x'(k)$ denote the inverse Fourier transform of $X'_N(j)$, then the C_0 complexity is defined as

$$C_0 = \frac{\sum_{k=1}^{N} |x(k) - x'(k)|^2}{\sum_{k=1}^{N} |x(k)|^2}. \quad (15.26)$$

Note that the C_0 statistic is invariant to the scaling of the signal $\{x(k)\}$, and it has the following properties [15]:

- $0 \leq C_0 \leq 1$ for any real-valued time series.
- $C_0 = 0$ if $\{x(k)\}$ is constant.
- C_0 approaches to zero with the length N if $\{x(k)\}$ is periodic.

15.5.4 Detrended Fluctuation Analysis

A time-dependent time series $\{x(t)\}$ is called self-similar if it satisfies $x(t) \stackrel{d}{=} c^\alpha x(t/c)$, where $\stackrel{d}{=}$ means that the statistical properties of both sides of the equation are identical. In other words, a self-similar process, $x(t)$, with a parameter α has the identical probability distribution as a properly rescaled process, $c^\alpha x(t/c)$. The exponent α is called the *self-similarity parameter* or *scaling parameter*.

The α-exponent can be estimated efficiently by a method called *detrended fluctuation analysis* (DFA) [23]. The advantages of the DFA method over other estimation methods (e.g., spectral analysis and Hurst analysis) are that it permits the detection of intrinsic self-similarity embedded in a seemingly nonstationary time series and also avoids the spurious detection of apparent self-similarity, which may be due to an artifact of extrinsic trends.

15.5.5 Quantitative Comparison Results

For qEEG analysis, the PFA and ICA procedures described in Sect. 15.3 were first applied to the EEG signals in an attempt to extract the brain waves. Based on the qualitative results obtained from the Fourier spectrum analysis, we selected some segments (10 s for each subject) of EEG recordings from the coma patient group. For the quasi-brain-death patient group, we randomly selected the recording segments with equal duration lengths. However, once the EEG recordings were selected, neither noise reduction nor ICA was applied before applying the above four complexity measures to the raw EEG signals (for all six channels).

First, we computed the *relative power ratio* (RPR) across six channels

$$\text{RPR} = \frac{\theta + \alpha + \beta(4\text{--}30\,\text{Hz})}{\text{total power }(1\text{--}30\,\text{Hz})},$$

where θ, α, β denote the power from θ, α, and β spectral bands, respectively. Here the relative power (ratio) is preferred to the absolute power of single spectral band because the latter directly depends on the signal amplitude, thereby also dependent on the scaling of the signal after ICA processing. The reason we exclude the low-frequency component (1–4 Hz) is that there always exist non-EEG slow waves in the recorded signals (including white noise) which is more difficult to distinguish based merely on power spectrum. For each subject, we computed the RPR values from six channels and only reported the *maximum* value (the motivation behind that is to emphasize the contribution from brain wave components). Comparison was further made between the subjects from two groups. It was our intention to investigate whether the simple relative power statistic can reveal any statistical difference with regard to the qualitative observations (Table 15.1). We applied the one-way ANOVA (*analysis of variance*) as well as the *Mann–Whitney test* (also known as *Wilcoxon rank sum test*) to evaluate the RPR statistics between two groups. The ANOVA is a parametric test (by assuming that the two group samples' distributions are both Gaussian) that compares the means for two groups and returns the p-value for the null hypothesis that the two groups have equal means; whereas the Mann–Whitney test is nonparametric (by assuming that the two group samples' distributions have similar shape) and tests the hypothesis if two groups have equal medians. From our experiments, statistical significance was found from both tests with our selected EEG data, and the null hypotheses were rejected. The quantitative results are shown in Table 15.2.

Next, based on the four complexity measures, we also computed the quantitative statistics of the selected EEG signals (see [13] for details). Summary of quantitative results is given in Table 15.3. The statistical tests were further applied to evaluate their statistical significance. Specifically, the one-way ANOVA or Mann–Whitney test could be applied to these quantitative

Table 15.2. Results of statistical tests on the maximum relative power ratio (RPR) of six channels for two groups: coma vs. quasi-brain-death

	Coma	Quasi-brain-death
No. subjects	19	18
Mean	0.8572	0.7469
Median	0.8515	0.7379
ANOVA	$p = 2.3 \times 10^{-5}$	
Mann–Whitney	$p = 2.5 \times 10^{-4}$	

Table 15.3. Summary of quantitative statistics applied to the raw and filtered EEG data for two groups: coma (C) vs. brain death (D)

Measure	Chan.	Raw data			Filtered data		
		Median (C)	Median (D)	p-Value	Median (C)	Median (D)	p-Value
ApEn	Fp1	0.227	0.598	**	0.103	0.217	*
	Fp2	0.267	0.727	**	0.123	0.362	**
	F3	0.302	0.836	**	0.115	0.362	**
	F4	0.314	0.853	**	0.135	0.331	**
	F7	0.232	0.755	**	0.097	0.263	**
	F8	0.274	0.798	**	0.097	0.305	**
NSSE	Fp1	0.427	0.659	**	0.236	0.336	**
	Fp2	0.453	0.732	**	0.256	0.412	**
	F3	0.474	0.755	**	0.264	0.411	**
	F4	0.464	0.748	**	0.270	0.398	**
	F7	0.402	0.718	**	0.223	0.363	**
	F8	0.419	0.715	**	0.228	0.393	**
C_0	Fp1	0.040	0.093	**	0.030	0.033	**
	Fp2	0.043	0.109	**	0.033	0.038	*
	F3	0.043	0.1191	**	0.030	0.042	*
	F4	0.049	0.137	**	0.034	0.043	0.063
	F7	0.041	0.096	**	0.028	0.040	**
	F8	0.031	0.112	**	0.023	0.041	**
α-exponent	Fp1	1.212	1.147	*	1.277	1.217	0.088
	Fp2	1.203	1.074	**	1.251	1.152	*
	F3	1.194	1.067	**	1.250	1.161	*
	F4	1.220	1.089	*	1.259	1.178	0.053
	F7	1.245	1.096	**	1.307	1.198	**
	F8	1.217	1.101	**	1.292	1.179	**

For the p-value column, * means $p < 0.05$ and ** means $p < 0.01$, and they both show statistical significance

measures within two groups (coma vs. quasi-brain-death) for each electrode channel. The associated p-values are computed, and we denote $p < 0.05$ (i.e., with 95% confidence interval) by * and $p < 0.01$ (i.e., with 99% confidence interval) by **. As seen from Table 15.3, statistical tests show significant differences in all complexity measures and all channels for the raw EEG data. For the filtered EEG data (0.5–100 Hz), significant differences between two groups are still found in all or most of the channels for all complexity measures; therefore, these complexity statistics are very informative for group (coma vs. quasi-brain-death) discrimination.

Among the four features, it seems from Table 15.3 that the ApEn and NSSE statistics lead to more robust difference between the two groups (coma vs. quasi-brain-death). Interestingly, these measures are both entropy-based statistics. The statistically significant difference still holds when applying these statistics to the filtered EEG data. We suspect that the entropy-based measures are less sensitive to the stationarity and noise conditions. Especially, NSSE achieves the overall best performance, which might be due to the fact that the calculation of NSSE measure is based on EVD (thereby insensitive to noise and any subspace rotation). In contrast, the C_0 complexity is based on the Fourier transform, which is restricted by the stationarity of the signal; and the calculation of α-exponent is known to be sensitive to the level of noise.

15.5.6 Classification

Upon computing the four complexity statistics for EEG signals per channel, we obtain $6 \times 4 = 24$ features for each subject. To further extract *uncorrelated* features, we used the PCA for dimensionality reduction [13]. This is done by projecting the data into a subspace that has the maximum or dominant variance. With the features at hand for the two groups, we can feed them into a linear or nonlinear binary classifier, such as the *Fisher linear discriminant analysis* (LDA) and the *support vector machine* (SVM) [26].

Let ℓ denote the total number of samples, we further define three performance indices:

$$\text{MIS} = \frac{\text{FP+FN}}{\text{Total}}, \quad \text{SEN} = \frac{\text{TP}}{\text{TP+FN}}, \quad \text{SPE} = 1 - \frac{\text{FP}}{\text{FP+TN}}$$

where the above nomenclature follows: false positive (FP, type I error), false negative (FN, type II error), true positive (TP), true negative (TN), sensitivity (SEN), and specificity (SPE). In addition, it is informative to compute the *receiver operating characteristic* (ROC) curve, which is a graphical illustration that shows the relation between the specificity ($1-\text{SPE}$ value in the abscissa) and sensitivity (SEN value in the coordinate) of the binary classifier. The *area under the ROC curve* (AUROC) reveals the overall accuracy of the classifier (with value 1 indicating perfect performance and 0.5 indicating random guess). In our experiments, since the available data set is rather small, thus far we only tested the classifier's performance accuracy using a *leave-one-out* cross-validation procedure (i.e., using $\ell - 1$ samples for training and the remaining one sample for testing, and repeating the procedure for the whole data set). The average leave-one-out misclassification (MIS) performance was 9.2% for SVM and 11.3% for LDA. For SVM, we used a Gaussian kernel function with a kernel width of 0.1 (chosen from cross validation). The optimal AUROC value, we obtained is 0.852 with the nonlinear SVM classifier. The results are summarized in Table 15.4. In addition, we also compared the classification performance using the raw features without PCA dimensionality reduction, the MIS results from SVM are similar, while the performance of LDA is slightly worse than the the one with PCA feature reduction. This is probably because LDA is a linear classifier whereas SVM is a nonlinear classifier, and the latter is less sensitive to the number of linearly correlated features.

Table 15.4. Summary of classification results on the misclassification (MIS) and AUROC indices

Performance index	LDA		SVM	
	MIS	AUROC	MIS	AUROC
With PCA	11.3%	0.804	9.2%	0.845
Without PCA	12.7%	0.798	9.2%	0.852

The MIS performance is based on the leave-one-out cross-validation procedure

15.6 Conclusion and Future Study

In this chapter, we proposed a practical EEG-based examination procedure for supporting the clinical diagnosis of brain death. The pre-diagnosis system integrates several features of EEG signal processing and complexity measure-based quantitative analysis. The proposed EEG preliminary examination procedure can be applied at the patient's bedside using a small number of electrodes. The robust PFA+ICA signal processing approach can be used to reduce the power of additive noise and to separate the brain waves and interference signals. Moreover, the qEEG analysis can be used to evaluate the statistical differences between the coma patient group and quasi-brain-death patient group. Our empirical data analysis showed that the complexity measures are quite robust to the presence of non-EEG sources such as the artifacts, noise, or power interference.

In the future study, we plan to investigate several advanced signal processing methods (such as the *complex empirical mode decomposition* [27]) that might be able to overcome the limitation of standard Fourier analysis for better distinguishing the low-frequency components of EEG signals. We also plan to collect more EEG data and build a *knowledge-based* classifier that discriminates the coma patient from the known breath death patients. It is hoped that the performance of the classifier will be improved by incorporating some prior knowledge and by collecting more training samples. Since all signal processing algorithms described here can be implemented efficiently in real-time (e.g., running the LabVIEW© on a portable computer), our proposed EEG-based preliminary diagnosis system (Fig. 15.1b) may be potentially used as a diagnostic and prognostic tool in clinical practice.

Although our results on brain death diagnosis reported here are still empirical and the solid confirmation of our claims requires further investigation and more data analysis, we believe that advanced EEG signal processing and qEEG analysis tools would shed a light on the real-time medical diagnosis in the future clinical practice.

Acknowledgments

The authors like to acknowledge Dr. Zhen Hong, Dr. Guoxian Zhu and Yue Zhang at Shanghai Huashan Hospital, Prof. Yang Cao and Prof. Fanji Gu at Fudan University, China for the EEG experiment and useful comments. This project was supported by the Japan Society for the Promotion Science (JSPS) and the National Natural Science Foundation of China (NSFC).

References

1. Akay, M. (ed.): Nonlinear Biomedical Signal Processing, vol. II Dynamical Analysis and Modeling. IEEE Press, New York (2001)
2. Amari, S.: Natural gradient works efficiently in learning. Neural Computation **10**, 251–276 (1998)
3. Amari, S.: Natural gradient for over- and under-complete bases in ICA. Neural Computation **11**(8), 1875–1883 (1999)
4. Amari, S., Cichocki, A., Yang, H.: Advances in Neural Information Processing Systems, vol. 8, chap. A new learning algorithm for blind signal separation, pp. 757–763. MIT, Cambridge, MA (1996)
5. Cao, J.: Lecture Notes in Artificial Intelligence, vol. 3973, chap. Analysis of the quasi-brain-death EEG data based on a robust ICA approach, pp. 1240–1247. Springer, Berlin Heidelberg New York (2006)
6. Cao, J., Murata, N., Amari, S., Cichocki, A., Takeda, T.: Independent component analysis for unaveraged single-trial MEG data decomposition and single-dipole source localization. Neurocomputing **49**, 255–277 (2002)
7. Cao, J., Murata, N., Amari, S., Cichocki, A., Takeda, T.: A robust approach to independent component analysis with high-level noise measurements. IEEE Transactions on Neural Networks **14**(3), 631–645 (2003)
8. Cardoso, J., Laheld, B.: Equivariant adaptive source separation. IEEE Transactions on Signal Processing **44**, 3017–3030 (1996)
9. Cardoso, J., Souloumiac, A.: Jacobi angles for simultaneous diagonalization. SIAM Journal of Matrix Analysis and Applications **17**, 145–151 (1996)
10. Carmeli, G., Knyazeva, G., Innocenti, G., Feo, O.: Assessment of EEG synchronization based on state-space analysis. NeuroImage **25**, 330–354 (2005)
11. Chen, F., Xu, J., Gu, F., Yu, X., Meng, X., Qiu, Z.: Dynamic process of information transmission complexity in human brains. Biological Cybernetics **83**, 355–366 (2000)
12. Chen, Z., Cao, J.: An empirical quantitative EEG analysis for evaluating clinical brain death. In: Processings of the 2007 IEEE Engineering in Medicine and Biology 29th Annual Conference, pp. 3880–3883. Lyon, France (2007)
13. Chen, Z., Cao, J., Cao, Y., Zhang, Y., Gu, F., Zhu, G., Hong, Z., Wang, B., Cichocki, A.: An empirical EEG analysis in brain death diagnosis for adults. Cognitive Neurodynamics (under review)
14. Eelco, F., Wijdicks, M.: Brain death worldwide. Neurology **58**, 20–25 (2002)
15. Gu, F., Meng, X., Shen, E.: Can we measure consciousness with EEG complexities? International Journal of Bifurcation and Chaos **13**(3), 733–742 (2003)

16. Hyvärinen, A., Oja, E.: A fast fixed-point algorithm for independent component analysis. Neural Computation **9**, 1483–1492 (1997)
17. Lee, T., Girolami, M., Sejnowski, T.: Independent component analysis using an extended infomax algorithm for mixed sub-gaussian and super-gaussian sources. Neural Computation **11**, 417–441 (1998)
18. Lin, M., Chan, H., Fang, S.: Linear and nonlinear EEG indexes in relation to the severity of coma. In: Proceedings of the 2005 IEEE Engineering in Medicine and Biology 27th Annual Conference, vol. 5, pp. 4580–4583 (2005)
19. Makeig, S., Bell, A., Jung, T., Sejnowski, T.: Advances in Neural Information Processing System, vol. 8, chap. Independent component analysis of electroencephalographic data, pp. 145–151. MIT, Cambridge, MA (1996)
20. Marks, S., Zisfein, J.: Apneic oxygenation in apnea tests for brain death: a controlled trial. Neurology **47**, 300–303 (1990)
21. Molgedey, L., Schuster, H.: Separation of a mixtures of independent signals using time delayed correlations. Physical Review Letters **72**(23), 3634–3637 (1994)
22. Niedermeyer, E.: Electroencephalography: Basic Principles, Clinical Applications, and Related Fields. Lippoincott Williams & Wilkins, Baltimore, MD (1991)
23. Peng, C., Buldyrev, S., Havlin, S., Simons, M., Stanleyn, H., Goldberger, A.: Mosaic organization of DNA nucleotides. Physical Review E **49**, 1685–1689 (1994)
24. Pincus, S.: Approximate entropy (apen) as a complexity measure. Proceedings of National Academy of Science **88**, 110–117 (1991)
25. Roberts, J., Penny, W., Rezek, I.: Temporal and spatial complexity measures for EEG-based brain–computer interfacing. Medical and Biological Engineering and Computing **37**(1), 93–99 (1998)
26. Schölkopf, B., Smola, A.: Learning with Kernels: Support Vector Machines, Regularization, Optimization and Beyond. MIT, Cambridge, MA (2002)
27. Tanaka, T., Mandic, D.: Complex empirical mode decomposition. IEEE Signal Processing Letters **14**, 101–104 (2007)
28. Taylor, R.: Reexamining the definition and criteria of death. Seminars in Neurology **17**, 265–270 (1997)

16

Automatic Knowledge Extraction: Fusion of Human Expert Ratings and Biosignal Features for Fatigue Monitoring Applications

Martin Golz and David Sommer

A framework for automatic relevance determination based on artificial neural networks and evolution strategy is presented. It is applied for an important problem in biomedicine, namely the detection of unintentional episodes of sleep during sustained operations of subjects, so-called microsleep episodes. Human expert ratings based on video and biosignal recordings are necessary to judge microsleep episodes. Ratings are fused together with linear and nonlinear features which are extracted from three types of biosignals: electroencephalography, electrooculography, and eyetracking. Changes in signal modality due to nonlinearity and stochasticity are quantified by the 'delay vector variance' method. Results show large inter-individual variability. Though the framework is outperformed by support vector machines in terms of classification accuracy, the estimated relevance values provide knowledge of signal characteristics during microsleep episodes.

16.1 Introduction

Besides the great progress in expert knowledge in several areas of science and technology, there are still several areas which suffer from complexity of the processes making them difficult to describe. This is mainly due to lack of methodology, to unsolved theoretical problems, or simply to the relatively short span of time for research. Several areas in biomedicine are faced with such problems. One example is the research of central fatigue and of its extreme end, the so-called microsleep episode (MSE). A unifying theory of both is lacking and the expert knowledge in this area is not free of disagreement. Several recent contributions in this field claimed that large inter-individual differences may be one reason [6, 10, 11].

Automatic knowledge extraction is very helpful in such cases because it can potentially handle much more data than human experts. Furthermore, both humans and machine learning have limited methodological access to the problem and hopefully, they are able to complement each other.

Here we describe results of applied research to the important problem of driver fatigue. It will be shown that experts are necessary to score critical episodes in fatigue driving and that a fusion of different features extracted from different biosignals is successful to get high detection accuracies for such episodes.

Afterwards, these accuracies will be utilized to evaluate empirically a variety of optimizations. The aim of the first optimization is to find out which biosignal or which combination of different signals performs better. Therefore, we compare three types of biosignals: electroencephalography (EEG) as a measure of brain activity, electrooculography (EOG) as a measure of eye and eyelid movements, and eyetracking signals (ETS) as a measure of eye movements and of pupil size alterations. Second, automatic relevance determination (ARD) is performed to get knowledge whether particular features are important or not. This should be important to stimulate discussion on the underlying theory of the psychophysiology of driver fatigue, and moreover, to reduce complexity of signal processing. This potentially leads to better robustness and less consumption of computational power. ARD is realized by a feature fusion framework based on evolution strategies and on optimized learning vector quantization networks (OLVQ1) which will be introduced (Chap. 3). In addition, ARD also optimizes classification accuracy. Their results will be compared to the non-extended OLVQ1 and to support vector machines (SVM). For the latter method, it is well known that they head several benchmarking contests.

From the practical point of view, the aim of our research is to search for reference standards of extreme fatigue. They are often also called gold standards and the aim is to measure this as precisely as possible. They are not compulsorily linked to the field use but rather to the laboratory use. In case of driver fatigue, lab methods allow us to get non-contactless biosignals with high quality, whereas industry is developing contactless operating sensors. The most successful solutions for field use seem to be based on image processing. Despite noticeable progress in this development, their sensitivity, specificity, and robustness has to be further improved.

A vital issue for reference standards in biomedicine is their intra- and inter-subject variability, whereby the first has to be small as possible. The latter should also be preferably low otherwise much more effort is requested to adapt such standards to particular groups of subjects or in the extreme to particular subjects.

As mentioned above, several authors have reported inter-individual differences. In a recent paper, Schleicher et al. [19] investigated oculomotoric parameters in a dataset of 82 subjects. The parameter most correlating to independent fatigue ratings was the duration of eye blinks. In addition to correlation analysis, this parameter was investigated in detail immediately before and after an MSE which they defined as overlong eye blinks. The mean duration of overlong eye blinks is substantially longer (269 ms) than of blinks immediately before (204 ms) and after (189 ms) which suggests that blink duration alone is not robust enough for MSE detection and not at all suited for predictions of upcoming episodes. In addition to the reported inter-individual differences their duration of overlong eye blinks seems to be much lower than the reported 0.7 s of Summala et al. [23]. Ingre et al. [10] also reported large inter-individual variability of blink duration in a driving simulation study of ten subjects after working on a night shift. In conclusion, only gradual changes and a large inter-subject variability are observable in this important parameter which is mostly used in fatigue monitoring devices. The same is reported of other variables, e.g., delay of lid reopening, blink interval, and standardized lid closure speed [19].

A similar picture of inter-individual differences, of non-unique parameter values and of non-specific patterns turned out in EEG analysis. In their review paper, Santamaria and Chiappa [18] stated: "There is a great deal of variability in the EEG of drowsiness among different subjects." In a large normative study with 200 male subjects the EEG of drowsiness was found to have "infinitely more complex and variable patterns than the wakeful EEG" [17]. Åkerstedt et al. [1] showed that with increasing working time subjectively rated sleepiness strongly increases and EEG shows a significant but moderate increase of hourly mean power spectral density (PSD) only in the alpha band (8–12 Hz) but not in the theta band (4–8 Hz). In contrast, Makeig and Jung [15] concluded from their study that the EEG typically loses its prominent alpha and beta (12–30 Hz) frequencies as lower frequency theta activity appears when performance is deteriorating due to strong fatigue. Sleep-deprived subjects performing a continuous visuomotor compensatory tracking task [16] showed increasing PSD in the upper delta range (3–4 Hz) during episodes of poor performance. Other studies stated a broad-band increase of PSD [2, 9, 13].

In summary, it is obvious how important it will be for further research to investigate inter-subject variability. We will give here some first results based on cross validation. The data of each subject have to be held out for testing whereas training is performed on data of all other subjects. This way we get an estimate how large the inter-subject as well as the intra-subject variability is.

16.2 Fatigue Monitoring

16.2.1 Problem

Since the 1970s, and notwithstanding continued traffic growth by factor three, the number of fatalities in Germany has declined by 75%, but the number of injuries has been virtually unchanged. The German In Depth Accident Study (GIDAS) has found that 93.5% of all roadway crashes are caused by misconduct, i.e. driver limitations and driver or pedestrian error. Driver behaviour is the most complex and least understood element in the roadway system. One of the most important human factors leading to accidents is driver fatigue which was estimated to play a role in about 25% of all casualties [14].

A problematic issue is the measurement of fatigue where a reliable and largely accepted indicator has not been found up to the now. Some authors prefer subjective ratings, i.e., self-assessment, others utilize performance measures, e.g., reaction time. A lot of authors have investigated physiological parameters, e.g., blink duration, galvanic skin response, or electroencephalography (EEG). An overview of several methods of objective and subjective measurement of fatigue can be found in [4].

At the extreme top of fatigue are episodes of microsleep (MSE) which are brief periods of involuntarily ongoing sleep under the requirement of sustained attention. In case of automobile drivers they are typically lasting between 0.5 and 5 s, whereas for train drivers and for aircraft pilots much longer durations have been observed.

16.2.2 Human Expert Ratings

In contrast to fatigue, it is relatively easy for observers to judge ongoing MSE during driving simulation experiments. Among others the following signs are typical:

- Overlong eye lid closure (longer than ca. 250 ms)
- Burst pattern in the EEG, e.g., theta (4–8 Hz), alpha (8–12 Hz) burst
- Slow roving eye movements
- Head nodding
- Major driving incident, e.g., lane departure

Each of them is not unique, but when more than one sign arise then it is likely to have MSE ongoing. Note that the investigator observes not only subject behaviour by means of video cameras but also by means of several biosignals and of driving performance measures. He has the past and present of them in mind and observes subject's fight against sleep. So he can judge the temporal development of each sign of MSE. We therefore believe that in case of extreme fatigue he should be better able to judge subject's state than the subject itself.

Nevertheless, there exist a small amount of events which are vague to judge. They are typically characterized by an occurrence of only one of the above-mentioned signs of MSE. For example, if strong burst activity in the EEG or large lane deviation is observed but the subject eyes are still open. Such events are sometimes described in the literature as stare blanks or as driving without awareness. Other vague events are characterized by e.g., short phases with extremely small eyelid gap, inertia of eyelid opening or slow head down movements. Such vague events have been discarded, because our aim is to discriminate between samples of clear MSE and of clear non-MSE[1] based

---

[1] Non-MSE was selected at all times outside of clear and of vague MSE.

on feature extracted from simultaneously acquired biosignals. We assume that, if the biosignals contain patterns which are characteristic for clear MSE then it should be also possible to detect such patterns also for vague events. Unfortunately, there exist up to now no independent method to detect such problematic cases and a validation of this assumption can therefore not yet be provided.

16.2.3 Experiments

Many biosignals that are more or less coupled to drowsiness do not fulfill temporal requirements. Electrodermal activity and galvanic skin resistance, for example, are too slow in their dynamics to detect such suddenly occurring events. EEG and EOG are relatively fast and direct measurements. EEG reflects mainly cortical and are to some low degree also subcortical activities; it should be the most promising signal for MSE detection. The endogenous components of eye and eyelid movements measured by EOG are coupled to the autonomic nervous system which is affected during drowsiness and wake–sleep transitions. EEG (C3, Cz, C4, O1, O2, A1, A2, common average reference) and EOG (vertical, horizontal) was both recorded by an electropolygraph at a sampling rate of 128 Hz. The first three locations are related to electrical activity in somato-sensoric and motoric brain areas, whereas O1, O2 are related to electrical activity in the primary and secondary visual areas, and A1, A2 are assumed to be functionally less active and often serve as reference electrodes.

Disadvantageously, EEG/EOG are non-contactless and are corrupted by large noise which is originated by other simultaneously ongoing processes. This is not the case with ETS, which is contactless working and features by high temporal resolution. For each eye of the subject it outputs time series of pupil diameter and of horizontal and vertical coordinate of eyegaze location in the projection plane of the driving scene. Therefore, ETS consisting of another six signals (sampling rate 250 Hz) was not strictly synchronized to EEG and EOG. This is not problematic for later fusion at the feature level.

Twenty-two young adults started driving in our lab (Fig. 16.1) at 1:00 a.m. after a day of normal activity and of at least 16 h of incessant wakefulness. All in all, they had to complete seven driving sessions lasting 35 min, each followed by a 10-min long period of responding to sleepiness questionnaires and of vigilance tests, and then followed by a 5-min long break. Experiments ended at 8:00 a.m. Driving tasks were chosen to be intentionally monotonous to support drowsiness and occurrence of MSE. As mentioned earlier, MSE were scored by two observers utilizing three video camera streams: (1) of subjects left eye region, (2) of her/his head and of upper part of the body, and (3) of driving scene.

This step of online scoring is critical, because in addition to the above-mentioned non-unique signs of MSE, their exact beginning is sometimes hardly to define. Therefore, all events were checked offline and were eventually corrected by an independent expert. All in all, we have found 3,573 MSE (per

Fig. 16.1. Real car driving simulation lab which is specialized for recording of overnight-driving. A small city car in conjunction with an interactive 3D driving simulation software is utilized to present a monotonic lane tracking task to the subjects. Subject behaviour is recorded by infrared video cameras. In addition, driving performance and biosignals of the subject are recorded

1. Electropolygraphy (EEG, EOG, ECG, EMG)
2. Eyetracking system (ETS)
3. Equipment control and synchronization
4. Video capture
5. Driving simulation
6. ETS multi camera unit
7. Steering acquisition and force feedback
8. Electropolygraphy head box
9. Video projector
10. Real car (GM Opel "Corsa")
11. Infrared video cameras
12. Projection plane
13. Intercom system

subject: mean number 162 ± 91, range 11–399) and have picked out the same amount of non-MSE to have balanced data sets.

All in all, 15 different signals were recorded. In subsequent preprocessing stages three steps are performed: signal segmentation, artifact removal, and missing data substitution.

Segmentation of all signals was done with respect to observed temporal starting points of MSE/non-MSE using two free parameters, the segment length and the temporal offset between first sample of segment and starting point of an event. The trade-off between temporal and spectral resolution is adjusted by segment length and the location of the region-of-interest on the time axis is controlled by the temporal offset. Therefore, both parameters are of high importance and have to be optimized empirically. We reported earlier that offset is crucial and the optimal value was about 3 s, whereas segment length can vary between 5 and 12 s [22].

Artifacts in the EEG are signal components which are presumably originated extracerebrally and often exhibit as transient, high-amplitude voltages. For their detection a sliding double data window is applied, in order to compare PSD in both windows. When the mean-squared difference of them is higher than a defined threshold value, then the condition of stationarity is violated and as a consequence this example of MSE/non-MSE is excluded from further analysis.

Missing data occurred in all six ETS during eyelid closures. This is caused by the measuring principle. They are substituted by data generated by an autoregressive model which is fitted to the signal immediately before the eyelid closure. So, artificial data replace missing data under the assumption of stationarity. Nevertheless, this problem should be of minor significance. For instance, periods of missing data are in the size of 150 ms which is small compared to the segment length of 8 s.

16.2.4 Feature Extraction

After having segmented all signals, features have to be extracted which should characterize signals generally and should preferably support discrimination between MSE and non-MSE. Ideally, features which support sufficiently compact regions in the feature space are sought. This can be achieved in the original time domain, or in the spectral or wavelet or some other transform domain.

The periodogram as a spectral domain transform has been widely used in biosignal analysis. It is assumed that signals are an output of a linear, stochastic process which has to be stationary. The fact that the periodogram is an asymptotically unbiased estimator of the true power spectral density (PSD) does not mean that its bias is necessarily small for any particularly sample size N. It is only then largely unbiased for autoregressive processes, if their PSD values have low dynamic range. Here, we apply the modified periodogram which uses data tapering to control between bias and variance. After linear trend removal the PSD is directly estimated.

This step is commonly followed by a feature reduction step of simple summation of PSD values over equidistantly frequency intervals (spectral bands). As a consequence, three further parameters have to be optimized, namely the lower and upper cut off frequency and the width of the bands. Finally, PSD values have been logarithmically scaled. It was shown that both operations, summation in spectral bands and logarithmic scaling, are of high value to improve classification accuracy [7, 22].

Having in mind that irregularities in signals have at least two possible sources: stochasticity and nonlinearity of the underlying signal generating system then PSD features should be complemented by such which describe both characteristics. PSD estimators generally rely solely on a second-order statistics. This is not the case with the following state space method.

The recently introduced method of delay vector variance (DVV) [5] provides an estimate to indicate to which extent a signal has a nonlinear or a stochastic nature, or both. The stochasticity is estimated by the variance of time-delayed embedding of the original time series, whereas nonlinearity is estimated by relating the variance of delay vectors of the original time series to the variance of delay vectors of surrogate time series.

In the following, we want to give a short summary of three important steps of the DVV method:

1. Transformation from the original space into the state space by time-delay embedding:

Given a segment of a signal with N samples $s_1, s_2, \ldots, s_N$ as a realization of a stochastic process. For each target s_k generate delay vectors

$$s(k) = (s_{k-m}; \ldots; s_{k-1})^{\mathrm{T}}$$

where m is the embedding dimension and $k = m+1, \ldots, N$.

2. Similarity of states of the generating system:
For each target s_k establish the set of delay vectors

$$\Omega_k(m, r_d) = \{s(i) \mid \|s(k) - s(i)\| \leq r_d\}$$

where r_d is a distance uniformly sampled from the interval

$$[\max(0, \mu_d - n_d \sigma_d), \mu_d + n_d \sigma_d].$$

The free parameter n_d controls the level of details if the number of samples over the interval N_r is fixed (here, we have chosen $N_r = 25$). All delay vectors of $\Omega_k(m, r_d)$ are assumed to be similar. The mean μ_d and standard deviation d have to be estimated over the Euclidian distances of all pairs of delay vectors $\|s(i) - s(j)\| \; \forall \; i \neq j$.

3. Normalized target variances:
For each set $\Omega_k(m, r_d)$ compute the variances $\sigma_k^2(r_d)$ over the targets s_k. Average the variances $\sigma_k^2(r_d)$ over all $k = m+1, \ldots, N$ and normalize this average by the variance of all targets in state space ($r_d \to \infty$).

In general, target variances are monotonically converging to unity as r_d increases, because more and more delay vectors are belonging to the same set $\Omega_k(m, r_d)$ and its target variance tends to the variance of all targets which is almost identical to the variance of the signal. If the signal contains strong deterministic components then small target variances will result [5]. Therefore, the minimal target variance is a measure of the amount of noise and should diminish as the SNR becomes larger. If the target variances of the signal are related to the target variances of surrogate time series then implications on the degree to which the signal deviates from linearity can be made. For linear signals, it is expected that the mean target variances of the surrogates are as high as the original signal. Significant deviations from this equivalence indicate that nonlinear components are present in the signal [5].

For each segment of a signal, the DVV method results in N_r different values of target variances. They constitute the components of feature vectors **x** which feed the input of the next processing stages and represent a quantification to which extent the segments of the measured signals has a nonlinear or a stochastic nature, or both.

16.3 Feature Fusion and Classification

After having extracted a set of features from all 15 biosignals, they are to be combined to obtain a suitable discrimination function. This feature fusion step can be performed in a weighted or unweighted manner. OLVQ1 [12] and SVM [3] as two examples of unweighted feature fusion and one method of weighted fusion [21] are introduced in this section. Alternative methods earlier applied to MSE detection problems (e.g., Neuro-Fuzzy Systems [24]), were not further pursued due to limited adaptivity and lack of validated rules for the fuzzy inference part. We begin with OLVQ1 since it is also utilized as central part of the framework for weighted feature fusion. SVM attracts attention because of their good theoretical foundation and their coverage of complexity as demonstrated in different benchmark studies.

16.3.1 Learning Vector Quantization

Optimized learning vector quantization (OLVQ1) is a robust, very adaptive and rapidly converging classification method [12]. Like the well-known k-means algorithm it is based on adaptation of prototype vectors. But instead of utilizing the calculation of local centres of gravity LVQ is adapting iteratively based on Riccati-type of learning and aims to minimize the mean squared error between input and prototype vectors.

Given a finite training set S of N_S feature vectors $x^i = (x_1, \ldots, x_n)^{\mathrm{T}}$ assigned to class labels y^i:

$$S = \left\{ (x^i, y^i) \subset \mathbb{R}^n \times \{1, \ldots, N_C\} \,|\, i = 1, \ldots, N_S \right\}$$

where N_C is the number of different classes, and given a set W of N_W randomly initialized prototype vectors w^j assigned to class labels c^j:

$$W = \left\{ (w^j, c^j) \subset \mathbb{R}^n \times \{1, \ldots, N_C\} \,|\, j = 1, \ldots, N_W \right\}$$

where n is the dimensionality of the input space. Superscripts on a vector always describe the number out of a data set, and subscripts on a vector describe vector components.

The following equations define the OLVQ1 process [12]: For each data vector x^i, randomly selected from S, find the closest prototype vector w^{j_C} based on a suitable vector norm in $\mathbb{R}^n$:

$$j_C = \arg\min_{j} \left\| x^i - w^j \right\| \forall j = 1, \ldots, N_W \tag{16.1}$$

Adapt w^{j_C} due to the following update rule, whereby positive sign has to be used if w^{j_C} is assigned to the same class as x^i, i.e., $y^i = c^{j_C}$, otherwise the negative sign has to be used:

$$\Delta w^{j_C} = \pm \eta_{j_C} \left(x^i - w^{j_C} \right). \tag{16.2}$$

The learning rates η_{j_C} are computed by

$$\eta_{j_C}(t) = \frac{\eta_{j_C}(t-1)}{1 \pm \eta_{j_C}(t-1)}, \tag{16.3}$$

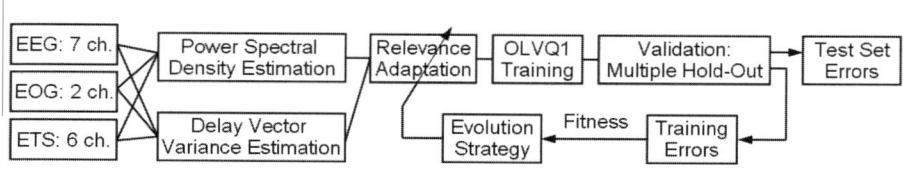

Fig. 16.2. Proposed OLVQ1+ES framework for adaptive feature fusion

whereby the positive sign in the denominator has to be used if $y^i = c^{j_C}$ and hence η_{j_C} is decreasing with iteration time t. Otherwise, it is increasing because the negative sign has to be used if $y^i \neq c^{j_C}$. It remains to be stressed that the prototype vectors and assigned learning rates are both updated, following (16.2) and (16.3), respectively, if and only if the prototype is closest to the data vector.

16.3.2 Automatic Relevance Determination

A variety of methods for input feature weighting exists not only for classification tasks, but also for problems like clustering, regression, and association, to name just a few. If the given problem is solved satisfactory then the weighting factors are interpretable as feature relevances. Provided that a suitable normalization of all features was done a priori, features which are finally weighted high have large influence on the solution and are relevant. On the contrary, features of zero weight have no impact on the solution and are irrelevant. On one hand such outcomes constitute a way for determining the intrinsic dimensionality of the data, and on the other hand, features ranked as least important can be removed and thereby a method for input feature selection is provided. In general, an input space dimension as small as possible is desirable, for the sake of efficiency, accuracy, and simplicity of classifiers.

To achieve MSE detection both estimated feature sets are fused by an adaptive feature weighting framework (Fig. 16.2). The mean training error[2] is used to optimize feature weights; they also serve as fitness function in an evolution strategy (ES) which is the essential part for updating weight values. Evolutionary algorithms are heuristic optimization algorithms based on the principles of genetic selection in biology. Depending on the concepts used they can be subdivided into Genetic Programming, Genetic Algorithm, Evolution Strategy [20], and others. For signal processing purposes mainly Genetic Algorithms and Evolution Strategies (ES) have been utilized. The first is based on binary-valued gene expressions and fits more to combinational optimizations (e.g., feature selection or model selection of neural networks). ES is based on real-valued gene expressions.

ES adapts a weighted Euclidean metric in the feature space [21]

$$\|x - w\|_\lambda^2 = \sum_{k=1}^n \lambda_k |x_k - w_k|^2 \qquad (16.4)$$

---

[2] Test errors were not used, directly and indirectly.

where λ_k are weighting scalars for each space dimension. Here, we performed standard (μ,λ)-ES with Gaussian mutation and an intermediary recombination [20].

OLVQ1+ES was terminated after computation of 200 generations; population consisted 170 individuals. This was repeated 25 times to have an estimate of the variability in optimal gene expressions, i.e., in the relevance values. As will be seen (Sect. 16.4), up to 885 features were processed, because 35 PSD and 24 DVV features had been extracted from 15 signals. It is not known a priori which features within different types of features (PSD, DVV) and of different signal sources (EEG, EOG, ETS) are suited best for MSE detection. Intuitively, features should differ in their relevance to gain an accurate classifier.

The proposed data fusion system allows managing extensive data sets and is in general a framework for fusion of multivariate and multimodal signals on the feature level, since individual methods can be exchanged by others. Feature fusion is advantageous due to fusion simplicity and ability to fuse signals of different types which are often non-commensurate [21]. But, it processes only portions of information of the raw signal conserved in features. Therefore, raw data fusion has the potential to perform more accurate as feature fusion.

16.3.3 Support Vector Machines

Given a finite training set S of feature vectors as introduced above, one wants to find among all possible linear separation functions $wx + b = 0$, that one which maximizes the margin, i.e., the distance between the linear separation function (hyperplane) and the nearest data vector of each class. This optimization problem is solved at the saddle point of the Lagrange functional:

$$L(w,b,\alpha) = \tfrac{1}{2} \|w\|^2 - \sum_{i=1}^{N_S} \alpha_i \left(y^i \left[(wx^i) \right] + b \right] - 1 \right) \quad (16.5)$$

using the Lagrange multipliers α_i. Both the vector w and the scalar b are to be optimized. The solution of this problem is given by

$$\bar{w} = \sum_{i=1}^{N_S} \alpha_i y^i x^i, \quad \text{and} \quad \bar{b} = -\tfrac{1}{2}\bar{w}\left(x_+ + x_-\right), \quad (16.6)$$

where x_+ and x_- are support vectors with $\alpha_+ > 0$, $y_+ = +1$ and $\alpha_- > 0$, $y_- = -1$, respectively. If the problem is not solvable error-free, then a penalty term

$$p(\xi) = \sum_{i=1}^{N_S} \xi_i$$

with slack variables $\xi_i \geq 0$ as a measure of classification error has to be used [3]. This leads to a restriction of the Lagrange multipliers to the range

$0 \leq \alpha_i \leq C \ \forall i = 1, \ldots, N_S$. The regularization parameter C can be estimated empirically by minimizing the test errors in a cross-validation scheme. To adapt nonlinear separation functions, the SVM has to be extended by kernel functions $k\left(x^i, x\right)$:

$$\sum_{i=1}^{N_S} \alpha_i y^i k\left(x^i, x\right) + b = 0. \tag{16.7}$$

Recently, we compared results of four different kernel functions, because it is not known a priori which kernel matches best for the given problem:

1. Linear kernel: $k\left(x^i, x\right) = x^i x$
2. Polynomial kernel: $k\left(x^i, x\right) = \left(x^i x + 1\right)^d$
3. Sigmoidal kernel: $k\left(x^i, x\right) = \tanh\left(\beta x^i x + \theta\right)$
4. Radial basis function kernel (RBF): $k\left(x^i, x\right) = \exp\left(-\gamma \left\|x^i - x\right\|^2\right)$

for all $x^i \in S$ and $x \in \mathbb{R}^n$. It turned out that RBF kernels are most optimal for the current problem of MSE detection [7].

16.4 Results

16.4.1 Feature Fusion

It is important to know which type of signal (EEG, EOG, ETS) contains enough discriminatory information and which single signal within of one type is most successful. Our empirical results suggest that the vertical EOG signal is very important (Fig. 16.3) leading to the assumption that modifications in eye and eyelid movements have high relevance, which is in accordance to the results of other authors [1]. In contrast to the results of EOG, processing of ETS led to lower errors for the horizontal than for the vertical component. This can be explained by the reduced amount of information in ETS compared to EOG. Rooted in the measurement principle, ETS measures eyeball movements and pupil alterations, but cannot acquire signals during eyelid closures and cannot measure eyelid movements. Both aspects seem to have large importance for the detection task, because errors were lower for EOG than for ETS. It turns out that also the pupil diameter (D) is an important signal for MSE detection. Despite the problem of missing ETS data during eye blinks, their performance for MSE detection is in the same shape as the EEG.

Compared to EOG, EEG performed inferior; among them the single signal of the Cz location came out on top. Relatively low errors were also achievable in other central (C3, C4) and in occipital (O1, O2) electrode locations, whereas both mastoid electrodes (A1, A2), which are considered as electrically least active sites, showed highest errors, as expected. Similarities in performance between symmetrically located electrodes (A1–A2, C3–C4, O1–O2) meets also expectancy and supports reliance on the chosen way of signal processing.

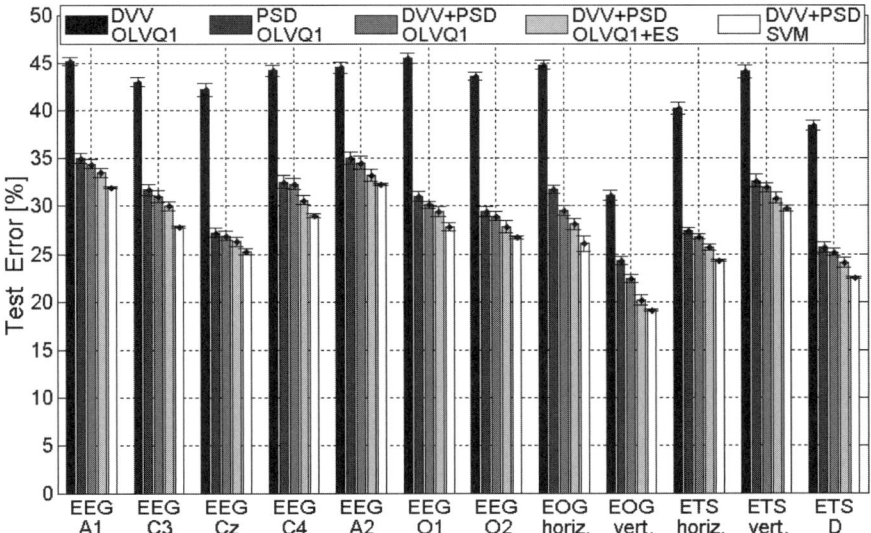

Fig. 16.3. Test errors (mean, standard deviation) of 12 single biosignals; a comparison of two different types of features and of three classification methods

Features estimated by DVV showed low classification accuracies (Fig. 16.3) despite additional effort of optimizing free parameters of the DVV method, e.g. embedded dimension m and detail level n_d. This is surprising because DVV was successfully applied to sleep EEG [22]. Processing EEG during microsleep and drowsy states and, moreover, processing of shorter segments seems to be another issue. PSD performed much better and performance was only slightly improved by fusion of DVV and PSD features (DVV+PSD).

Introducing an adaptively weighted Euclidean metric by OLVQ1+ES, our framework for parameter optimization, yielded only slight improvements (Fig. 16.3). SVM outperformed OLVQ1 and also OLVQ1+ES, but only if Gaussian kernel functions were utilized and if the regularization parameter and the kernel parameter were optimized previously.

Considerable decrease in errors was gained if features of more than one signal were fused. Compared to the best single signal of each type of signals (three left-most groups of bars in Fig. 16.4), the feature fusion of vertical EOG and central EEG (Cz) led to more accurate classifiers, and was better than fusion of all EOG features. Feature fusion based on all seven EEG signals was inferior. But, if features of nine signals (all EOG + all EEG) or, moreover, of all 15 signals (all EOG + all EEG + all ETS) were fused, then errors were considerably lowered. This is particularly evident if OLVQ1+ES or if SVM has been utilized. Both methods seem to suffer scarcely from the so-called curse of high dimensionality, but at the expense of much more computational load than the non-extended OLVQ1 [7].

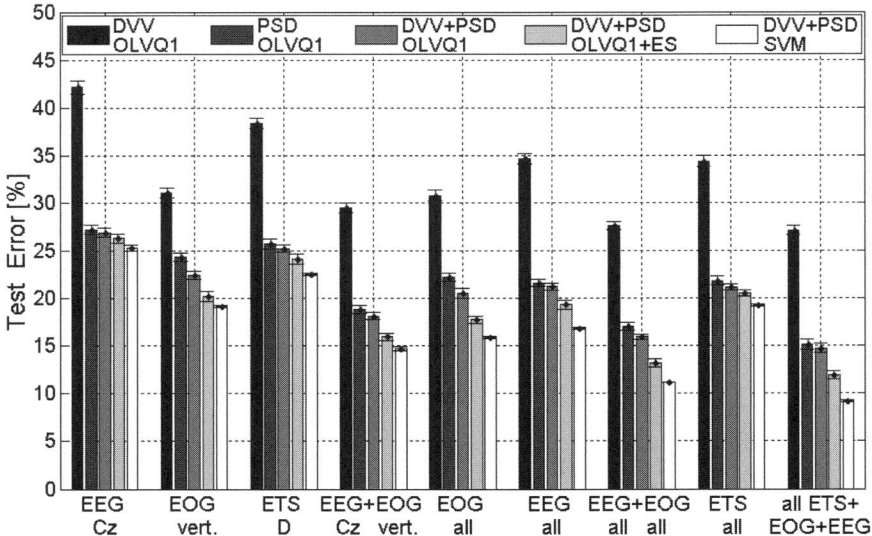

Fig. 16.4. Test errors (mean, standard deviation) for feature fusion of different signals; a comparison of two different types of features and three classification methods

16.4.2 Feature Relevance

The above introduced OLVQ1+ES framework for the adaptation of a weighted Euclidean metric in the input space resulted in much higher classification accuracies than OLVQ1 networks without metric adaptation, in particular when many features have to be fused. Nevertheless, OLVQ1+ES was outperformed by SVM. On the other hand, OLVQ1+ES has the advantage to return relevance values for each feature which is important for further extraction of knowledge. Relevance values of EOG have larger differences than of EEG, at a glance (Fig. 16.5).

PSD features of EEG are relevant in the high delta – low theta range (3–6 Hz) and in the high alpha range (10–12 Hz), and to some degree in the very low beta range (13–15 Hz) and low delta range (<1 Hz). They are less relevant in the high theta – low alpha range (7–9 Hz) and in the middle beta range (16–24 Hz) and are slightly higher in the high beta range (25–35 Hz). PSD features of EOG are relevant at very low and low frequencies (<1 Hz, 4–7 Hz); they are less relevant at 2–3 Hz and in the 8–13 Hz and in the 16–30 Hz range. A discussion of these results will be given later (Chap. 5). DVV features of EEG are roughly equal relevant with a decreasing tendency. This is not the case with DVV features of EOG where at high target variances (σ_n^{*2}, $n > 19$) relevance values are much higher than at low target variances, especially at very low n (1–3, 5–12). In summary, it can be stated that no feature relevance is marginal and no one is huge. Attempts to prune least relevant features led

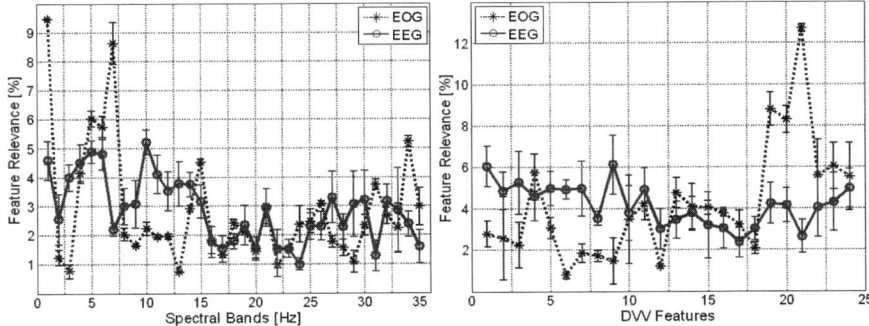

Fig. 16.5. Feature relevance values for MSE detection estimated by OLVQ1+ES. Relevances of PSD features (*left*) and of DVV features (*right*) are separated for each type of signal (EEG, EOG)

to remarkable increases in MSE detection errors, which are not reported here in detail.

16.4.3 Intra-Subject and Inter-Subject Variability

To estimate the stability of MSE classification and its independence to individual data the following subject-hold-out validation has been performed. The data set of subject k is used as test set whereas data sets of all other subjects, but subject k excluded, are used as training set. This is done for all k ($k = 1, \ldots, 22$). Doing so, we simulate that subject k did not took part in the study and is now a new subject, unknown for the classifier. Ideally, a single subject should not have too much influence on the classification performance; the classifier should represent biosignal patterns typical for all subjects. But this is basically not a prerequisite. It is also quite conceivable that groups of subjects have different patterns in their biosignals during MSE. Up to the present, nothing is known in the literature open to the public. Recently, several authors claimed large inter-individual variability in the EEG and EOG of drowsy subjects [6, 10, 19].

Our results show high inter-individual variability (Fig. 16.6) indicating that patterns common to all subjects were rarely found. For example subject five cannot be explained having the feature vector distribution of all other subjects, because mean errors of 47% are nearly as high as completely random classifications. In contrast, it was possible to explain several subjects (3, 12–14, 16, 21) nearly as good as for the case of no subject-hold-out (dashed line). But for all subjects except one (#16), errors are much lower if the classifier has been adapted solely on the data of the subject itself. For this, the training as well as the test set was restricted to the single subject itself. This clearly demonstrates that biosignals during MSE are very different from subject to subject. It is most reasonable to construct MSE recognition sensors based on individual characteristics.

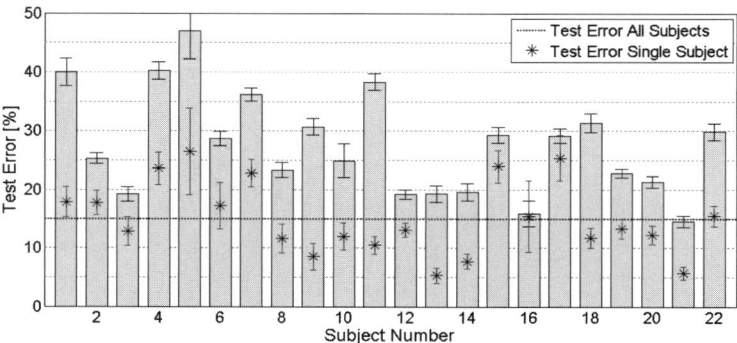

Fig. 16.6. Test errors (mean, std.) of subject-hold-out validation (*bars*) and of multiple hold-out validation: single subjects (*stars*) and all subjects (*dashed line*)

16.5 Conclusions and Future Work

In this chapter, a real-world problem of high importance for future developments in operator and dispatcher safety is addressed. Sensors for extreme fatigue with high sensitivity and high specificity are needed to counteract large problems caused by humans operating long-lasting under demands of sustained attention.

To the authors' knowledge, more than 20 different fatigue monitoring devices are commercially available at the present time. Most of them are contactless operating and have limited access to psychophysiological state of the subject. To validate such devices and to point to specific aspects of them to improve, a reference standard is needed. For the present, this has to be based on laboratory measurements, like e.g., EEG, EOG, and also eyetracking, because they are considered as most reliable. Quantitative biosignal analysis in combination with modern computational intelligence methods offers the feasibility of processing a great deal of different quantitative features as extracted before from the different biosignal sources. For this, we have presented three different methods. SVMs involving Gaussian kernel function outperformed both other methods, particularly in case of large number of features. The recently introduced OLVQ1+ES framework also performed well. Moreover, it automatically extracts knowledge of how relevant each individual signal feature is.

Relevance values showed reasonable differences and relatively low standard deviations. EEG frequencies in the region between the delta and theta band and in the alpha band, but not in the beta band are important for MSE detection. These results are in line with those in fatigue research [1, 15, 16], but somehow contradict other authors [2, 9, 13]. Also, the reported downshift from alpha to high theta is in contrast to our results [18]. Otherwise, results on large inter-individual variability reported here and also by others [6, 10, 19] give rise to doubt such considerations. Relatively simple rules, however, coarse they might be, seem to be not valid.

It was shown that human ratings are necessary and at the same time not free of vagueness to score episodes of extreme fatigue. Therefore, we processed only such episodes which appeared clearly to the observers. Vague episodes were ignored. It is in our expectation that classifiers accurately detecting clear episodes are to some extent also able to detect vague episodes.

Such problems notwithstanding, extensive signal acquisition and processing should support human scoring or eventually pass it. Like EEG and EOG also the ETS, in particular pupillography, turned out to be suitable measure of fatigue on a second-by-second basis [8]. Our results of all three types of signals showed that, based on a large data set, feature fusion techniques reduce classification errors down to an average of 9%. To our knowledge, there exist no comparable results in the scientific literature. Large inter-individual variability is a serious matter in this field for many scientists. We reported here first quantitative results of our data based on cross validation.

Future work has to validate intra-subject variations based on more data of single subjects and has to show if groups of subjects with similar biosignal characteristics can be found. Moreover, we are optimistic that new methods in higher-level signal processing and computational intelligence will come up against the large variability and will extract knowledge in more detail.

References

1. Akerstedt, T., Kecklund, G., Knutsson, A.: Manifest sleepiness and the spectral content of the EEG during shift work. Sleep **14**, 221–225 (1991)
2. Cajochen, C.: Power density in theta/alpha frequencies of the waking EEG progressively increases during sustained wakefulness. Sleep **18**, 890–894 (1995)
3. Cortes, C., Vapnik, V.: Support vector networks. Mach Learn **20**, 273–297 (1995)
4. Dinges, D.: An overview of sleepiness and accidents. J Sleep Res. **4**(Suppl 2), 4–14 (1995)
5. Gautama, T., Mandic, D., VanHulle, M.: The delay vector variance method for detecting determinism and nonlinearity in time series. Physica D **190**, 167–176 (2004)
6. Golz, M., Sommer, D.: The performance of LVQ based automatic relevance determination applied to spontaneous biosignals. In: Gabrys et al. (eds.) Proceedings of the 10th International Conference KES-2006, *LNAI*, vol. 4253, pp. 1256–1263. Springer, Berlin Heidelberg New York (2006)
7. Golz, M., Sommer, D., Chen, M., Trutschel, U., Mandic, D.: Feature fusion for the detection of microsleep events. J VLSI Signal Proc **49**, 329–342 (2007)
8. Golz, M., Sommer, D., Seyfarth, A., Trutschel, U., Moore-Ede, M.: Application of vector-based neural networks for the recognition of beginning microsleep episodes with an eyetracking system. In: Kuncheva et al. (eds.) Computational Intelligence: Methods and Applications, pp. 130–134. ICSC Academic, Canada (2001)

9. Horne, J., Baulk, S.: Awareness of sleepiness when driving. Psychophysiology **41**, 161–165 (2004)
10. Ingre, M., Åkerstedt, T., Peters, Anund, Kecklund: Subjective sleepiness, simulated driving performance and blink duration. J Sleep Res. **15**, 47–53 (2006)
11. Knipling, R.: Individual differences in commercial driver fatigue susceptibility: Evidence and implications. In: Proceedings of the International Conference on Fatigue Management in Transportation Operations. Seattle (2005)
12. Kohonen, T.: Self-Organizing Maps, 3rd edn. Springer, Berlin Heidelberg New York (2001)
13. Lal, S., Craig, A.: Driver fatigue: Electroencephalography and psychological assessment. Psychophysiology **39**, 313–321 (2002)
14. Langwieder, K., Sporner, A., Hell, W.: Einschlafen am Steuer: Hauptursache schwerer Verkehrsunfälle. Wien Klin Wochenschr **145**, 473 (1995)
15. Makeig, S., Jung, T.: Changes in alertness are a principal component of variance in the EEG spectrum. Neuroreport **7**, 213–216 (1995)
16. Makeig, S., Jung, T., Sejnowski, T.: Awareness during drowsiness: Dynamics and electrophysiological correlates. Can J Exp Psychol **54**, 266–273 (2000)
17. Maulsby, R., et al.: The normative electroencephalographic data reference library. Final report, National Aeronautics and Space Administration (1968)
18. Santamaria, J., Chiappa, K.: The EEG of drowsiness in normal adults. J Clin Neurophysiol **4**, 327–382 (1987)
19. Schleicher, R., Galley, N., Briest, S., Galley, L.: Looking tired? Blinks and saccades as indicators of fatigue. Ergonomics, in press (2008)
20. Schwefel, H., Rudolph, G.: Contemporary evolution strategies. In: Proceedings of the 3rd European Conference on Artificial Life, vol. 929, pp. 893–907. Springer, Berlin (1995)
21. Sommer, D., Golz, M., Trutschel, U., Mandic, D.: Fusion of state space and frequency-domain features for improved microsleep detection. In: W. Duch, et al. (eds.) Proceedings of the ICANN, *LNCS*, vol. 3697, pp. 753–759. Springer, Berlin Heidelberg New York (2005)
22. Sommer, D., Hink, T., Golz, M.: Application of learning vector quantization to detect drivers dozing-off. In: Proceedings of the 6th World Multiconf Systemics, Cybernetics and Informatics (ISAS-SCI2002), vol. XI, pp. 95–98. Orlando, USA (2002)
23. Summala, H., Häkkänen, H., Mikkola, T., Sinkkonen, J.: Task effects on fatigue symptoms in overnight driving. Ergonomics **42**, 798–806 (1999)
24. Trutschel, U., Guttkuhn, R., Ramsthaler, C., Golz, M., Moore-Ede, M.: Automatic detection of microsleep events using a neuro-fuzzy hybrid system. In: Proceedings of the EUFIT-98, vol. 3, pp. 1762–1766. Aachen, Germany (1998)

Index

Adaptive noise cancellation, ANC, 56, 59, 62, 65, 67
  demonstrative study, 60
Ambient intelligence, 185, 189, 198
Assisted living, 181, 195

Behaviour interpretation, 195
  eyetracking, 300
  microsleep, 302
Behaviour modelling
  fatigue, 300
  inter-subject variability, 300
Biomarker, 235
Blind source separation, 56, 62, 132
  ACMA, 134
  convolutive BSS, 59
  fusion of ICA and ANC, 56, 66, 69, 72
  independent component analysis, 283
  limitations, 71, 72
  parameter setting, 67
  performance evaluation, 64
  reference signal based, 70
  speech separation, 62
Brain computer interface, 262, 272
Brain modelling, 221
  brain death diagnosis, 275
  brain imaging, 223, 224, 228
  diffusion, 226
  EMDsonic, 262
  Fisher linear discriminant analysis, 295
  independent component analysis, 225
  Mann–Whitney test, 293
  relative power ratio, 293
  sonification, 262
  steady-state visual evoked potential, 262, 263
  Wilcoxon rank sum test, 293
Camera calibration
  Bayer pattern, 165
Car navigation systems
  dead reckoning, 142
  GPS, 141, 144
  guidance, 141
  location, 143
  map matching, 142, 146
  navigation, 141, 152
  positioning, 141, 151
Car onboard sensors
  accelerometer, 162
  adaptive cruise control, ACC, 161
  global positioning system, GPS, 161
  gyroscope, 162
  inertial measurement unit, IMU, 162
  odometer, 162
  speedometer, 162
Cardiopulmonary activity, 121
Classification, 78, 81, 94
  artifacts, 304
  evolution strategy, 300, 308
  false nearest neighbour method, 25, 28
  learning vector quantization, LVQ, 300, 307
  missing data, 304
  nearest neighbor, NN, see NN

support vector machines, SVM, 295, 309
support vectors, 79
validation, 303
Collaborative filtering, 6
  distributed learning, 79, 92–94
  hybrid filtering, 7
Collaborative processing, *see* Human pose estimation
Convex optimisation, 6

Data fusion, 309
  component selection, 235
  data fusion via fission, 246, 248, 263
  feature extraction, 305
  feature fusion, 182, 185, 186, 190, 222, 310
  high level fusion, 163
  intermediate level fusion, 163
  low level fusion, 163
  multimodal fusion, 222, 223, 226, 228, 231, 233
  opportunistic fusion, 185
  spatiotemporal fusion, 186
  waterfall model, 244
Dead reckoning, *see* Kalman filter
DECT, 116
Deterministic modelling
  $\psi\phi$ method, 30
  direct method, 30
  iterative method, 30
Distributed learning, *see* Collaborative filtering
Distributed vision networks, 185, 198
Doppler radar, 121
  Doppler shift, 121
DPM, *see* Signal modelling
Driving simulation, 302

E-cosmetics, 201, 212
  E-make, 201
  just noticeable difference, 215
Electroencephalogram, EEG, 224, 227, 264, 268, 275, 300
  EEG modelling, 11
  EEG preliminary examination, 276
  fusion of fMRI and EEG, 235
  ocular artifacts, 230, 248
  visual stimuli, 262, 272

Electrooculography, EOG, 300
Empirical mode decomposition, 244, 264, 265
  complex EMD, 251, 296
  ensemble EMD, 247
  fixed point iteration, 257
  rotation invariant EMD, 254
  sifting algorithm, 245
Expert knowledge, 299

fMRI, *see* Magnetic resonance imaging
Fuzzy logic, 141
  angle classification, 156
  fuzzy guidance, 154
  membership functions, 155

Hardware implementation
  field-programmable gate array, FPGA, 177
Heart rate variability, 122
Hierarchical filters, 38
  biased estimates, 46
  disjoint inputs, 41
  local cost function, 43
  structure, 39
  Wiener solution, 44
  overlapping, 40
HRV, *see* Heart rate variability
Human computer interface, 191
Human pose estimation, 181
  3D human body model, 190
  appearance attributes, 190
  gesture analysis, 182, 184–187
  gesture interpretation, 190
  interaction with environment, 195
  passive gestures, 195
  posture analysis, 187
Human posture estimation, 187, 188
  3D body model, 189
  3D views, 194
  assisted living, 195
  collaborative posture estimation, 192
  fall detection, 196
Hybrid filtering, *see* Collaborative filtering

Image processing
  data-driven, 212, 213
  Laplacian pyramid, 211, 212
  physiologically based, 202, 212, 213

Independent component analysis, ICA, 202, 203, 205, 207, 210, 222, 225, 229
  parallel ICA, 235

Kalman filter, 141, 143, 163
  computational complexity, 146
  dead reckoning improvement, 145
  decentralized Kalman filter, 144
  extended Kalman filter, EKF, 163
  trajectory tracking, 146, 151
  unscented Kalman filter, UKF, 163
Kohonen, see SOM

Least squares algorithms, 79, 84, 85
  iteratively reweighted least-squares, IRLS, 173
  kernel regression, 78, 79, 82, 83, 85, 86
  kernel subspace algorithm, 82
  minimum mean square error, see MMSE
  regression, 86
  subspace problems, 78
  support vector machines, 81, 83
LM, see Signal modelling

Magnetic resonance imaging
  functional MRI, fMRI, 222–224, 230
    auditory oddball task, 232
    multitask fMRI, 231
    preprocessing, 225
  structural MRI, sMRI, 223, 226, 228, 230, 232, 236
MAP, 103
Maximum likelihood estimation, MLE, 56–59, 67
  generalized likelihood ratio test, 127
  maximum a posteriori, see MAP
  maximum likelihood, ML, see ML
  normalised maximum likelihood, 29
ML, 103
MMSE, 103
Motion modelling
  canonical disparity plane, 174
  disparity plane, 173
  ground plane, 173
  independent motion, 165
  independent motion detection, 165

independently moving object, IMO, 161
LeNet, 167
normalized coordinates, 165
optical flow, 165
stereo disparity, 165
visual cues, 165
Multiple input multiple output, MIMO, systems, 123
MWM, see Signal modelling

Navigation, 141
Neural networks
  kernel function, 310
  kernel methods, 78, 82
  multi-layer perceptron, MLP, 163, 165
  radial basis function model, 28
  self-organizing map, see SOM
  support vectors, 81, 83, 84, 86
Neuronal spikes, 248, 252, 254
NN, 104

PLM, see Signal modelling

Radio frequency, 99
RF, see radio frequency
Room acoustics
  reverberation, 55
  reverberation noise, 57
  reverberation time, 57, 59, 60, 62
  room impulse response, 65, 67
  simulated room environment, 60
RSS, 97, 99

Sensor fusion, 142, 144
  automatic relevance determination, ARD, 300
  Kalman filter, 144
Sensor networks, 77–79, 87, 89
  localization, 78–82, 84, 85, 89, 90, 97, 103, 113
  and learning, see SLL
  mobile motes, 79, 90
  simultaneous localization and learning, see SLL
  WLAN, 116
Signal modality characterisation, 3
  delay vector variance, DVV, 305

fully-complex nonlinearity, 14
iterative amplitude adjusted Fourier
  transform surrogates, 27
non-uniform embedding, 27
random shuffle surrogates, 26
signal nonlinearity, 8, 13
sparsity detection, 38
split-complex nonlinearity, 14
surrogate data analysis, 25
Signal modelling
  C_0 complexity, 291
  analysis of variance, 293
  approximate entropy, 289
  complexity measure, 288
  constant modulus, 133
  detection, 301
  detrended fluctuation analysis, 292
  dominant path, 102
  eigenvalue decomposition, 281
  Fourier spectrum analysis, 285
  Hilbert transform, 133
  index of orthogonality, 268
  linear model, 99
  measurement noise, 109
  multi wall, 100
  non-parametric model, 102
  normalized singular spectrum
    entropy, 290
  parametric modelling, 99
  periodogram, 305
  piecewise linear, 100
  power spectral density, 305
  prediction, 301
  principal component analysis, 280
  propagation, 99
  singular value decomposition, 290
  Wigner–Ville distribution, 285
Skin appearance, 201, 202, 212
  chromophore, 203, 206, 208, 211
  hemoglobin, 202, 206, 210

melanin, 202, 206, 210–212
shading, 202, 211
shading removal, 205, 208
SLL, 104, 106, 107, 116
Smart camera networks, 184, 185
Smart environments, 184, 195
SOM, 105, 107
Sonification
  MIDI, 262
  pianoroll, 268, 269
State machines, 147
  feature generation, 147, 150
Statistical modelling
  kernel regression, 87–89, 92
  received signal strength, see RSS
  signal strength, 78–80, 84–89
  Wayland statistic, 26
Stochastic gradient, 8
  complex LMS, 12
  generalised normalised gradient
    descent, GNGD, 9
  equivalence between hierarchical and
    FIR filters, 41
  hierarchical gradient descent, HGD,
    45
  hierarchical least mean square,
    HLMS, 43
System identification, 46
  sparse channel, 47

Traffic control
  charge-coupled device, CCD, 164
  differential GPS, DGPS, 161
  light detection and ranging, LIDAR,
    161
  radar, 161

Wind modelling, 17
  complex EMD, 254
  polar coordinate system, 30
  rectangular coordinate system, 30

Printed in the United States of America